The Ancient Origins of Consciousness

The Ancient Origins of Consciousness

How the Brain Created Experience

Todd E. Feinberg and Jon M. Mallatt

The MIT Press
Cambridge, Massachusetts
London, England

© 2016 Massachusetts Institute of Technology

Illustrations © Mount Sinai Health System, reprinted with permission (unless otherwise noted).

All rights reserved. No part of this book may be reproduced in any form by any electronic or mechanical means (including photocopying, recording, or information storage and retrieval) without permission in writing from the publisher.

This book was set in ITC Stone by Toppan Best-set Premedia Limited. Printed and bound in the United States of America.

Library of Congress Cataloging-in-Publication Data

Names: Feinberg, Todd E., author. | Mallatt, Jon, author.
Title: The ancient origins of consciousness : how the brain created experience / Feinberg, Todd E., and Jon M. Mallatt.
Description: Cambridge, MA : The MIT Press, [2015] | Includes bibliographical references and index.
Identifiers: LCCN 2015038381 | ISBN 9780262034333 (hardcover : alk. paper)
Subjects: LCSH: Consciousness. | Brain.
Classification: LCC QP411 .F45 2015 | DDC 612.8/233--dc23 LC record available at http://lccn.loc.gov/2015038381

10 9 8 7 6 5 4 3 2 1

Contents

Preface   vii
Acknowledgments   xi
List of Figures and Tables   xiii

1 The Mystery of Subjectivity   1
2 The General Biological and Special Neurobiological Features of Conscious Animals   17
3 The Birth of Brains   37
4 The Cambrian Explosion   51
5 Consciousness Gets a Head Start: Vertebrate Brains, Vision, and the Cambrian Birth of the Mental Image   69
6 Two-Step Evolution of Sensory Consciousness in Vertebrates   101
7 Searching for Sentience: Feelings   129
8 Finding Sentience   149
9 Does Consciousness Need a Backbone?   171
10 Neurobiological Naturalism: A Consilience   195

Appendix: Table References   229
Notes   251
References   287
Index   349

Preface

How is consciousness created? When did it first appear on Earth? How did it evolve and which living animals have it? Consciousness researcher Todd Feinberg has been pondering these questions for over twenty years, while biologist Jon Mallatt has been fascinated by the early evolution of animals for an even longer time, so in 2013 we teamed up to see if we could find some answers. Today, consciousness studies are undergoing a great surge of interest, as long-standing philosophical questions are starting to be addressed scientifically. In this book we draw upon the diverse fields of neuroscience, evolutionary neurobiology, and philosophy to answer these questions. We focus on *primary* or *sensory consciousness*, which is the most basic type and means just having any kind of experience at all. Using as a starting point philosopher Thomas Nagel's proposal that conscious animals have experiences that constitute "something it is like to be," we seek the evolutionary origins of central *philosophical indices of consciousness* such as John Searle's "ontological subjectivity" and David Chalmers's "hard problem of consciousness" (how the physical brain produces personal experience). By applying a range of recent discoveries about the nervous systems of many different kinds of animals, about animal behavior, and about evolution and ancient life on Earth, we aim to identify the date when sensory consciousness and animal sentience first appeared and the "hard problem" was created.

This book is written for a wide range of readers: those interested in big philosophical questions about life and meaning, in consciousness and brain science, the vertebrate and invertebrate animals, or fossils and deep time. These fields are vast and varied, but we strive to cover and join them all, presenting at least their basics without demanding that the reader has an expert background in any of them. Still, we did not skimp as we worked to present clearly such topics as the full anatomy of vertebrate brains, the defining philosophical axioms of conscious states as well as the brain structures associated with consciousness, the theories and controversies of consciousness research, the first explosive diversification of animals half a billion years ago, the role of the senses in survival, neural pathways, the various kinds of vertebrates and their evolutionary histories, learning in animal behavior, the origins of pleasure and pain, and even some invertebrate biology.

In the first part of this book, we explain the basic philosophical puzzles of consciousness and then begin to assemble a list of features that seem responsible for consciousness: the "correlates" of consciousness. With this list, we start to formalize our own scientific theory of consciousness called *neurobiological naturalism*, previously proposed by Feinberg, systematically laying it out here so that it can be adjusted by the discoveries of the later chapters.

Next, we consider the fossil record of animal evolution, as well as the living groups of animals that descended from the ancient ancestors. We examine the great "Cambrian explosion" of animal diversity, which occurred between about 560 and 520 million years ago, and produced all the known animal phyla (vertebrates in the chordate phylum, arthropods, molluscs, etc.). Most importantly, this explosion also produced the first complex nervous systems and brains, the first complex behaviors in animals, and the earliest evidence of animal cognition. We deduce that in the first vertebrates, our fishlike early relatives, something wonderful occurred. That is, these neural advances were accompanied by the first appearance of consciousness, when simple reflexives evolved into a unified "inner world" of experiences filled with the most mysterious feature of consciousness, called "qualia" or the subjective feeling of things. The first qualia we consider are mapped *mental images* of the external world as sensed by vision, hearing, smell, and the other "distance senses." We refer to this as *mapped exteroceptive consciousness*. Challenging conventional wisdom and long-standing taboos, we then deduce that all vertebrates are conscious, not just humans and other mammals but also every fish, amphibian, reptile, and bird. All vertebrates have always been conscious, but we detect a large, memory-enhanced advance in sensory consciousness in the first mammals and first birds, when the end-site of consciousness switched from the lower midbrain (tectum) to the higher cerebrum.

While the book's first half is about exteroceptive consciousness of the distance senses—experiences that can exist without any emotion—we next cover another fundamental aspect of consciousness called *sentience* that includes affect and entails positive and negative feelings. By surveying which animals living today show the behaviors and brain structures known to be associated with affective feelings, we deduce that all vertebrates past and present have affective consciousness, as well as mapped exteroceptive consciousness.

Then, having worked out the markers that identify both aspects of consciousness in the vertebrates, we apply these same criteria to the invertebrates, and find that the arthropods (including insects and crabs) and cephalopods (like the octopus) meet many of the criteria for exteroceptive and affective consciousness. This would mean that consciousness evolved simultaneously but independently in the first arthropods and first vertebrates over half a billion years ago.

The final chapter summarizes our findings about which animals are conscious, what brain regions are involved in its creation, and how consciousness first evolved. Consciousness in fact turns out to be a more diverse and widespread evolutionary adaptation than most workers in the field have realized. Our analysis leads us to update our theory of neurobiological naturalism. This updated version, still based fully on known biological laws and principles, is then used to tackle the most fundamental philosophical question of the nature of consciousness: how does the material brain create subjective experience? We find an answer in the transitions from reflex-to-image and from reflex-to-affect, as they occurred over 520 million years ago.

We gain even more insight by subdividing the problem, that is, by tracing the origins of four different aspects of consciousness: not only qualia (1), but (2) *unity*, or why the conscious experience is unified; (3) *referral*, or why conscious brains focus experience on the outer world and inner body, but never ever experience the workings of their brain neurons; and (4) *mental causation*, or how immaterial consciousness can cause changes in the material world.

Overall, our analysis of the neural origins of sensory consciousness attempts to cut through the Gordian knot of the mind–brain problem and provide a path of reconciliation between a philosophy of personal subjectivity and the objective structure and functions of the brain. It does so by chronicling the *evolution* of the first "conscious brains," while also incorporating the *neurobiological* and *philosophical* aspects of consciousness.

The book's special contribution is its finding that consciousness can be understood if we combine the evolutionary, neurobiological, and philosophical approaches. In fact, combining the three approaches is essential to solving the problem. Each approach has limitations that cannot be recognized unless someone considers the full picture from all three points of view. For example, the biological approach works to solve problems by reducing them to their most basic parts and then exploring the complex interactions between these parts, but the philosophical approach shows that such traditional scientific reductionism cannot solve the hard problem of subjectivity. Then evolutionary history resolves it by explaining why consciousness is both irreducible and natural. Our discovery that the triple approach is needed has important implications not just for consciousness science, philosophy, and paleobiology, but also for understanding ourselves in relation to the natural world and the animal kingdom.

Acknowledgments

Numerous authors kindly provided insights and knowledge from their respective fields, including Ann Butler, Bud Craig, Sten Grillner, Brian Hall, Martin Heisenberg, Nicholas Holland, Jon Kaas, Harvey Karten, Thurston Lacalli, Trevor Lamb, Bjorn Merker, Georg Northoff, Andrew Parker, Gerhard Schlosser, Gordon Shepherd, Barry Stein, Georg Striedter, Edgar Walters, and Mario Wullimann. While our work has greatly benefited from their assistance, none of these investigators of course is responsible for the opinions we express in this book.

We thank the artists at Mount Sinai Health Systems, Jill Gregory and Courtney McKenna, for their tireless and exceptional work on the many illustrations in this book.

Todd Feinberg thanks his wonderful wife Marlene, who—as always—is his best friend and ally. Jon Mallatt thanks his devoted and ever-young wife Marisa for her support at all times.

At the MIT Press, we especially thank Phil Laughlin, who first brought the book on board at MIT, and Chris Eyer, who shepherded the manuscript through the preparation process. Judith Feldmann, the copy editor, was very helpful, dedicated, and professional, and her "Goldilocks" approach of offering just the right number of suggestions steered us to clear, direct writing.

List of Figures and Tables

Figures

Figure 1.1
Three approaches to understanding consciousness.

Figure 1.2
The concept of mental unity, as shown by the phenomenon of cyclopean perception ("cyclopean" = as if with one eye).

Figure 1.3
A timeline of the history of life on Earth, based on fossil evidence and dating of rocks through rates of radioactive decay.

Figure 2.1
How embodiment and "bodies" became more elaborate during the early evolution of life on Earth.

Figure 2.2
A biological hierarchy drawn as a stack of rectangles, showing the concepts of different levels, emergence, and constraint.

Figure 2.3
Reflexes: simple and complex.

Figure 2.4
Different types of hierarchies.

Figure 2.5
Isomorphism, affects, and two kinds of consciousness.

Figure 3.1
Tree showing the relations among the chordates, including vertebrates and the nonvertebrate cephalochordates and urochordates.

Figure 3.2
Basic parts of the chordate nervous system.

Figure 3.3
The nonvertebrate chordates: tunicates and cephalochordates (amphioxus).

Figure 3.4
Brains of tunicates of different life stages and classes.

Figure 3.5
Brain of amphioxus as the platform on which chordate and vertebrate brains are built.

Figure 4.1
Timeline from the start of the Earth to the present.

Figure 4.2
Today's animal phyla, which originated in the Cambrian explosion, presented in a phylogenetic "tree of life" that shows their interrelations.

Figure 4.3
Ediacaran (A) versus Cambrian (B) seafloor communities.

Figure 4.4
Cambrian representatives of many bilaterian animal phyla, known from fossils, with some that are strange and of uncertain relationship.

Figure 4.5
Bilaterian ancestor reconstructed: complex (A) versus simple (B) versions.

List of Figures and Tables

Figure 4.6
Comparison of Cambrian arthropods with vertebrates.

Figure 5.1
Relations and evolution of the chordate and vertebrate animals.

Figure 5.2
Main subdivisions and parts of the vertebrate brain (forebrain, midbrain, and hindbrain).

Figure 5.3
Brains of various groups of vertebrates in side view.

Figure 5.4
The limbic system, seen as the shaded structures in this mid-sagittal view of a generalized vertebrate brain.

Figure 5.5
Stepwise evolution of the vertebrate eye.

Figure 5.6
The embryonic tissues that are unique to vertebrates: neural crest and placodes.

Figure 5.7
The brain of amphioxus (as a proxy for the ancestral, prevertebrate brain), compared to the brain of a lamprey (a proxy for the brain of the first true vertebrate).

Figure 5.8
Prevertebrates and early vertebrates.

Figure 5.9
Vision: Visual pathway in the human.

Figure 5.10
Touch: Somatosensory-touch pathway in the human.

Figure 5.11
Hearing: Auditory pathway in the human.

Figure 5.12
Smell: Olfactory pathway in the human.

Figure 6.1
The clades of vertebrates and their phylogenetic relationships, according to current understanding.

Figure 6.2
Fossil jawless vertebrates from the late Cambrian and after.

Figure 6.3
Cyclostomes: Modern lamprey and hagfish.

Figure 6.4
Lamprey brain and sensory pathways.

Figure 6.5
Simple diagram of a sensory neural hierarchy.

Figure 6.6
Optic tectum, showing its size in a teleost zebrafish and a bird, with its layered neuronal structure in a teleost.

Figure 6.7
Pallium of the cerebrum of the vertebrate brain.

Figure 6.8
A sampling of extinct synapsid and sauropsid amniotes, from 280 to 70 million years ago: mammal-like reptiles, a mammal, a dinosaur, and a bird.

Figure 6.9
Comparing the brains of different amniote vertebrates.

List of Figures and Tables   xvii

Figure 7.1
Interoceptive and pain pathways to the brain of humans.

Figure 7.2
Limbic system in vertebrate brain.

Figure 8.1
Phylogenetic tree showing that affective, interoceptive, and exteroceptive consciousness all existed in the first vertebrates of the Cambrian explosion.

Figure 9.1
Relations of the living animal groups.

Figure 9.2
Plan of the central nervous system of most protostome invertebrates, as reconstructed in a presumed worm ancestor.

Figure 9.3
Insect and arthropod nervous system.

Figure 9.4
The relation between brain weight and body weight in various animals: vertebrates with some invertebrates.

Figure 9.5
Gastropod nervous system.

Figure 9.6
Cephalopod nervous system.

Figure 10.1
Timeline showing the major events in the history of consciousness.

Figure 10.2
Animals in the Cambrian ocean (520–505 million years ago).

Figure 10.3
Animals in the seas of the Carboniferous Period (330 million years ago), the age of the great coal forests.

Figure 10.4
Animals on land in the Carboniferous Period (330 million years ago), the age of the great coal forests.

Figure 10.5
Animals in the seas in the Triassic Period (220 million years ago), the age of reptiles.

Figure 10.6
Animals on land in the Triassic Period (220 million years ago), the age of reptiles.

Figure 10.7
Phylogenetic relations of the animals that show evidence of consciousness.

Figure 10.8
Relationships among the three types of primary consciousness in vertebrates: exteroceptive, interoceptive, and affective.

Figure 10.9
The problem of auto- and allo-ontological irreducibilities of consciousness.

Tables

Table 1.1
The neuroontologically subjective features of consciousness (NSFC)

Table 2.1
The defining features of consciousness

Table 5.1
Parts of the central nervous system of vertebrates, emphasizing the brain

List of Figures and Tables

Table 5.2
Vertebrate CNS versus that of tunicates and amphioxus: A comparison of parts

Table 5.3
Summary of the major sensory pathways of vertebrates

Table 7.1
Three different aspects of sensory consciousness (qualia) and "something it is like to be"

Table 7.2
The interoceptive and nociceptive pathways in mammals

Table 7.3
Summary of theories of the origin of affective consciousness

Table 8.1
Behaviors that do not indicate pain/pleasure (or negative/positive affect)

Table 8.2
Criteria for operant learned behaviors that probably indicate pain/pleasure (or negative/positive affect)

Table 8.3
Distribution of the behavioral evidence for positive and negative affect across animals

Table 8.4
Comparative neuroanatomy of positive and negative affects: Features for reward, nociception, and fearlike responses

Table 8.5
Comparative neuroanatomy of positive and negative affects: Mesolimbic reward system (MRS), or mesolimbic dopamine (reward/aversion) system, with the functions listed for mammals

Table 8.6
Adaptive behavior network ("social behavior network"), which signals behaviors necessary for survival; linked to mesolimbic reward system, which motivates and rewards these adaptive behaviors

Table 9.1
Affective consciousness: Suggested behavioral evidence of positive and negative affect in protostome invertebrates

Table 9.2
Sensory (exteroceptive) consciousness: Evidence for protostome invertebrates

Table 10.1
The three postulates of neurobiological naturalism

1 The Mystery of Subjectivity

"Something It Is Like to Be"

But no matter how the form may vary, the fact that an organism has conscious experience at all means, basically, that there is something it is like to be that organism. ... Fundamentally an organism has conscious mental states if and only if there is something that it is like to *be* that organism—something it is like *for* the organism. We may call this the subjective character of experience.

—Thomas Nagel[1]

Have you ever wondered whether a fish or a frog, or a bat or a bee, is conscious? Supposing you could actually "get inside" the brain of a bird, what criteria would you use to decide whether it is "sentient"? How could you know, as Nagel put it in his classic 1974 paper,[2] if there is "something it is like to be a bat"? Nagel was saying that the way to judge whether an animal is conscious is not by how intelligent it is, or how big its brain is, but rather by whether that organism has "subjective experiences."

The aspect of consciousness that comes closest to "something it is like to be" and that poses the greatest scientific challenge is *sensory consciousness*. Sensory consciousness is also called *phenomenal consciousness, primary consciousness, perceptual consciousness,* or the experiencing of *qualia* (perceived qualities). As defined by Antti Revonsuo, sensory experiences do not have to be elaborate, lingering, or humanlike to be conscious:

The mere occurrence or presence of any experience is the necessary and minimally sufficient condition for phenomenal consciousness. For any entity to possess primary phenomenal consciousness only requires that there are at least *some* patterns—any patterns at all—of subjective experience *present-for-it*. It is purely about the *having* of *any* sorts of patterns of subjective experience, whether simple or complex, faint or vivid, meaningful or meaningless, fleeting or lingering.[3]

If we want to know how the brain creates subjective experiences, or whether an animal possesses "something it is like to be," then investigating the origins and basis of *sensory* consciousness is the best place to start. This is because sensory consciousness allows a brain to create an "inner world" of personal experience. If an animal creates subjective, qualitative, "single point-of-view" awareness, then that animal possesses "something it is like to be." Therefore, this book is not about "higher consciousness," "self-consciousness," "access consciousness,"[4] "reflective consciousness" of one's own thoughts,[5] "intelligent consciousness," or "theory of mind." It is about explaining the nature and origins of the *most basic, sensory* type of consciousness. How the brain creates this experience still eludes a satisfying scientific explanation. Why?

The "Hard Problem"

Many questions regarding the nature of consciousness remain unexplained, but perhaps the most puzzling is how the flesh-and-blood brain creates a subjective experience. Philosopher David Chalmers called this the "hard problem" of consciousness, which is the difficulty of objectively explaining the subjective aspects of experience:

It is undeniable that some organisms are subjects of experience. But the question of how it is that these systems are subjects of experience is perplexing. Why is it that when our cognitive systems engage in visual and auditory information-processing, we have visual or auditory experience: the quality of deep blue, the sensation of middle C? How can we explain why there is something it is like to entertain a mental image, or to experience an emotion? It is widely agreed that experience arises from a physical basis, but we have no good explanation of why and how it so arises. Why should physical processing give rise to a rich inner life at all? It seems objectively unreasonable that it should, and yet it does.[6]

The problem is that when it comes to explaining sensory consciousness in terms of physics, chemistry, or even neurobiology, certain aspects of subjectivity always appear to remain unexplained. As philosopher Joseph Levine put it, a mysterious "explanatory gap" remains between the physical properties of the brain as we know them and the subjective experiences that the brain thereby creates.[7] He argues that no matter how detailed the objective explanation of the neural pathways that create subjective experience is, something is always left out—namely, the personal "experience" itself.

The Blind Men and the Elephant

An ancient parable from the Indian subcontinent serves as an apt metaphor for the difficulties we confront when trying to explain scientifically how the brain creates subjective experience. The story tells of a group of blind men—the exact number differs depending on the version, but let us say four—who are set to the task of deciding what an elephant looks like. The problem is that, being blind, the men must make their individual judgments by touch alone, and to make matters even worse, because the elephant is so large, each man is able to feel only a small part of the entire animal.

The blind man who feels the trunk declares that the elephant must look like a snake, while the one who feels the tail thinks it must look like a rope; the one who feels the tusk thinks it must have the appearance of a spear, while the one who touches the ear declares it must appear like a hand fan. In nearly all versions of the parable, the whole matter ends badly, and in one version that is attributed to the Buddha, the men ultimately come to blows:

> O how they cling and wrangle, some who claim
> For preacher and monk the honored name!
> For, quarreling, each to his view they cling.
> Such folk see only one side of a thing.[8]

This story illustrates the mystery of sensory consciousness in two ways. First, it shows how each blind man is unable to escape his own subjectivity and personal point of view. But more importantly, it shows how different approaches to a difficult problem can lead to differing opinions, each of which is only a partial answer to an overall "truth." The logical path to a solution is to integrate multiple approaches and find a consilience among them.

Multiple Paths to Solving the Mystery of Consciousness

We propose that the hard problem of subjectivity remains mysterious because each field of study that is interested in its solution only addresses it from the perspective of its own set of questions and answers. In this book, we endeavor to tackle the hard problem with three different approaches, each important in its own way, yet each incomplete. These three approaches are the *philosophical*, *neurobiological*, and *neuroevolutionary* domains (figure 1.1).[9]

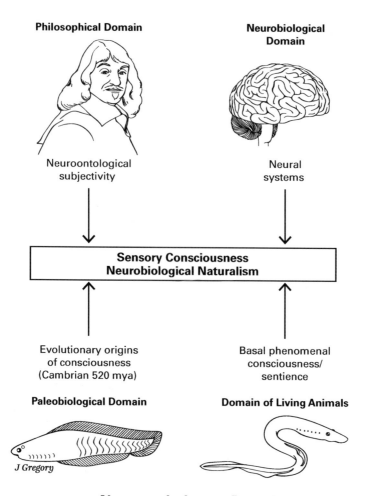

Figure 1.1
Three approaches to understanding consciousness. At the upper left is René Descartes, the famous seventeenth-century philosopher who pondered the mind–body problem of how the material brain relates to the immaterial mind. The neuroevolutionary domain is illustrated by the oldest known fossil fish, *Haikouichthys* (left), and by the lamprey (right), which represents the most basally evolved of the living fish.

The Philosophical Approach: The Gap between Subjectivity and Objectivity

The major philosophical question is whether and how *subjective* experience can ever be explained by any *objective* scientific theory. Philosopher John Searle ascribes this difficulty to the "first-person ontology" of consciousness (where "ontology" has to do with being, reality, essence, or fundamental nature):[10]

> Consciousness has a first-person or subjective ontology and so cannot be reduced to anything that has third-person or objective ontology. If you try to reduce or eliminate one in favor of the other you leave something out ... Biological brains have a remarkable biological capacity to produce experiences, and these experiences only exist when they are felt by some human or animal agent. You can't reduce these first-person subjective experiences to third-person phenomena for the same reason that you can't reduce third-person phenomena to subjective experiences. You can neither reduce the neuron firings to the feelings nor the feelings to the neuron firings, because in each case you would leave out the objectivity or subjectivity that is in question.[11]

The four neuroontologically subjective features of consciousness (NSFC)

Therefore, the most problematic characteristics of sensory or primary consciousness from the philosophical perspective are its first-person *experiential* aspects. But upon close analysis, we find that rather than there being a single "explanatory gap," there are in fact multiple gaps between subjective experience and the brain. To delineate these gaps as systematically as possible, we identified four *neuroontologically subjective features of consciousness (NSFCs)*.[12] The NSFCs have in common the following features: they are (1) ontologically subjective and (2) there is an explanatory gap between the way each is experienced by the brain versus the way each is observed or understood from the third person perspective. The phenomenon of *qualia* is the most frequently considered of the NSFCs, but we recognize three more: *referral, mental unity,* and *mental causation* (table 1.1). It is these features of consciousness that are the most resistant to scientific explanation. In attacking the problem of consciousness, this book will have to explain the neurobiology and evolution of these neuroontologically subjective features.

Table 1.1 The neuroontologically subjective features of consciousness (NSFC).*

Referral
Mental unity
Qualia
Mental causation

*Adapted from table 1 in Feinberg and Mallatt (2016a).

Referral

This feature is the "referral" or "projection" of neural states. No conscious experience refers to the brain itself. That is, no experience is perceived as the firing of the neurons (nerve cells) inside the brain that generate it, nor is one subjectively aware of these firings at all. Instead, the experience is referred to something out in the world or else to someplace on or within the body. Such referral characterizes *exteroceptive* experiences of the world that are projected outward into the world as "mental images," and also *interoceptive* and *affective* experiences of the body that are experienced partly or fully as internal bodily states.

First, consider exteroceptive sensory experiences, which are created by processing stimuli from the external environment. For the special senses of seeing, hearing, tasting, and smelling, Charles Sherrington called their referral *projicience*,[13] which means the projected externalization of sensation away from the body. One does not "feel" a visual stimulus in the eye or an auditory stimulus in the ear, nor does one experience the stimulus as arising in the brain. Instead, these sensations are experienced as if projected out to the source of the stimuli, as if in the outside world that is being viewed or heard.[14]

Some interoceptive feelings, like pain or the stretch of a full stomach, are referred or externalized away *from* the neurons in the brain *to* some other place in or on the body. Positive or negative affective states (feelings) are also referred away from the brain but not into the external world, as occurs with information picked up by the distance receptors. Affects are therefore experienced entirely within the self (being happy or unhappy, for example), but not as if within the brain. In other words, when I am sad, I do not feel like my brain's neural circuits are sad.

Mental unity

Though the nervous system objectively consists of billions of different neurons, consciousness is subjectively experienced as a unified field, the central stage of our consciousness. This is called the "grain argument," meaning that while the brain objectively appears like grains of sand, consciousness is subjectively experienced like the whole beach.[15] A classic example of the puzzle of subjective mental unity is "cyclopean perception" (figure 1.2): How is it that we experience a unified visual world that appears to emanate from a single visual point of view but is based on separate information coming from two eyes? Thus, objective characterization says the brain matter is divisible into parts and extended, while consciousness is normally unified into one

The Mystery of Subjectivity

Figure 1.2
The concept of mental unity, as shown by the phenomenon of cyclopean perception (cyclopean = as if with one eye). What we consciously see is unified as if from one central eye, although it actually comes from the information of our two separate eyes.

central experience.[16] How can we resolve the contradiction and bridge the gap?

Qualia

As explained above, qualia are the subjectively experienced felt qualities of sensory consciousness, such as a perceived color, sound, or smell, or a negative affect. Many investigators consider qualia to be the central puzzle of consciousness. Francis Crick and Christof Koch state:

> The most difficult aspect of consciousness is the so-called "hard problem" of qualia—the redness of red, the painfulness of pain, and so on. No one has produced any plausible explanation as to how the experience of the redness of red could arise from the actions of the brain. It appears fruitless to approach this problem head-on.[17]

In essence, neurons as objectively characterized are not "red," "painful," or "stinky," but the subjective states associated with these neurons are. This is, to most philosophers and neuroscientists, the most perplexing gap between subjective experience and the brain.

Mental causation

The problem of mental causation, according to philosopher Jaegwon Kim, is explaining "how it is possible for the mental to exercise causal influences in the physical world"; or, according to Revonsuo, how is it that "the mind or mental phenomena have causal powers to change some purely material (e.g., biological or neural) process in the brain"?[18] How can consciousness—an intangible, unobservable, and fully subjective entity—cause material neurons to direct behaviors that change the world? How can an immaterial process build the Great Wall of China or the Panama Canal?[19]

In summary, the four NSFCs all fall into the gap between sensory consciousness as experienced and sensory consciousness as objectively observed. In this book, we will show how to demystify these gaps, but this requires bringing in the other two approaches (see figure 1.1), which take into account the unique neurobiology of the brain as well as the early evolution of consciousness.

The Neurobiological Approach: The Gap between the Brain and Sensory Consciousness

Many writers from diverse fields have long suspected that there is something different about the biology of consciousness when compared to other domains of biology. They all see a gap between the biology of the brain and sensory consciousness, and seek a way to close this gap.

Two major explanatory approaches have been used to try to understand subjective experience in terms of biological neural functioning: *reduction* and *emergence*. First, let us consider reduction. While there are several types of scientific reduction, *ontological reduction* is the most pertinent for investigating consciousness.[20] Searle describes an ontological reduction between two entities as a "nothing but" relationship:

> The most important form of reduction is ontological reduction. It is the form in which objects of certain types can be shown to consist in nothing but objects of other types. For example, chairs are shown to be nothing but collections of molecules. This form is clearly important in the history of science. For example, material objects in general can be shown to be nothing but collections of molecules, genes can be shown to consist in nothing but DNA molecules.[21]

To clarify this quotation, Searle actually means "nothing but molecules *plus their fields of force*," which therefore includes the *interactions* between the parts.[22]

In biology, the most common form of ontological reduction is explaining how wholes are reduced to their parts, their functions, and their interactions. The digestive organs (stomach, intestines, etc.) and their interactions that break down food, taken together, explain how digestion occurs, and we can readily reduce digestion to its parts and their interactions. Similarly, as noted by Searle in the above example, the discovery by Watson and Crick of the microscopic structure of DNA led to the nonmysterious elucidation of the great role this molecule plays in heredity and life.

In neurobiology, we usually start with a macroscopic ("big picture"), definable property (such as sensory or movement processes, epilepsy, memory, or paralysis after brain injury) and then try to explain it based on more fundamental, known properties. This reductionist approach often works well. For instance, there is no ontological "mystery" of epilepsy or paralysis; we have discovered how an epileptic seizure is caused by abnormal electrical discharges in the brain[23] and how cutting nerve fibers interrupts the signals necessary for movement of muscles. But even though biologists do not yet fully understand how all the chemical reactions or electrical signals of life operate, they do not encounter—nor do they anticipate—any reductive gaps between the macroscopic biology of the nervous system and its more microscopic physiological substrate. Only consciousness seems to resist reduction and raise explanatory gaps.

Emergence is a second possible approach to explaining consciousness. If consciousness cannot be simply reduced to the brain, perhaps it is an "emergent feature" of the brain. Emergence theories come in *weak* and *strong* or *radical* varieties, which differ in how they treat reduction. The weak version of emergence simply says that complex systems have higher-order properties that are new relative to the component parts and processes that create them— yet the new properties are still explained by (reducible to) these components. A well-known example of a weakly emergent property is the fluidity of water. Water, as an aggregate of water molecules, is fluid, but its individual molecules are not. This presents no ontological mystery because science can (or should be able to) explain all the principles of water molecules and their interactions to reveal fluid dynamics. The biological processes of muscular contraction, hormonal regulation, blood circulation, and so on are all weakly emergent in that they can be explained by the component parts and their interactions. Some weakly emergent features are very complex and cannot be explained exactly at present (for example, in so-called chaotic systems such as the weather). This bothers some philosophers, but scientists are confident that these systems have natural causes that can be explained in principle with further knowledge.[24]

Unlike this weak version, the strong form of emergence theory says that complex systems may have emergent properties that cannot *in principle* ever be reduced to the system's constituent parts. If consciousness were a radically emergent process, then dissecting it would always leave something unexplained, thus creating the hard problem.

The philosophical obstacles posed by ontological subjectivity have led some philosophers and scientists to take this more radical path. They say consciousness is a radically emergent feature of the brain that can never be reduced to the brain or derived from the laws of physics as science now understands these laws. Nobel laureate Roger Sperry was an advocate of this view. He claimed that the "mysterious" features of consciousness are radically emergent nonmaterial properties of the brain and that the mind is more than the sum of the brain's material parts and physical processes: "Conscious phenomena are different from, more than, and not reducible to, neural events."[25]

Sperry is not alone in his suspicion that we will never in principle be able to reduce consciousness fully to the brain. The physicist Erwin Schrödinger, one of the early innovators of quantum mechanics, also said as much:

> The sensation of colour cannot be accounted for by the physicist's objective picture of light-waves. Could the physiologist account for it, if he had fuller knowledge than he has of the processes in the retina and the nervous processes set up by them in the optical nerve bundles and in the brain? I do not think so.[26]

Chalmers endorses a similar view, saying that we may as well just consider consciousness a fundamental feature of the world and leave it at that:

> I suggest that a theory of consciousness should take experience as fundamental. We know that a theory of consciousness requires the addition of *something* fundamental to our ontology, as everything in physical theory is compatible with the absence of consciousness. We might add some entirely new nonphysical feature from which experience can be derived, but it is hard to see what such a feature would be like. More likely, we will take experience itself as a fundamental feature of the world, alongside mass, charge, and space-time. If we take experience as fundamental, then we can go about the business of constructing a theory of experience.[27]

The position we will endeavor to prove in this book is that consciousness is explained by naturally occurring but unique biological features and that positing any new "fundamental" or "radically emergent" features of the brain is unnecessary.

The Neuroevolutionary Approach: How Old Is Sensory Consciousness, How Did It Evolve, and Which Animals Have It?

So far we have identified two major approaches to the problem of sensory consciousness, the philosophical and the neurobiological, and both appear to face insurmountable roadblocks and unbridgeable "explanatory gaps" when trying to solve the hard problem of ontological subjectivity. A third avenue is the neuroevolutionary approach, which can be divided into the paleobiological domain of fossil animals and deep time, and the modern domain of living animals (see figure 1.1). Some thinkers have started to include evolution[28] in their models of consciousness,[29] mostly by using humans and other mammals as their point of reference. Throughout the book, we will build on this neuroevolutionary foundation by adding much more hard data from the literature.[30] These data focus less on mammals and more on other animals, including important new findings on the early evolution of animals with brains, on the comparative biology of brains and sensory systems across the animal phyla, and on the developmental genetics of nervous systems.[31] These findings let us peer even further back in evolutionary history and take a closer look at the very early origins of sensory consciousness, to the time when neurobiological innovations spurred the transition from the nonconscious to the conscious. This goldmine of data provides a whole new perspective on the problem.

In this book, along with trying to resolve the philosophical gap between first-person subjective experiences and third-person explanations, or the neurobiological gap between subjective experience and the brain, we approach the mystery of consciousness from another direction: we search for *the evolutionary origins of the gaps themselves*. In essence, we use the origin of the hard problem as a "marker" for the origins of sensory consciousness. We reason that if we can explain the evolutionary origins and neurobiological basis of sensory consciousness at the ancient date when the hard problem first appeared—and can do it with conventional biological principles—then we can find a way to resolve the philosophical mystery of consciousness.

The paleobiological domain: How ancient is consciousness?

We do not know when consciousness first evolved on Earth. Is it relatively recent or was it a primordial feature of the first nervous systems? Where did it come from? Consciousness surely did not leap onto the world stage out of nowhere. At some point in the history of life, there existed only simpler animals that did not have the neural apparatus to produce sensory consciousness. And at some later time, animals evolved that were capable of integrating the nuts and bolts of their working neurons into a fabric of experience. What—or maybe we should say "who"—were the first animals to create an inner world of subjective experience?

Of course, how one judges which animals are conscious will depend on one's approach and the criteria used. Some would say that only the "smartest" living species of animals are conscious, which would mean consciousness is of relatively recent origin. That is, certain theories claim that a large cognitive capacity is the best measure of the presence of consciousness.[32] Using such indices of animal intelligence has the advantage that they can be measured objectively and make cross-species comparisons relatively easy. However, they have the disadvantage that they do not directly address the "hard problem" issues of basic consciousness, including the nature and origin of subjectivity and qualia. Further, while it has been proposed that complex cognition and consciousness roughly correlate in evolution with advancing brain complexity, there is no a priori reason why this must be so. For instance, while humans as a species have higher intelligence than other primates, we have no reason to assume that a human "sees" a tree more acutely or in some way differently vis-à-vis primary consciousness than an orangutan does. A human baby, its intelligence not yet developed, has conscious sensory experiences very much like an adult's.[33] Also, cognitive measures, such as the ability to reason, communicate, categorize, learn and remember, do not address the nature and origin of the *affective* qualities of consciousness, such as the unpleasantness of being cold, in pain, or hungry—all important to the question of the nature of qualia. We want to know when and how this feeling aspect of consciousness evolved as well.

In this book we develop the idea that consciousness has been around a long, long time, far longer than many have supposed. Figure 1.3 shows a timeline of the periods of Earth history since life first evolved billions of years ago. We will build the case that 560 to 520 million years ago (mya) was a critical period in the evolution of the first vertebrate animals and conscious brains. We will even give evidence, tentatively, that some *invertebrates* evolved consciousness at this time. This idea, we believe, cuts the Gordian knot at least in the vertebrate line and allows one to uncover the most basic neural architecture of sensory consciousness (e.g., of qualia) and reconcile it with the philosophy of ontological subjectivity and the "hard problem." If this can be accomplished, it would be the first time anyone has dated, explained the architecture, and found the most basal forms of consciousness on Earth all in one fell swoop.

The Mystery of Subjectivity

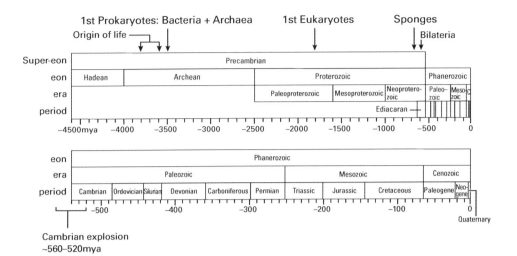

Figure 1.3
A timeline of the history of life on Earth, based on fossil evidence and dating of rocks through rates of radioactive decay. The bottom line is an enlargement of the right end of the top line. The time of the Cambrian explosion (bottom left) is especially important to our ideas. mya = million years ago.

The domain of living animals: What animals possess sensory consciousness?
The other part of our neuroevolutionary approach is the *domain of living animals* (see figure 1.1), which asks, what are the most basal *living* descendants of those first conscious pioneers from so long ago? What animals on Earth today have crossed the evolutionary Rubicon into sensory consciousness? For some organisms, for instance those without a brain at all, it seems obvious, to most biologists at any rate, that there cannot be anything that "it is like" to be that creature. The single-celled amoeba responds to its environment, approaching chemical gradients in its immediate vicinity that come from a food source, and withdrawing from potentially damaging noxious chemicals that contact its cell membrane. The amoeba can even respond to a light source, and it reacts differentially to hot and cold temperatures. But it has no nervous system where complex information processing could occur, no neural substrate where an integrated subjective experience could be created.

So unless one believes that no organisms on Earth are conscious except humans—a position that many, many researchers have come to deny in recent years—then somewhere between the brainless and mindless amoeba and the brainy and conscious human are other animals with conscious minds or sentience. But which ones are these? Most everyone now attributes consciousness to chimpanzees, but are there others? All mammals and birds? Invertebrates with complex behaviors, such as octopuses and honeybees?[34] To approach these questions, we will reframe them as "What type of neuronal organization and how much complexity must a nervous system have in order to have subjectivity?"

Searching for a Theory of Neurobiological Naturalism

Skeptical of exotic or new fundamental causes, we start from Searle's theory of biological naturalism, in which he proposed that consciousness results from purely natural biological principles. He begins his book *The Rediscovery of the Mind* with this explanation and definition:

> The famous mind–body problem, the source of so much controversy over the past two millennia, has a simple solution. This solution has been available to any educated person since serious work began on the brain nearly a century ago, and, in a sense, we all know it to be true. Here it is: Mental phenomena are caused by neurophysiological processes in the brain and are themselves features of the brain. To distinguish this view from the many others in the field, I call it "biological naturalism." Mental events and processes are as much part of our biological natural history as digestion, mitosis, meiosis, or enzyme secretion.[35]

We tend to agree with Searle that mind is based on matter and physical processes as we normally understand them, so that a theory of the "radical" emergence of consciousness from the brain is incorrect. However, a critical neuroscientific limitation of his theory of biological naturalism remains: namely, *exactly why* can't we reduce subjective experience to objective neural processes in the same way as other biological processes can be reduced to their constituent parts and their interactions? This limitation is the greatest challenge to any theory that tries to naturalize consciousness. Indeed, were it not for this unexplained reduction barrier, there would be no difficult mind–body problem, and no hard problem.

To solve these problems, we offer a theory of *neurobiological naturalism*.[36] This theory claims that while Searle is correct in saying "mental events and processes are as much part of our biological natural history as digestion, mitosis, meiosis, or enzyme secretion," there are in fact some crucial differences between those biological processes and the unique neural processes that lead to consciousness; and that these very differences both generate experiences and provide the solution to the hard problem of consciousness. Thus, although we believe that Searle's theory of biological naturalism is fundamentally sound, it is nonetheless missing some critical details that are required for a unified scientific theory of consciousness and subjectivity.

Our theory will reveal that the reason why the hard problem is so hard, ontological subjectivity is so confounding, and the mind–body problem persists to this day is that the solution requires integrating philosophy, neurobiology, and neuroevolution (see figure 1.1). Each of these overlapping and interconnected domains provides specific and unique answers to the mysteries of consciousness, together producing a single solution to the problem.

Our strategy for presenting the three domains and tackling the hard problem is as follows. Chapter 2 starts to link the neurobiological to the philosophical. It lays out the unique features of complex nervous systems that could lead to consciousness by filling the gap between simple, nonconscious reflexes and conscious brains. Chapters 3 through 8 explore and date the evolution of exteroceptive, interoceptive, and affective consciousness in the vertebrate animals (fish, amphibians, reptiles, birds, and mammals), thereby integrating the evolutionary and neurobiological domains. At this stage, we will have assembled numerous criteria for recognizing consciousness, so chapter 9 uses these criteria to see if any *invertebrate* animals are conscious. In chapter 10, informed by the criteria for, distribution of, and dates of consciousness in animals, we unite all three domains (see figure 1.1) to upgrade our theory of neurobiological naturalism and offer a solution to the philosophical hard problem. What distinguishes this study from previous investigations of consciousness is that it thoroughly addresses the philosophical problems through a detailed and modern treatment of evolutionary neurobiology.

2 The General Biological and Special Neurobiological Features of Conscious Animals

Critical to our theory of neurobiological naturalism is the idea that consciousness emerged in steps or stages, with the features of lower stages still represented in the stages that evolved later.[1] Based on the timing of their emergence in evolution from earlier to later, we divide the critical biological features on the road to consciousness into three categories (table 2.1).

The first are *general biological features* that apply to all living things. These are necessary precursors to consciousness, but none of them taken separately or collectively can account for consciousness. The second is the level of *reflexes*, which exist in all animals that have nervous systems. Reflexes occur in animals who have the general biological features plus the added dimensions that nervous systems bring to an animal's life, but they operate without creating sensory consciousness. Finally, from these basal neural systems emerged the *special neurobiological features*, the properties of consciousness alone. We propose that although consciousness is not "radically" different from other emergent biological properties, nonetheless the special neurobiological features—in association with the general features and reflexes—provide the advanced and unique properties that make consciousness possible. The special features are the missing ingredients that evolved from and built on the general neurobiological features.

Level 1: General Features of All Living Things

Life, Embodiment, and Process
Life
It is a fact that the only things we know are conscious are alive, and however and whenever consciousness begins, it depends on life and thus ends with the death of the brain. That said, even though explaining life in scientific terms poses no mysterious obstacles for scientists or philosophers, defining what

Table 2.1 The defining features of consciousness*

Level 1: General biological features that apply to all living things
- *Life: embodiment and process*
- *System and self-organization*
- *Hierarchy, emergence, and constraint*
- *Teleonomy and adaptation*

Level 2: Reflexes that apply to animals with nervous systems
- *Rates and connectivity*

Level 3: Special neurobiological features that apply to animals with sensory consciousness. (For more, see table 9.2.)
- *Complex neural hierarchies; a brain*
- *Nested and non-nested hierarchical functions*
- *Neural hierarchies create isomorphic representations and mental images and/or affective states*
- *Neural hierarchies create unique neural-neural interactions*
- *Attention*
- *Sensory consciousness may be created by diverse neural architectures*

*An earlier and simpler version of this table appears in Feinberg and Mallatt (2016a).

constitutes "life" is less clear. Evolutionary biologist Ernst Mayr summarizes the problem in his wonderful book *The Growth of Biological Thought*:

> Attempts have been made again and again to define "life." These endeavors are rather futile since it is now quite clear that there is no special substance, object, or force that can be identified with life. The process of living, however, can be identified. There is no doubt that living organisms possess certain attributes that are not or not in the same manner found in inanimate objects. Different authors have stressed different characteristics, but I have been unable to find in the literature an adequate listing of such features.[2]

Although the task is a difficult one, defining life may not be as difficult as Mayr said. Scientists do agree that all cells and organisms made of cells are alive, and that the simplest known life forms are bacteria and the bacteria-like *archaeans*. And all living, cellular organisms share the distinct traits of using energy to sustain and organize themselves, responding, reproducing, having genes, and evolving by natural selection. All are chemically complex enough to have emergent features. In addition, all cells have a membrane that separates them from the environment outside, this being the environment from which they get their nutrients and the other molecules needed for their life processes.[3] With a membranous boundary enclosing it, even the tiniest cell is *embodied*.

Embodiment

Embodiment means that a living thing is a distinct entity with an interior, and it is separate from the external surroundings. To illustrate this concept, we will show how embodiment became more complex and elaborate over time (figure 2.1). Earth's oldest fossils reveal that the earliest known living organisms were indeed the one-celled bacteria and archaeans, which lived in the oceans about 3.5 billion years ago.[4] Later, the more-complex *eukaryote* cells evolved,[5] when a large archaean engulfed a type of bacteria that is highly efficient at producing energy. The resident bacteria survived and became the power plants (mitochondria) in the host cell. This energy boost allowed the new, composite eukaryote cell to evolve many more functions,[6] so obtaining mitochondria is a good example of an *emergent property* of life. The early eukaryotes gave rise to many kinds of single-celled organisms such as amoebas and the *choanoflagellates* in figure 2.1, and later to the many-celled organisms with bodies of greater complexity. By amassing, the first simple groups of eukaryote cells illustrated a general tendency for life to increase in size (as protection against predation). Although the colonies began as loose clusters of identical cells, they were the precursors of the real animals and plants whose different cells interact and specialize for different functions for efficiency. The animals included the sponges and our own wormy ancestors. This is how complex multicellular organisms with *real bodies* emerged, so that embodiment reached its highest level.

The concept of embodiment also has important philosophical implications for our understanding of consciousness. Both life in general and consciousness in particular are embodied, meaning that they depend on a material body and a material brain, respectively, for their existence. Because they can only occur in a body with borders, all the features of life and consciousness are internal and "ontologically individual." That is, the individual life of an organism is specific to that particular organism. Evan Thompson argues that this contrasts with the traditional Cartesian way to think about consciousness as "mind versus matter" dualism, because it forces us to change our idea of what is internal versus what is external:

My point is rather that to make headway on the problem of consciousness we need to go beyond the dualistic concepts of consciousness and life in standard formulations of the hard problem. In particular, we need to go beyond the idea that life is simply an "external" phenomenon in the usual materialist sense. Contrary to both dualism and materialism, life or living being is already beyond the gap between "internal" and "external." A purely external or outside view of structure and function is inadequate for life. A living being is not sheer exteriority (*partes extra partes*) but instead embodies a kind of interiority, that of its own immanent purposiveness.[7]

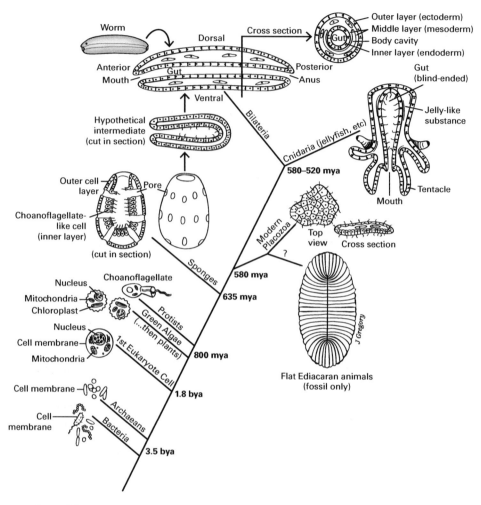

Figure 2.1

How embodiment and "bodies" became more elaborate during the early evolution of life on Earth. Going up the tree shows how small cells evolved into more complex cells, which joined to become the multicellular animals. The Bilateria worm at top is our own ancestor. bya = billion years ago. mya = million years ago.

Process

Second, life and consciousness belong together because they are both embodied *processes*. They are not simply structures or things. Here is how Mayr said it:

As far as the words "life" and "mind" are concerned, they merely refer to reifications of activities and have no separate existence as entities. "Mind" refers not to an object but to mental activity and since mental activities occur throughout much of the animal kingdom (depending on how you define "mental"), one can say that mind occurs whenever organisms are found that can be shown to have mental processes. Life, likewise, is simply the reification of the processes of living. Criteria for living can be stated and adopted, but there is no such thing as an independent "life" in a living organism. The danger is too great that a separate existence is as assigned to such "life" analogous to that of a soul. ... The avoidance of nouns that are nothing but reifications of processes greatly facilitates the analysis of the phenomena that are characteristic for biology.[8]

In fact, William James, many years before, already realized that consciousness is a process, not an object:

To deny plumply that "consciousness" exists seems so absurd on the face of it—for undeniably "thoughts" do exist—that I fear some readers will follow me no farther. Let me then immediately explain that I mean only to deny that the word stands for an entity, but to insist most emphatically that it does stand for a function. There is, I mean, no aboriginal stuff or quality of being contrasted with that of which material objects are made, out of which our thoughts of them are made; but there is a function in experience which thoughts perform, and for the performance of which this quality of being is invoked. That function knows. "Consciousness" is supposed necessary to explain the fact that things not only are, but get reported, are known. Whoever blots out the notion of consciousness from his list of first principles must still provide in some way for that function's being carried on.[9]

What is essential here is that consciousness, like all known life, is physiologically an embodied process. Consciousness is part of what a living brain *does*. All the general features of biology we now consider, and all the special features of conscious brains, are functional features of particular embodiments.

System and Self-Organization

Living organisms are self-organizing systems. As stated by Camazine and colleagues:

In biological systems self-organization is a process in which pattern at the global level of a system emerges solely from numerous interactions among the lower-level components of the system. ... In short, the pattern is an emergent property of the system, rather than a property imposed on the system by an external ordering influence.[10]

Indeed, self-organization is a characteristic of all *complex adaptive systems*.[11] Self-organization therefore must be involved in the emergence of consciousness, for which it is a necessary although insufficient condition.

Hierarchy, Emergence, and Constraint

Complex biological systems have a hierarchical structure, becoming more complex toward the top of the hierarchy (figure 2.2).[12] In these living hierarchical systems, most evolutionary change tends to occur at the higher levels. For example, the genetic code of DNA is the same in virtually all organisms, the same organelles occur in the cells of all animals, but the kinds of tissues and organs vary more across the animal phyla. Then at the top of the hierarchy, the *bodies* of animals differ markedly, from worms to shrimps to iguanas. We have proposed a reason for this tendency: "Although natural selection acts on every level of the hierarchy it is strongest at the top level, the individual organism, which has the most direct interaction with the challenges and changes of particular external environments."[13] Lower levels, on the other hand, are more buffered within the body.

The interrelated concepts of *emergence* and *constraint* apply to all hierarchical systems (figure 2.2). Emergence was briefly introduced in chapter 1, and will be considered more fully here. The close relationship between emergent features and hierarchical structure was explained by Jaegwon Kim, among others:[14]

Although the fundamental entities of this world and their properties are material, when material processes reach a certain level of complexity, genuinely novel and unpredictable properties emerge, and ... this process of emergence is cumulative, generating a hierarchy of increasingly more complex novel properties. Thus, emergentism

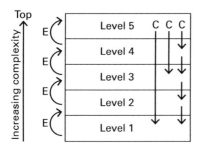

Figure 2.2
A biological hierarchy drawn as a stack of rectangles, showing the concepts of different levels, emergence and constraint. E = emergent features of increasing complexity. C = constraints imposed by higher levels on lower levels.

presents the world not only as an evolutionary process but also as a *layered structure*—a hierarchically organized system of levels of properties, each level emergent from and dependent on the one below.[15]

The favored example of emergence is that our body's most complex physiological processes emerge from its cellular and molecular interactions. Other common examples have more of a time dimension: single-cell organisms evolving into the multicellular animals and plants (see figure 2.1)[16] and a fertilized egg developing into a fully formed body via an embryo, then a fetus, then the newborn. Then there is the evolutionary elaboration of the vertebrate forebrain.[17] In this example, new, large and complex forebrain centers evolved in the highest part of the neuraxis (*neuraxis* = the brain atop the spinal cord) in reptiles, birds, mammals, and some fish, from the simpler forebrains of their fish and amphibian ancestors. The best-known instance of this is the large cerebral cortex of humans and other mammals.[18]

Whereas emergence is when the parts of a hierarchy produce new properties at the higher levels, constraint is when the upper levels exert control over the lower levels.[19] According to the biologist Howard H. Pattee, hierarchical constraint is essential for the emergence of complex biological systems:

If there is to be any theory of general biology, it must explain the origin and operation (including the reliability and persistence) of the hierarchical constraints which harness matter to perform coherent functions. This is not just the problem of why certain amino acids are strung together to catalyze a specific reaction. The problem is universal and characteristic of all living matter. It occurs at every level of biological organization, from the molecule to the brain. It is the central problem of the origin of life, when aggregations of matter obeying only elementary physical laws first began to constrain individual molecules to a functional, collective behavior. It is the central problem of development where collections of cells control the growth or genetic expression of individual cells. It is the central problem of biological evolution in which groups of cells form larger and larger organizations by generating hierarchical constraints on subgroups. It is the central problem of the brain where there appears to be an unlimited possibility for new hierarchical levels of description. These are all problems of hierarchical organization. Theoretical biology must face this problem as fundamental, since hierarchical control is the essential and distinguishing characteristic of life.[20]

In a living system such as the human body, the cells constrain their subunits (organelles) to work together, the tissues and organs constrain their cells to cooperate, and the entire body constrains its organs to team up, all to perform the many physiological functions needed for the body to survive. If the constraints were to fail at any level, the body would disassemble and die.

This is not to imply that the higher levels always fully and rigidly control the lower levels. In many hierarchies, the lower levels interact to perform important functions with a minimum of top-down influence. For example, cells in the pancreas that produce insulin successfully secrete this hormone into nearby blood vessels without much influence from the rest of the pancreas. Also, the nature of its lower levels limits what a hierarchical system can achieve. For instance, our bodies and brains are made mostly of water molecules and soft elements, so they cannot freeze solid, boil, or withstand the impact of cannon balls and survive. Below, we will see how *neural* hierarchies have special features that are critical for the creation of consciousness.

Teleonomy and Adaptation
Biological structures have functional roles. This is obvious, but it is remarkably easy to go too far and say that these structures were "designed" for their roles, as in "the limbs evolved for locomotion" and "the spleen evolved to clean the blood of infection." Although biologists sometimes talk this way for brevity, they do not literally mean it. They realize that accepting a "predesigned" purpose is *teleological thinking*—invoking the end result of a process as the cause of that process, which is a logical impossibility and a big error. Evolution adapts to the now, and though it can proceed step by step over time, it seeks no future goal.

There is a way to avoid the trap of teleological thinking. Biological structures do have adaptive functions—a tooth's function is to chew, for example. To recognize this kind of purpose without any implication of planning or design, the biologist Jacques Monod introduced the term *teleonomy*. Sometimes we can and will use "teleonomic" as a synonym for "adaptive."[21]

From the start, we will consider sensory consciousness, like many other biological processes, to be adaptive and beneficial to the survival of the animals that have it. We will give all the reasons for this in chapter 10, after we have explored animal consciousness more thoroughly. But a good initial argument that consciousness is an adaptation is that of Shaun Nichols and Todd Grantham. They point out that consciousness is built on the great *structural complexity* of the brain, and that "according to evolutionary theory, the structural complexity of a given organ can provide evidence that the organ is an adaptation, even if nothing is known about the causal role of the organ."[22] Such complexity implies adaptation because complexity is expensive to evolve and maintain, a waste of energy, and selected against if not of value. Nichols and Grantham demonstrate that consciousness is *structurally* complex because its neural circuits combine many diverse types of sensory input into one experience, on the central stage or unitary scene of consciousness.

Level 2: Reflexes, Which Occur in All Animals with Nervous Systems

Rates and Connectivity

Reflexes are fast, automatic responses to sensory stimuli. Although reflexes are not conscious, their neuronal chains are the ingredients from which evolved the complex circuits of consciousness, and they bridge the gap between the general and special features that create consciousness (table 2.1, Level 2). The simplest and fastest type of reflex is *monosynaptic*, in which just two neurons join at a single communicating synapse. An example of this is the knee-jerk reflex that helps us maintain our balance (figure 2.3A). Being involuntary, it operates even if a person is in a coma. In addition, more complex, multineuron reflexes exist and still operate without consciousness. These are also called polysynaptic or multisynaptic. An example is the pupillary light reflex, whereby light shone in an eye causes the pupil to get smaller so that the eye's retina is not hurt by strong light (figure 2.3B).

Over evolutionary time, some of the simple reflexes elaborated into increasingly complex reflexes and then into long neuronal hierarchies. This was done by adding more neurons, like adding links to the center of a chain (but with lots of cross-talk between the links). It yielded more hierarchical

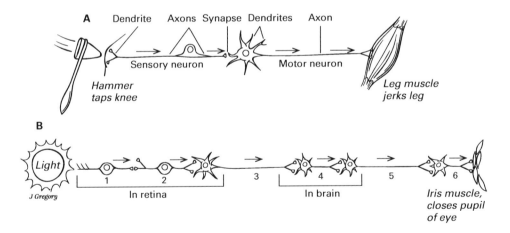

Figure 2.3
Reflexes, simple and complex. A. Simple knee-jerk reflex with a chain of just two neurons (nerve cells). B. Complex pupillary-light reflex, with a chain of more neurons: six neurons are present, in which the first is the sensory neuron and the sixth is the motor neuron. This figure also previews the anatomy of neurons with their processes, the dendrites and axon.

levels of information processing. And the advances were not all in *sensory* processing: in some of the hierarchies their *motor*-control arm evolved into basic motor programs and central pattern generators (CPGs) for the repetitive and rhythmic actions of locomotion, feeding, and breathing.[23] Simple sensory cues and basic sensory feedback influenced these stereotyped and nonconscious actions of the body. Complex reflexes, expanding neural hierarchies, and basic motor programs all had to be in place before consciousness evolved.

Therefore, reflexes were the vital "royal road" to the evolution of consciousness. Contributing to this is their speed. Some kinds of living hierarchical systems, such as a saguaro cactus growing for centuries in the desert, operate slowly, but others, such as nervous systems, operate much faster. Reflexes communicate at the fast rates necessary for the later evolution of consciousness. Also essential was the tight (synaptic) connectivity between neurons, a connectivity that first evolved for reflexes. Tight and rapid connections made the neural hierarchy possible, by permitting many electrical signals to flash back and forth throughout this system. All the unique characteristics of consciousness followed from this.

Reflexive responses foreshadow the phenomenon of conscious referral (defined in chapter 1). The blinking reflex automatically shields the eye from dust blowing in the air, and a harmful cut provokes unconscious withdrawal by a worm or a person. In these primitive reflexes, the neural processing is internal (in-body), but the response already involves, or is an adaptive reaction to, the external world.

Level 3: Special Neurobiological Features That Apply to Consciousness

The special neurobiological features (table 2.1) mark the transition from reflexes and basic motor programs to consciousness. They define this junction both literally, as a historic evolutionary event, and conceptually, as when we compare different groups of animals and compare conscious with nonconscious sensory systems. Thus, these special features of consciousness are either completely absent or only exist in primordial form in nonconscious animals and nonconscious (reflexive) neural systems. This critical junction also creates the "explanatory gaps" and the "hard problem."

Complex Neural Hierarchies

Because harm to our *complex* brain interferes with it, consciousness is unlikely to be "fundamental" (p. 10) but instead requires a large amount of neural complexity (p. 24). This seems especially true for complex processing of *sensory* information. Let us turn to Herbert Simon, an originator of complexity theory, for his statement that nature builds complexity with hierarchical systems:

> Complexity frequently takes the form of hierarchy, and ... hierarchic systems have some common properties that are independent of their specific content. Hierarchy, I shall argue, is one of the central structural schemes that the architect of complexity uses.[24]

Hierarchies gain their complexity in several ways: by gaining more levels, by increasing the number and specialization of components in each level, and by more interaction between the levels. From there, new properties emerge. For neural hierarchies specifically, this increase meant evolving more of these hierarchies in the brain, adding more levels of neurons to the existing hierarchies, and adding more of the speedy neuronal processing within and between levels. From this growing complexity emerged more-sophisticated behaviors, greater learning ability,[25] and, we claim, consciousness. Similarly, Giulio Tononi says that the amount of consciousness in a system is the quantity of *integrated information* generated by the system's elements and their interactions beyond the quantity generated by the individual parts of the system. More of this complex information equals more consciousness. Along with Christof Koch, Tononi especially emphasizes that organized interactions and feedback between neuronal centers are important for consciousness, with which we agree.[26]

Also contributing to brain complexity and consciousness is *cellular* diversity. Many structural classes of neurons, with specific roles, exist both across the animal kingdom and within each animal's brain.[27] The trend toward differentiated cell types began at the level of neural reflexes, but it exploded when the first complex brains evolved (see chapters 5 and 6). Neuron types differ in the numbers and branching patterns of their dendrites (information-receiving processes) and their axons (information-sending processes). Neurons also vary by function, that is, by how their axons carry signals—for instance, either *tonically* by firing the whole time the neuron is stimulated, or *phasically* by firing only briefly at the very start of a stimulus.[28]

Especially important for the evolution of neural complexity and sensory consciousness is that all the vertebrate animals have a very wide variety of sensory *receptors* associated with their sensory neurons. Most of these receptors develop from embryonic tissues that are unique to vertebrates, tissues known as *neural crest* and *ectodermal placodes*.[29]

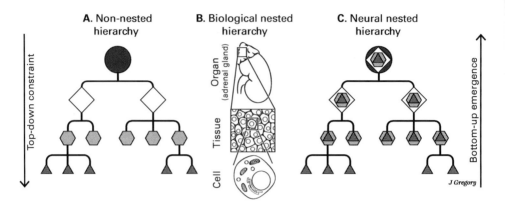

Figure 2.4
Different types of hierarchies. All hierarchies have emergent properties that are created through the interaction of lower and higher levels and display constraint of higher on lower levels. A. In a non-nested hierarchy, the entities at higher levels are physically separate from the entities at lower levels, and there is strong constraint by higher on lower levels; but lower levels can inform upper levels. B. In a nested hierarchy, higher levels are physically composed of lower levels, and there is no central control of the system, resulting in weak constraint of higher on lower levels. C. A *neural* nested hierarchy is a unique biological system that displays features of both non-nested and nested hierarchies.

No other organ system of the body remotely approaches the nervous system in its cellular diversity. The mammal brain has hundreds of kinds of neurons, whereas the muscles, liver, kidney, ovary, and other organs consist of tissues with under a dozen cell types.[30] The great diversity of neurons, which are the communicating units of the nervous system, allows the complex information-processing of consciousness.

Special Nature of Neural Hierarchies: Nested and Non-nested Functions

Exploration of the neurohierarchical basis of consciousness starts with recognizing two basic kinds of hierarchies (figure 2.4), *non-nested* and *nested*.[31] In a non-nested hierarchy (figure 2.4A) the different levels are physically separate, though they communicate with each other and can be physically interconnected. The lower levels *converge* upon higher levels in a "bottom-up" fashion or else the higher levels *control* the lower ones in a "top-down" fashion. Non-nested hierarchies are often shown as having the shape of a pyramid with the highest levels controlling the lowest (figure 2.4A). Think of an army in which information about enemy movements flows from the many soldiers on the ground up through the hierarchy of command until the general at the top gets briefed. Then the general, acting as central commander, uses this information to direct and control the lower ranks of command below.

In a nested hierarchy, on the other hand (figure 2.4B), the lower levels *physically make up* (compose) the higher levels, into which the lower levels are combined to create increasingly complex wholes. For this reason, a nested hierarchy is also called a *compositional* hierarchy. An organ of the body like the adrenal gland is composed of its constituent tissues and cells; and the entire body, which is the highest level, is composed of all its organs. Much more interaction occurs within and between the levels, from which new properties emerge on the way up the hierarchy. This interaction still allows the higher levels to exert influence—and some control—over the lower levels, but in a more holistic manner, with no physical "central command."

The sensory neural hierarchies within the brain are at once both non-nested and nested (figure 2.4C). Physically, they are non-nested because their levels have different locations in the nervous system and brain; each level is not physically contained within the next. A major *function* is also non-nested. That is, non-nested functions emerge via upward flow through a pyramid in a process called "topical convergence." In our visual processing, for example, the low-level neurons have less-differentiated response characteristics (responding, for example, to viewed points or short lines) and they project to higher neurons that have increasingly specific responses. Finally, in the cerebrum's temporal lobe at the top of this pyramid, certain neurons respond to the most highly integrated visual stimuli. Some neurons respond only to a hand, or a specific face, or a specific building like the Leaning Tower of Pisa.[32] To qualify this statement a bit, although these top neurons do receive unified inputs like this, each neuron probably responds to a few, similar objects instead of just one.[33] These specialized neurons at the top of sensory processing streams are whimsically called "grandmother cells" because one of them might only respond to the face of your grandmother, or "pontifical neurons," as if it were the pope being briefed by the college of cardinals at the Vatican. Although in this instance the lower-order neurons are not strictly controlled by the higher-order neurons, the higher-order neurons do sit at the top of the sensory neural pyramid in a physically non-nested manner.

Such non-nested, topical convergence of the "many onto the one" helps to unify the brain's functional properties, but the necessary streamlining means that a lot of specific information is lost along the way (such as the spatial location of the face being viewed). To compensate, a nervous system with a unified awareness must have another aspect that uses and integrates all the sensory information at various hierarchical levels. This is achieved through the nesting of lower levels within higher levels. That is, sensory consciousness also has functional features of a *nested* hierarchy, wherein the upper levels *functionally* contain the lower levels. Consider a person seeing an apple as it falls off a tree. The earliest part of the visual-processing hierarchy (in the retina, etc.) represents the apple as many individual line segments. Then, higher in the hierarchy, in our brain's cerebrum, the apple's outline is constructed by binding the short line segments into longer segments; and small patches of the apple's red color are bound into larger patches of red. Assembled parts are becoming increasingly more complex. Then, still higher in the cerebrum, the apple's stem is bound to—or nested within—the overall outline of the apple, the color is bound to the apple's shape, and these higher-order representations are bound to the apple's movement as it falls.

In the end, a unified, subjective experience of the apple's journey has been built from anatomically dispersed representations[34] without there being any *central place where all the information converges.* In this case, there is nothing analogous to a "grandmother region" or a "pontifical zone." As Daniel Dennett put it, there is no physical "Cartesian Theater" where all the sensory information comes together to create a unified consciousness. The information is integrated because it is *functionally nested* in consciousness.[35]

Thus, any complex nervous system that creates consciousness combines properties of both non-nested and nested hierarchies (figure 2.4C), making it special and perhaps unique in all of Nature. It seems that all conscious things have this dual property, and all things that have it could be conscious. This dual kind of hierarchy is necessary for the creation of a unified consciousness.

Neural Hierarchies Create Isomorphic Representations and Mental Images or Affective States

Another special feature of brains that are known to have complex neural hierarchies and sensory consciousness is the creation of topographic or isomorphic maps. A topographic map means that *spatial* ordering is preserved from the lowest level of the sensory field to the higher levels in the central nervous system (*topo* = spatial map). That is, the same, precise organization of neurons and their signals characterizes multiple levels of the hierarchy, matching the spatial arrangement of the original sensory receptors. A classic topographic map is the *somatotopic* map for the touch-related senses, which is located in the cerebral cortex (parietal lobe) of mammals. In this map, the skin and the rest of the body surface are represented point by point, although in a distorted way (figure 2.5A). Especially large are the representations of the most

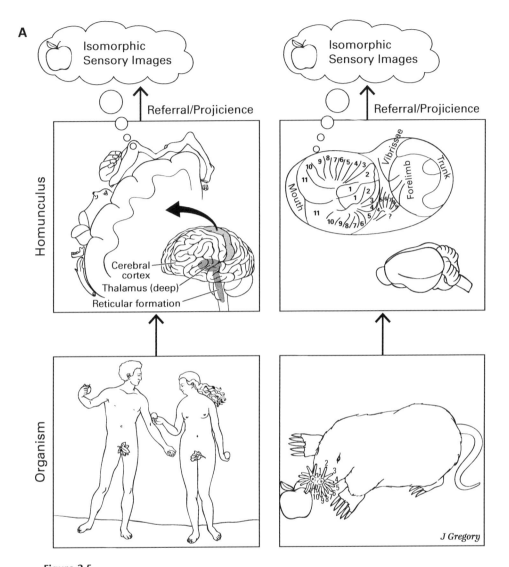

Figure 2.5

Isomorphism, affects, and two kinds of consciousness. A. Isomorphic consciousness, somatotopically mapped, in which sensory images are experienced in the brain's cerebral cortex and referred outward to an apple, according to the body part touched by that apple. *Homunculus* means "little man," with the cortical map of the whole body resembling a distorted little person. The animal at right is a star-nosed mole whose snout tentacles are highly sensitive to touch. The numbered map in the mole's brain, above, reflects the somatotopic arrangement of the tentacles (Catania & Kaas, 1996). B. Affective consciousness, without isomorphic images. Here it involves the affective feeling of shame. Part B is adapted from a famous painting by Masaccio from 1425: *The Expulsion from the Garden of Eden.*

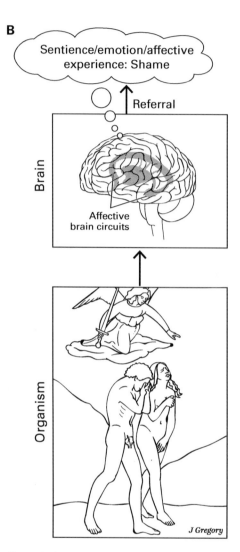

Figure 2.5 (continued)

touch-sensitive parts of the body—the face and hand—where the most sensory processing occurs. Another example of topography is the *retinotopic* map in the back of mammals' cerebral cortex (occipital lobe), where the visual field is mapped point by point.[36] The vertebrate brain has many topographical maps, with some of the sensory maps encoding aspects of the external environment and others representing aspects of the body.

However, not all these brain maps reflect the locations of stimuli in physical space. The smell map is instead arranged by odor combinations and the hearing map by tones.[37] For such senses without an obvious topography of input, we will use the term "isomorphic" alone. But ultimately the hearing and smell maps are based on the spatial arrangements of their receptors, in the ear's receptor organ (the cochlea) and in the brain's olfactory bulb, respectively. So they are still mainly topographical. Still, we prefer to use "isomorphic" as our general term for mapping.

Isomorphic neural representations are important because they relate to sensory consciousness. They are physical and biological entities, but conscious organisms can experience these maps as *sensory mental images* in mapped-out format. In short, objective *representations* mirror subjective *images* at the heart of the brain–mind divide. We use this term, "sensory mental image," to describe the isomorphic aspect of consciousness that directly and immediately arises from the brain's complex processing of sensory information.

Many other authors have also emphasized isomorphic maps as important for sensory consciousness.[38] To start, Sherrington said that the evolution of maps of the world, as created by the distance receptors, was critical to the ancient origin of the brain. In fact, he rather boldly claimed that "the brain is always the part of the nervous system which is constructed upon and evolved upon the 'distance receptor' organs."[39] Gerald Edelman made sensory maps and mental images a critical feature of what he called primary consciousness, concluding with: "Primary consciousness is the state of being mentally aware of things in the world—of having mental images in the present."[40] Antonio Damasio discusses how "mapped neural patterns" can result in "mental images" that are a part of what he calls "core consciousness," which is fairly similar to Edelman's "primary consciousness."[41] We take it as axiomatic in this book that mental images are a part of primary, sensory consciousness and therefore contribute to "something it is like to be."

But isomorphism is not in itself sufficient for consciousness, nor is it a part of every kind of sensory consciousness. We know this because there are "maps" in the lower levels of the sensory pathways—in the retina, cochlea of the ear, spinal cord, and so on—that do not in and of themselves create unified, conscious images. At the other end, as mapped information moves to the highest levels of the sensory processing paths, its isomorphic organization becomes increasingly indistinct so that it grows difficult even to find where the "map" is situated in the brain.[42] The point here is that advanced sensory consciousness, such as when we deeply reflect on what we have sensed, is not so obviously isomorphic.

Finally, *affective* consciousness—emotionally positive and negative feelings—has its own brain circuits, it does not require isomorphic mapping, and it may be experienced as mental states rather than mental images (figure 2.5B; chapters 7 and 8). Thus, isomorphic maps are only one part of the creation and evolution of subjectivity and "something it is like to be"; many other special and general features (table 2.1) are required to create sensory consciousness and ontological subjectivity.

Neural Hierarchies Create Unique Neural-Neural Interactions
Many theories see widespread and extensive neural *interactions* as the key to consciousness. These interactions are called "reciprocal," "recurrent," or "reentrant." They involve both communication within local neural circuits and extensive feedback between distant neuronal centers; and they are organized and integrative. They are considered essential for many conscious features including unity, qualia, and attending selectively to certain qualia. For example, Crick and Koch[43] said unified consciousness arises from a "dominant coalition" or coordinated "assembly" of specific, connected, and interacting neurons. They and many other investigators assign this patterned cross-talk of consciousness only to the highest brain regions of mammals (and perhaps birds), specifically between different regions of the mammalian cerebral cortex, or between the cortex and another forebrain region called the thalamus (pictured in figure 2.5A, the upper left square). While we heartily agree that complex neural interactions are essential to consciousness, we see this cortex- and thalamus-centered view as too restrictive. Extensive neural-neural interactions are known to occur in many other regions of the central nervous systems of vertebrates and invertebrates, and some of these regions could be associated with consciousness (chapters 5–9). In chapter 10 we will show how these unique interactions within and between complex chains of directly linked, hierarchically arranged, sensory class-specific neurons could create the unique aspects of sensory consciousness and the neuroontologically subjective features discussed in chapter 1, including qualia.

Attention

Attention is an intuitively obvious concept, but it is rather difficult to define scientifically. William James gave a commonly accepted, practical definition:

> It is the taking of possession by the mind, in clear and vivid form, of one out of what seem several simultaneously possible objects or trains of thought. Focalization, concentration, of consciousness are of its essence. It implies withdrawal from some things in order to deal effectively with others. [44]

For our sensory considerations we will center on *selective attention*, which *selects* important stimuli, filters out unimportant ones, and *shifts* from one important stimulus to the next.

While attention and consciousness are closely related, they do not seem to be the same thing.[45] Not every detail or aspect of objects that are attended to is necessarily in consciousness, and unattended objects are not entirely excluded from consciousness (people are aware of things in their peripheral vision, for example).[46] Actually, this is our middle-of-the-road conclusion about a highly debated topic, of whether attention and consciousness are the same, strongly associated, only weakly associated, or dissociated.[47] We take the view that consciousness involves or depends on attention, because the objects that are the focus of attention are much more likely to be most conscious, and vice versa. Thus, these two phenomena would tend to evolve together, or coevolve.

Consciousness-associated attention has several subtypes, including bottom-up (exogenous) versus top-down (endogenous) attention.[48] Bottom-up attention is driven by the importance of the incoming stimuli and leads to the animal orienting to things that happen suddenly in the environment. Top-down attention, on the other hand, involves proactive anticipation, maintaining attention by concentration and focusing on goals. We will deal primarily with bottom-up attention because it is simpler, seems not to demand as much brain complexity, and thus is more likely to have evolved first.[49] Top-down attention will receive little consideration because it relates to the higher forms of consciousness that are mostly outside the scope of this book.

In any case, attention is very likely to be adaptive. Inattentive animals are easy prey. Attention allows an animal to focus on particular elements in the environment that are most important for its survival and to screen out the others.

Sensory consciousness may be created by diverse neural architectures

Neuroscientists seeking to explain the neurobiology of sensory consciousness almost always treat the nature of qualia as a single neurobiological issue.

Recall from chapter 1 that Crick and Koch wrote of the "redness of red" and the "painfulness of pain," as though these two types of qualia are the same and come from the same brain mechanisms. But might the exteroceptive aspect of consciousness differ from the affective aspect, and both differ from the interoceptive aspect? Or might the neural mechanisms of consciousness differ for the three aspects, and differ in different groups of animals, such as a fish versus a human versus an octopus? In vertebrates, as we will see, the olfactory (smell) pathway to consciousness has some unique features in which it differs from the pathways for all other senses. Is that a diversity of consciousness across the different senses? And finally, if consciousness is diverse, are there ultimately some common features shared by all kinds of consciousness? The Australian neurophysiologist Derek Denton has already begun to ask these questions and to see signs of diversity, especially between affective and exteroceptive consciousness.[50] We will continue this inquiry in depth in later chapters of the book.

In this book, the general and special features of consciousness as shown in table 2.1 will be explored most thoroughly in the vertebrate animals. To lay the groundwork for this vertebrate story, the next chapter explores the simpler nervous systems of vertebrates' closest relatives among the invertebrates.

3 The Birth of Brains

To trace the evolution of sensory consciousness in vertebrates and the origins of the special neurobiological features that made consciousness possible, the first question to ask is "When, how, and why did complex brains evolve?" Essential to our hypothesis of neurobiological naturalism is that just before the first, explosive radiation of animals in the Cambrian Period, a tiny brain had evolved in the invertebrate ancestor of the vertebrates. This unimpressive brain had only begun to advance in complexity, having the general biological features that characterize all biological systems and some polysynaptic reflexes and basic motor programs, but few or none of the special neurobiological features that make sensory consciousness possible (see table 2.1).

Vertebrates are in the *chordate* phylum,[1] so the gist of this chapter is an exploration of how the simpler brain of ancient, non-vertebrate chordates could have evolved into the complex vertebrate brain. But first we should provide the necessary background information. All chordates have a notochord, a defining feature of the phylum. It is not a nervous structure but a supporting rod along the length of the back (see figure 3.3, parts B, D, E, and F). In vertebrates the notochord forms the core of the backbone—not the bony vertebrae of the backbone but the spongy discs between the vertebrae (these discs are involved in the human back injury called a "slipped disc").

Chordates are a phylum of bilaterian animals; that is, they are in the Bilateria (figure 3.1). Bilateria include almost all groups of animals except for sponges, the jellyfish and their relatives, and comb jellies. Other, nonchordate, groups of bilaterians are the arthropods like insects and lobsters; the nematode roundworms; the annelid worms like earthworms; the molluscs that include snails, squids, and clams; the echinoderms such as sea stars and sea urchins; and many more invertebrate groups.[2] As shown earlier in figure 2.1, the bodies of bilaterians are characterized by, at least primitively, an anterior and a posterior end, a top and a bottom (that is, a dorsal and a ventral

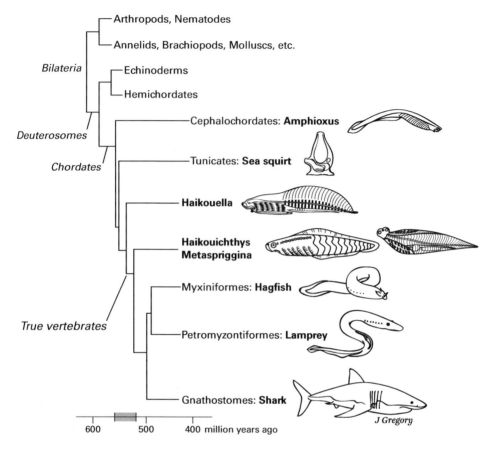

Figure 3.1
Tree showing the relations among the chordates, including vertebrates and the non-vertebrate cephalochordates and tunicates. All are modern, living animals except for *Haikouella*, *Haikouichthys*, and *Metaspriggina*, which are extinct fossil chordates.

side), and right and left sides that are mostly mirror images of one another (this condition is called bilateral symmetry). As biologist Peter Holland says of bilaterians, "They (primitively) have a through-gut with two openings, ingesting food through a mouth and passing food unidirectionally before ejection of waste through an anus. Either side of the gut are muscle blocks capable of contorting [and moving] the body ... [for] very active locomotion in a clear direction."[3] The bodies of bilaterians develop from three embryonic germ layers, all shown at top right in figure 2.1: the inner endoderm, the outer ectoderm, and the middle mesoderm. Bilaterians have a nervous system, although there is debate how complex their nervous system originally was, and whether or not the first bilaterians had a brain (see chapter 4).

The Birth of Brains

But all living chordates do have a brain, which is why we claim a simple brain had evolved by the stage of the early, nonvertebrate chordates. This chapter reconstructs that early brain, to set the stage for the later innovations that could have led to vertebrate consciousness. To reconstruct the ancestral chordate brain, we will describe the simple brains of vertebrate's closest living relatives in the chordate phylum, the fishlike cephalochordates (or amphioxus, or lancelets) and the urochordate tunicates (or sea squirts). The relation of the vertebrates to these ocean-dwelling invertebrate chordates is shown in figure 3.1. Before turning to the tunicates and amphioxus, however, we must present the basic principles of chordate neuroanatomy.

Basic Chordate Neuroanatomy

The nervous system of every chordate has two fundamental parts, central and peripheral (figure 3.2). The *central nervous system* (CNS) includes the brain in the head and the nerve cord or spinal cord in the neck and trunk of the body. The *peripheral nervous system* (PNS) consists of nerves that run throughout the body.[4] The CNS of chordates is dorsal to the notochord on the back side of the body, as opposed to ventral (the belly side).

As labeled in the figure, terms of direction help us navigate around the nervous system: "rostral" means toward the higher brain centers, whereas "caudal" refers to the opposite direction. In nonhuman chordates whose bodies are horizontal, "rostral" and "caudal" are the same as "anterior" and "posterior," respectively.

As indicated in chapters 1 and 2, the nervous system consists of many communicating nerve cells, or *neurons*, that form chains and information-processing networks, especially in the central nervous system and brain. Neurons are excitable cells that respond to sensory stimuli as well as to signals from other neurons. Two typical, communicating neurons are shown at the far left of figure 3.2. A neuron has thin *processes* that carry electrical signals. These processes are usually several signal-receiving *dendrites* and one signal-sending *axon*, all attached to the neuron's *cell body*. Any long axon is called a nerve *fiber*. Inside the cell body is the cell *nucleus*, the neuron's control center with the genes.

The synapses, where neurons communicate with one another, usually occur where an expanded ending of an axon releases a *neurotransmitter* chemical onto a dendrite or cell body of another neuron. This neurotransmitter drives the second neuron toward generating an electrical signal (or else, away from signal generation if the synapse is inhibitory instead of excitatory).

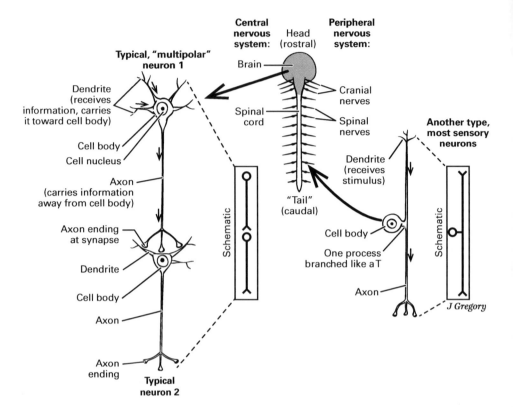

Figure 3.2
Basic parts of the chordate nervous system. Mostly based on vertebrates, with the human nervous system shown at top center as viewed from the front. Basic types of neurons in the nervous system are shown, as are synaptic connections. Neurons and synapses are also shown in figure 2.3.

Most neurons are multipolar ("many-processes") with several dendrites and an axon. These are the two neurons at left in figure 3.2. Another important kind of neuron is on the right side of the figure. This represents most of the sensory neurons, in which just one stout process attaches to the cell body.

Evolution and Origin of the Earliest Brains

Brains almost never fossilize, which is why we must use living chordates to reconstruct the earliest stages of brain evolution that occurred prior to the Cambrian Period. All of the nonvertebrate chordates—tunicates and amphioxus (figure 3.3)—have a sense of touch and can probably sense chemicals in the environment, but unlike the vertebrates they lack the elaborate sensory

The Birth of Brains

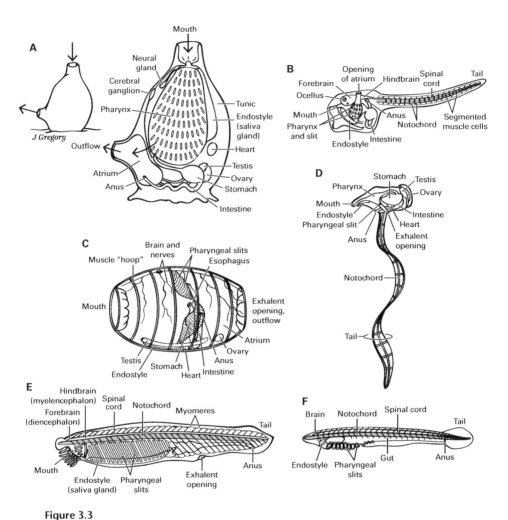

Figure 3.3
The nonvertebrate chordates: tunicates and cephalochordates (amphioxus). A. Ascidian tunicate, adult. External view of the animal at left, and with the body sectioned in half at right. The arrows show how water moves through the ascidian pharynx. B. Ascidian tunicate, larva. C. Thaliacean tunicate (allied to the salps). D. Larvacean tunicate. E. Adult amphioxus. F. Larval amphioxus. Rostral direction, which is headward and mouthward, is to the left in all parts except A.

organs of image-forming eyes, olfactory noses, ears, and so on. Although we must seek the ancestral, prevertebrate brain by studying the less elaborate brains of the nonvertebrate chordates, this is by no means an easy task. Scientists find the brains of tunicates and amphioxus difficult to interpret, perhaps because these brains are small, secondarily simplified, and evolutionarily specialized.

Tunicates

The tunicates[5] are a diverse and abundant group of ocean animals, almost all of which are filter feeders. They have a tough body covering, or tunic, that contains proteins and also cellulose, a substance otherwise present only in plants. Their main subgroup is the ascidians (figure 3.3A). Ascidians have a tiny, free-swimming larval stage (figure 3.3B), which looks like a tadpole, lasts only a few days, and does not feed but soon settles on the ocean floor and changes into the immobile, filter-feeding adult. The baglike adult body is mostly a large, water-straining pharynx plus the other visceral organs, but with no locomotory structures or locomotory muscles. Apparently evolved from one subgroup of ascidians are the swimming thaliaceans (figure 3.3C), the most familiar of which are the salps. The body resembles a barrel that is open at both ends, and it contracts to move the thaliacean through the water by jet propulsion.

The other tunicate group is the larvaceans, or appendicularians (figure 3.3D). These are free-swimming and tadpole-like throughout life, but extremely small (less than a centimeter long) as they feed on the tiniest plankton and bacteria in the water column.

For most of the twentieth century, scientists believed that vertebrates evolved either from larvacean tunicates or from the tadpole larvae of ascidians that gained gonads and became sexually mature, eliminating the former adult stage from the life cycle (a process called *neoteny*).[6] But genetic studies by Linda Holland and others now show that larvaceans have lost many of the genes that were kept by other bilaterian animals, and many larvacean traits are specializations for fast development and brief life spans. So larvaceans seem "weird," not primitive. In fact, *all* the tunicates have experienced much gene loss and rapid gene evolution, larvaceans being only the extreme example of this. And a closer look at their anatomy also suggests that all tunicates are specialized and make poor candidates for the ancestor of vertebrates. For example, the swimming muscles in the larval-tunicate trunk are not true, fishlike muscle segments but just single muscle cells, and the trunk region has no gut because the anus opens in the pharynx (neck) region (figure 3.3B).[7]

This "oddness" is ironic, because the same genetic studies that show that tunicates are the most unusual chordates also show that they are the closest relative of vertebrates—closer than the more fishlike cephalochordates (see figure 3.1). Most of these gene studies were by Frederic Delsuc, Herve Philippe, and their colleagues.[8] The evidence says that tunicates are just highly divergent, like a sister ("sister group" is the official term) with whom a person has little in common.

Tunicate brains are always small, with a few hundred neurons at most.[9] The biologists Thurston Lacalli and Linda Holland have studied the brains of the various tunicate subgroups, and seem to have identified the same three regions that are present in the vertebrate brain: a forebrain, midbrain, and hindbrain—but almost never all together.[10] That is, tunicates of different subgroups and life stages lack different subsets of these three brain regions (figure 3.4). It is as if tunicates' chordate ancestors once had a full, three-part brain that became differentially reduced in the different lineages of tunicates.[11] The diversity and apparent regression of brain parts across tunicates is shown in the figure, as is how all tunicate brains might derive from an ancestral pattern of forebrain-midbrain-hindbrain (upper right). The shaded part allows comparison across these different tunicates. Despite this hint of a common plan, it is all rather obscure as if seeing the tripartite brain "through a glass darkly." With the tunicate brains proving to be rather unhelpful, perhaps the cephalochordates can tell us more about the ancestral chordate brain. For a long time it was debated whether cephalochordates even have a brain because it is thin, unbulged, and almost uniformly tube-shaped, like the spinal cord behind it. But new methods show that this brain exists and is quite informative, as we will now see.

Cephalochordates (Amphioxus)

The cephalochordates (see figure 3.3E, 3.3F),[12] though less closely related to vertebrates than tunicates are, seem to have a more conservative body plan and to be better indicators of both the prevertebrate and primitive chordate conditions.[13] This is because their genomes (all their genes and DNA) are more like those of vertebrates, as are the way their body develops embryonically and their resultant fishlike anatomy. Cephalochordates are not a diverse group, and most species look like those shown in figures 3.3E and 3.3F.

The cephalochordate life cycle has a larval and an adult stage, in both of which the animal swims well and filter feeds. The swimming larvae feed high in the water column on ocean plankton, and when they grow about 1 cm long they change into adults and live burrowed in the sandy ocean bottom.

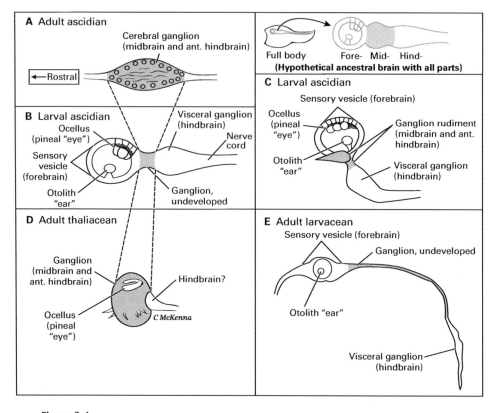

Figure 3.4
Brains of tunicates of different life stages and classes. The brains are shown from the side, with rostral to the left (for orientation, see the tadpole-shaped "Full body" at upper right). A. Adult ascidian, *Ciona*. B. Larval ascidian, *Ciona*. C. Larval ascidian of the group, aplousobranchs, that has the most elaborate brains (*Diplosoma*). D. Thaliacean, *Thalia*. E. Larvacean, *Oikopleura*. Shading and labeling mark the comparable regions of the different brains.

There, they eventually reach lengths of 4 to 6 cm. The adult brain of amphioxus (figure 3.5A), though not widened or bulged and certainly simpler than any vertebrate brain, still has many thousands of neurons.[14] This is too complex for neurobiologists to work out all the neurons and their connections. But the brain of the young larva is tiny enough that most of its neurons can be localized and traced (figure 3.5B), by reconstructing them from successive thin sections (slices) taken through the brain and viewing them with an electron microscope. In this way, the simpler larval brain has allowed Thurston Lacalli and his colleagues to identify the brain regions of amphioxus and to compare these with those of vertebrates.[15] Once their significance was worked out in the larva, the basic landmarks could then be used to decipher the brain of adult amphioxus, because these landmarks are also present there.

In the front of the brain of larval amphioxus is a frontal "eye" (figure 3.5B), with just a few light-receptor neurons (photoreceptors) and no ability to form an image, but still corresponding to the vertebrate retina.[16] In vertebrates, the retina grows from the part of the developing forebrain called the diencephalon, the other part being the telencephalon (or cerebrum: see figure 3.5C). The fact that the amphioxus "eye" is at the very front of its brain means that amphioxus has a diencephalon but probably no telencephalon.[17] Posterior to its frontal eye, the amphioxus brain has an *infundibular organ* made of special, secretory neurons resembling those in the pituitary gland of vertebrates (the *neurohypophysis*). In vertebrates, these cells are in the ventral part of the brain's diencephalon, that is, in the hypothalamus, so they show that amphioxus too has a hypothalamus. Farther dorsally, the amphioxus diencephalon has a *lamellar body*, with photoreceptors resembling those in the pineal organ (the "third eye") of vertebrates. Finally, near the front of the infundibular organ of adult amphioxus is a ring of cells that corresponds to the *suprachiasmatic nucleus* in the diencephalon of vertebrates ("clock cells" in figure 3.5A and 3.5C). This structure was positively identified in amphioxus because it contains "clock genes" as in vertebrates, where it generates circadian rhythms or bodily changes during the daily light-dark cycle.[18] In summary, four structures indicate that amphioxus has a diencephalon forebrain as do the vertebrates: frontal eye, infundibular organ, lamellar body, and the circadian-rhythm generator.

Posterior to the lamellar body is the *primary motor center* (figure 3.5B), whose neurons control the locomotion of amphioxus. Vertebrates have a comparable locomotor center in their ventral midbrain (figure 3.5C), where it is part of the "reticular formation" that sends motor instructions to the spinal cord to signal contraction of the body muscles. Thus, amphioxus must have a ventral midbrain. Posterior to this is amphioxus' hindbrain. This was identified because early in development it expresses *Hox* genes,[19] which are also characteristic of the vertebrate hindbrain.[20]

Although amphioxus has a primitive forebrain, midbrain, and hindbrain, Lacalli saw that these correspond almost exclusively to the *ventral* parts of these brain divisions in vertebrates (the shaded and stippled areas in figure 3.5C).[21] These ventral parts—the hypothalamus and the midbrain locomotor center—serve only the most basic survival behaviors of vertebrates and they control the functions that maintain homeostasis (that is, constant and optimal conditions within the body). In larval amphioxus, these brain regions

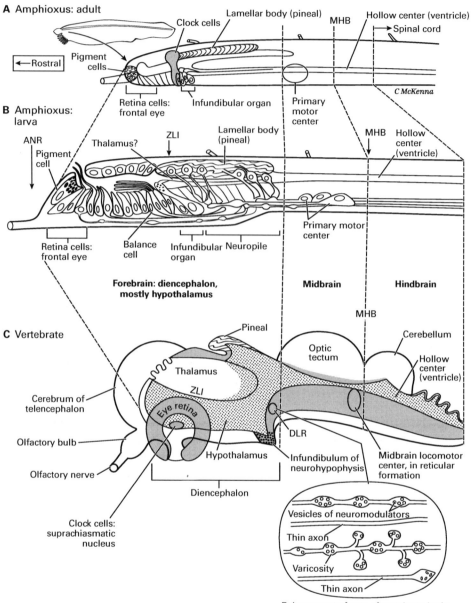

Figure 3.5
Brain of larval amphioxus as the platform on which chordate and vertebrate brains are built (hypothesis of Thurston Lacalli). Brains are shown from the side and cut in half in the midbody plane (sagittally). Notice the orientation diagram at upper left, showing the whole animal. A. Adult amphioxus brain, drawn small because it is not as well known as the larval brain. B. Larval amphioxus brain, with its neurons and subdivisions. C. Vertebrate brain (generalized). Corresponding structures in the three brains are labeled with similar names. The shaded and stippled areas of the vertebrate brain are the only parts that exist in amphioxus. The core of the vertebrate brain (lower right) corresponds to the neuropile in amphioxus ("Neuropile" in part B), because both have wide varicosities on thin axons. Also labeled are landmarks for defining the brain regions in both amphioxus and vertebrates: ANR = anterior neural ridge; MHB = midbrain-hindbrain border; ZLI = zona limitans intrathalamica; DLR = diencephalic locomotor region.

receive sensory information, but mostly just for choosing and signaling the larva's few, simple motor actions. What amphioxus is missing is the *dorsal* part of the vertebrate brain, which is for more extensive sensory processing. Specifically, Lacalli saw that the amphioxus brain lacks the large cerebrum, optic tectum, and cerebellum (figure 3.5C), processing centers that are associated with the more elaborate senses of vertebrates.

What do the existing parts of the amphioxus brain say about the evolutionary origin of vertebrate brains and vertebrate behaviors? In the 1980s and 1990s, Rudolph Nieuwenhuys and his colleagues developed the idea that the vertebrate brain has a primitive core region,[22] to which the rest of the brain lies peripheral and dorsal. This central core region has been called the limbic core, paracrine core, and other names. It includes the hypothalamus and the gray-shaded parts of the midbrain and hindbrain in figure 3.5C. In all vertebrates, this core controls routine homeostatic adjustments and the basic motivated movements that are necessary for vertebrate survival: adjusting blood pressure and heartbeat, maintaining a constant body chemistry, drinking, breathing, mating, aggression, chewing and swallowing, adjusting to day versus nighttime, and escape behaviors. Many of these are basic motor programs from central pattern generators (chapter 2), primitive in origin but essential to the vertebrate brain. One begins to sense a correspondence between this vertebrate core and most of the larval-amphioxus brain. We'll now present more evidence for this correspondence.

This core region has some unusual neuronal features, by which it differs from the rest of the vertebrate brain. As shown at bottom right in figure 3.5C, its neurons communicate less through fast-acting synapses and neurotransmitters and more by releasing a wide variety of "neuromodulatory" chemicals, which prime other neurons to become more active or less active over longer periods of time. This prepares the vertebrate for long-lasting behavioral states. The neuromodulatory chemicals are stored and released from saclike vesicles in varicosities (swellings) along the axons, and the core's axons in general are very thin.

The above applies to vertebrates, but Lacalli showed that these core features comprise most of the brain of larval amphioxus, in its hypothalamus and its locomotory regions that resemble the reticular formation of vertebrates.[23] Especially informative is the postinfundibular *neuropile* of the amphioxus hypothalamus (figure 3.5B). Here we find the classic features of Nieuwenhuys's paracrine core: thin axons with varicosities that contain diverse populations of vesicles, and few classical synapses. This neuropile receives every kind of sensory input that amphioxus picks up, and after working out its neuronal connections, Lacalli deduced that it is a center for the few behavioral programs that this larva can execute. That is, the neuropile's circuits switch from signaling the larva to swim quickly away from danger, to signaling slow upward swimming to reach the best level in the water column for filter feeding, to signaling the larva to stop moving so it can sink through the food-rich water and passively trap tiny food particles in its mouth.[24] So the amphioxus neuropile seems to be for basic survival behaviors, as is the corresponding paracrine core in vertebrates. Lacalli deduced that this neuropile also corresponds to a major motor-control center in the hypothalamus of lampreys and other vertebrates called the *diencephalic locomotor region* (DLR in figure 3.5C).[25]

After deciphering the larval-amphioxus brain, Lacalli used it as a model for the simple brain of the ancient ancestral chordates, as well as the brain of the earliest prevertebrates.[26] Although it received information from light, touch, and other environmental stimuli, such a simple ancestral brain did not allow extensive sensory processing or many different behavioral decisions. These functions came later, when the lineage leading to vertebrates evolved their elaborate eyes, nose, and ears (see chapters 5 and 6), along with their dorsal-brain regions that constructed sensory images of the external world.

According to Lacalli, the larval-amphioxus brain has a circuitry too simple to permit consciousness.[27] In agreement, when we traced out this simple circuit from Lacalli's descriptions, we found that most of the sensory pathways have only three or four levels of neurons between the external stimulus and the motor response. Put another way, this chain has just one or two neurons before the primary motor center, so it has very few links for sensory processing along the way. Such a short chain is not the long processing hierarchy needed for sensory consciousness, but is just a reflex arc (see figure 2.3) with a simple central pattern generator in the form of its primary motor center. In fact, larval amphioxus neatly fits the hypothesized "Level 2" stage (p. 25) that we called unconscious. In chapter 5, we will explore whether the new and more complex circuitry in the first true vertebrates allowed consciousness. But first, we will describe the important geological epoch during which that complexity—and complex brains—evolved: the Cambrian Period.

4 The Cambrian Explosion

Now we will explore the early years of animal history that led to a riot of animal diversification, of evolution in hyperdrive, called the Cambrian explosion.[1] This unique event lasted from around 560 or 540 million to 520 million years ago (mya). Vital to our study of consciousness, this was when animals first evolved complex senses, nervous systems, and behaviors. The vertebrates first appeared during this explosion, emerging from it as the most complex animals on Earth, with, as we will see, the most complex nervous systems. It is a story full of drama, of how the first fish evaded and competed successfully against their arch rivals, the early, predatory arthropods. This chapter asks why the Cambrian explosion occurred, and if it really set the stage for consciousness in vertebrates. Box 4.1 defines some useful terms on animal evolution for this and later chapters.

Evolution and Darwin's Dilemma

Though Charles Darwin assembled much evidence for evolution by natural selection in his 1859 book *On the Origin of Species*, he saw that something about the fossil record did not seem to support it. That is, the fossils of animals, consisting of their mineralized skeletons, appeared abruptly and for many different animal groups in rocks of the Cambrian Period, which we now know was from 541 to 488 mya (figure 4.1), with no evidence of earlier ancestors from older rocks.[2] This riddle of the sudden appearance of animals is called Darwin's dilemma.[3] Some scientists, including Darwin, claim that this is just an artifact, meaning that animals existed long before the Cambrian Period but were not preserved at first because, for instance, they were small and had only soft parts, which rarely fossilize. Other scientists, however, emphasize that soft bodies and small animals sometimes do preserve as

Box 4.1
Terms of Evolution and Animal Relationships

Chordate features   All animals of the chordate phylum (figures 4.2, 3.1, 3.3) share the unique features of a notochord, pouches in their pharynx region (gill pouches and gill slits of fish), a tail that extends posterior to the anus of the digestive tube, and a hollow nerve cord (spinal cord and brain) on the dorsal side of body. In nonchordate animals, by contrast, the nerve cord is usually solid and ventral (on the belly side).

Clade   A natural group of organisms, all united by being descended from a common ancestral species. For example, mammals are a clade because all the later mammals, and only mammals, evolved from the first mammals. Reptiles are not a clade, however, because birds evolved from reptiles, yet the word "reptile" is not typically used to include birds.

Convergent evolution   Independent evolution of similar structures by distantly related organisms, usually in response to similar evolutionary pressures. A classic example is the wings of insects, birds, and bats. These structures are said to be analogous or homoplasious, not homologous (next term).

Homology, homologues, homologous   Corresponding structures in different organisms, due to descent from a common structure in the ancestral species. The limbs of land vertebrates are homologues of, or homologous to, the paired fins of fish; feathers of birds are homologous to the scales of snake skin; and the eyes of all vertebrates are homologues (evolved from the ancestral vertebrate eye).

Invertebrates   All the animals that are not vertebrates. Every group shown in figure 4.2 except for the vertebrate clade. As shown in the figure, some invertebrates are closely related to vertebrates and others are not, so invertebrates definitely do *not* form a true clade (defined above).

Phylum   A group of organisms that share the same basic body plan. Thirty to thirty-five phyla of animals are said to exist (for example, arthropods, molluscs, and cnidarians, as the jellyfish and their kin), and all but the most minor phyla are shown in figure 4.2. Many of the phyla are clades (defined above as natural groups), although a few are not because they have recently been found to have evolved from another phylum, to which they must therefore belong. For example, the peanut worm and spoon worm "phyla" really belong to the annelid worms. Vertebrates are not a phylum but instead belong to the chordate phylum (figure 4.2).

Phylogenetics   Study of the evolutionary relationships among different groups (clades) of organisms.

Selective pressure   Strong natural selection.

Sister groups   The two most closely related clades of organisms. Humans and chimpanzees are sister groups. So are horses and donkeys.

Vertebrate features   All vertebrate animals have a head and distance senses for vision, smell, and hearing. But their entirely unique features are their vertebrae and vertebral column, a distinctive skull, and neural crest and ectodermal placodes (see chapter 5).

The Cambrian Explosion

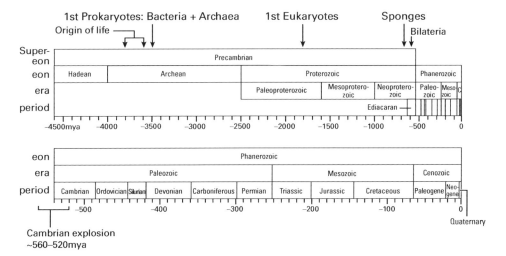

Figure 4.1
Timeline from the start of the Earth to the present. Note the period called the Cambrian (541–488 mya), and the Ediacaran before it (635–542 mya). Lower row expands the right side of upper row.

fossils, yet animal fossils are still absent until about 580 to 555 mya, in the period that came immediately before the Cambrian named the Ediacaran (635–542 mya). These scientists hold that the Cambrian explosion of animal diversity in the fossil record marks a real historical event.

The fossil record of soft-bodied animals from Pre-Cambrian and Cambrian times has improved since the days of Darwin, as has our knowledge of the modern animals whose phyla originated in those ancient times (figure 4.2).[4] Thus, today's scientists can better judge whether the Cambrian explosion was real or an artifact.

Figure 4.2 shows a phylogenetic tree of the history of animal life, topped by pictures of modern members of the different phyla and showing that all the phyla originated during or just before the Cambrian explosion. It also shows the interrelationships of the animal phyla. Scientists reconstruct these relations by comparing the anatomic traits and genes of the different animals.[5] About thirty-five phyla exist, but we left out a few minor ones and only drew animals from the major ones. Drawn at some of the basal branches of the tree are our attempts to reconstruct the ancestors of major groups of phyla, all of which reconstructions are worms.

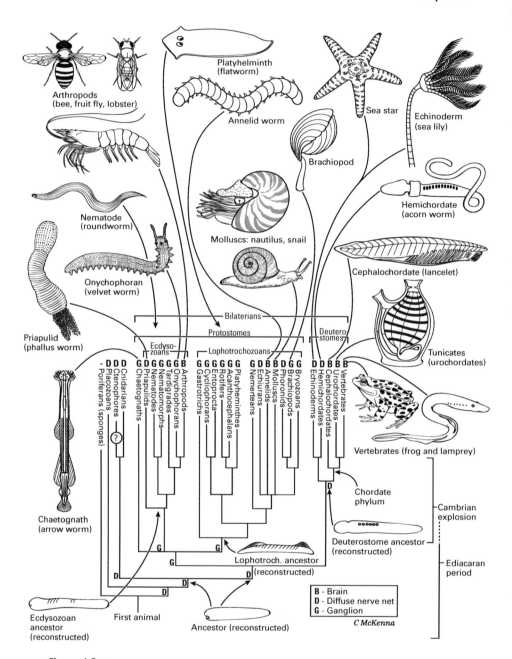

Figure 4.2
Today's animal phyla, which originated in the Cambrian explosion, presented in a phylogenetic "tree of life" that shows their interrelations. Notice how few of the phyla have true brains (B) versus simpler neural ganglia (G) or a diffuse nerve net (D). This implies that their ancestor lacked a brain. Tree is modified from the work of R. G. Northcutt (2012a). The position of the ctenophores (sea gooseberries or comb jellies) is the most uncertain (Moroz et al., 2014).

The Cambrian Explosion

Explosion's Eve

Our strategy for investigating the explosion is to reconstruct the ocean floor of the Ediacaran, then of the Cambrian (figure 4.3), and then compare these two to find what happened in the intervening years. The ocean floor—especially the floor of the *shallow* ocean—is crucial because for billions of years it was the center of evolution of life on Earth.

The older, Ediacaran world is reconstructed from soft-bodied fossils found mainly in southern Australia; Newfoundland, Canada; Namibia, Africa; and the White Sea region of Russia. The subsequent world of the Cambrian is reconstructed from soft- and hard-bodied fossils from two famous fossil formations, the Chengjiang shale of south China (520 mya) and the Burgess Shale of west Canada (505 mya).[6]

The Ediacaran ocean bottom of 555 mya looks strange, covered by a thick, green mat of microbes, and its soft-bodied animals were mostly of an unfamiliar pleated or flat type with no mouth or gut (figure 4.3A). Some of them looked like leaves. Some species could not move and others moved only slowly, so not much activity is evident: not on, nor above, nor in the sediment. Animals fed on the mat microbes, not on each other. Early bilaterians are represented by creeping worms. The Cambrian ocean of 520 mya, by contrast, is more familiar and diverse (figure 4.3B), with the mats mostly gone and containing early members of most of the modern phyla of animals. These included arthropods, fish, and animals with hard shells. As in today's oceans, many animals were continually moving, on, above, and deeper within the sediment.

The first of the multicellullar animals were probably the sponges, which may have evolved around 635 mya (see figure 4.1). Sponges are filter feeders. They have no nerve cells and show little movement or "behavior." One candidate for the first sponge had the size and shape of Wiffle golf balls, with many pores to let in water and the suspended food particles (see figure 2.1).[7]

Ediacaran sponges attached to the thick microbial mats on the ocean floor (figure 4.3A, animal E). Because these mats fed the later Cambrian explosion, we should consider how they got so thick. Actually, bacteria have always grown as films on surfaces (think of the plaque on your teeth), including the ocean floor, but in Ediacaran times algae cells joined these films and grew in the sunlight.[8] The mats became so lush that scientists call them the "Garden of Ediacara," a pun on the "Garden of Eden."

Figure 4.3

Ediacaran (A) versus Cambrian (B) seafloor communities. About 555 mya versus 520–505 mya. Note the great increase in animal diversity and activity. The projecting ledges of sediment show that worms did not burrow very deeply in the Ediacaran but burrowed deeper and more extensively in the Cambrian.

Key for A: A. Leaflike animal, *Charniodiscus*. B. Leaflike "rangeomorph," *Charnia*. C. Seaweed. D. Tubular body fossils, *Funisea dorthea*. E. Sponge, *Thectardis*. F. Microbial mat. G. Worm burrows on underside of the microbial mat. H. Segmented animal, *Spriggina*. I. Possible mollusc, *Kimberella*. J. Bilaterian worm foraging on microbial mat. K. Mobile segmented animal, *Dickensonia*.

Key for B: A. Sponge, *Vauxia*. B. Anomalocarid arthropod-relative, *Amplectobelua*. C. Jellyfish of family Narcomedusidae. D. Sea gooseberry, ctenophore *Maotianoascus*. E. Near-vertebrate *Haikouella*. F. Sponge, archaeocyathid. G. Brachiopod, *Lingulella*. H. Sponge, *Chancelloria*. I. Arthropod trilobite, *Ogygopsis*. J. Priapulid phallus worm, *Ottoia*. K. Mollusc relative, hyolithid. L. Arthropod, *Habelia*. M. Arthropod, *Branchiocaris*. N. Brachiopod, *Diraphora*. O. Hemichordate worm, *Spartabranchus*. P. Annelid worm, polychaete *Maotianchaeta*. Q. Vertebrate, *Haikouichthys*. R. Arthropod, *Sidneyia*. S. Varied worm trails within the sediment. T. Arthropod trilobite, *Naraoia*. U. Sea anemone, *Archisaccophyllia*. V. Lobopodian worm, *Microdictyon*.

Figure 4.3 (continued)

The ancestral bilaterian worms, perhaps having evolved from some elongated sponge (see figure 2.1), crept along to feed on the mat or on its sediment particles that were glued together with nutritious bacteria (figure 4.3A, animal J). The earliest trackways date to over 585 mya and were made by a small worm less than 1 cm long, with simple actions. The only "rules" for these actions were: follow the food gradient, do not dig deeply or vertically, and plow straight ahead or in shallow waves with no exploratory side trips. Fascinatingly this is "fossilized behavior," of a very primitive sort.[9]

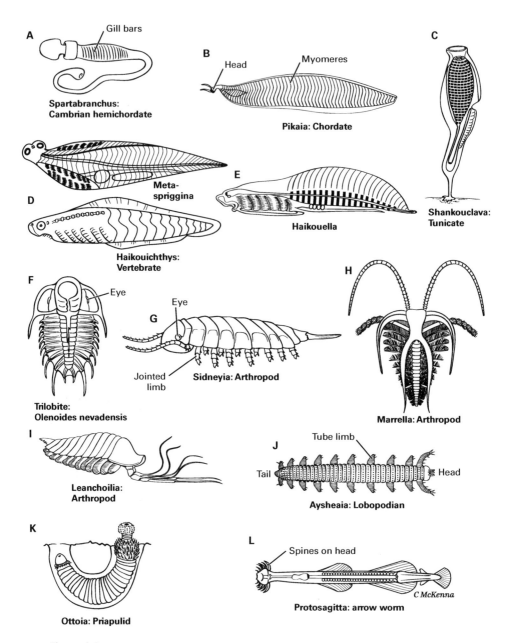

Figure 4.4
Cambrian representatives of many bilaterian animal phyla, known from fossils, with some that are strange and of uncertain relationship. A. Hemichordate acorn-worm *Spartabranchus*. B. Chordate *Pikaia*. C. Tunicate *Shankouclava*. D. Vertebrate fish *Haikouichthys* with its relative *Metaspriggina*, which is drawn as twisted to show the top of the head. E. Near-vertebrate *Haikouella*. F. Arthropod *Olenoides* (a trilobite). G. Arthropod *Sidneyia*. H. Arthropod *Marrella*. I. Arthropod *Leanchoilia*. J. Lobopodian *Asheaia*. K. Priapulid *Ottoia*. L. Chaetognath (arrow worm) *Protosagitta*.

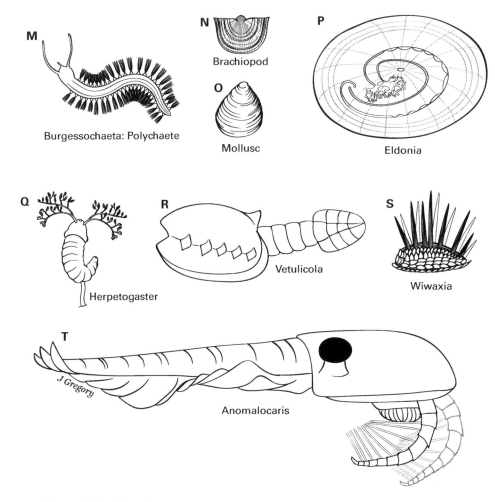

Figure 4.4 (continued)
M. Polychaete *Burgessochaeta*. N. Small brachiopod. O. Small bivalve mollusc. P. *Eldonia*, strange C-shaped worm in a floating umbrella. Q. *Herpetogaster*, odd worm with tentacles and a stalk. R. *Vetulicolia*. S. *Wiwaxia*: worm possibly related to molluscs, annelids, and brachiopods. T. arthropod relative, *Anomalocaris*.

The Explosion's Products: Animal Tree of Life

Ediacaran animals were well nourished by the microbial mats, but were much less diverse than Cambrian animals. To explain the increase in diversity, we must look more closely at the many animal phyla that emerged from the explosion, both the modern and Cambrian members of these phyla. The cnidarian jellyfish and their kin played a role, but we will focus on the bilaterians.

Bilaterians are divided into two groups of phyla, the protostomes and deuterostomes (see figure 4.2). Deuterostomes include our own chordate phylum. They also include two of the invertebrate phyla: the echinoderms (e.g., sea lilies, sea stars, sea urchins) and hemichordates (mostly acorn worms). The protostome group has two subdivisions, the ecdysozoans (e.g., arthropods) and lophotrochozoans (e.g., molluscs). *This means that we vertebrates are only distantly related to arthropods and molluscs.* Many other protostome phyla are shown in figure 4.2.[10]

Figure 4.4 shows that these phyla had Early Cambrian members.[11] Most of these are similar enough to their modern relatives that scientists can deduce their ancient lifestyles, and thus can deduce what their nervous systems could do. The chordates of the Cambrian, though not very common or diverse, have become well known to science in the past twenty years (figure 4.4B–E). Among these chordates are a fossil tunicate (figure 4.4C); the puzzling, laterally flattened eel-like *Pikaia*, with a tiny head (figure 4.4B); and vertebrates that include jawless fish (figure 4.4D). The Cambrian arthropods are hyperdiverse (figures 4.4F–I), and many have sensory antennae and intricate compound eyes. Stunningly similar to their modern relatives are some of the Cambrian worms and brachiopods (figure 4.4A, 4.4J, 4.4K, 4.4L, 4.4N).

Odd animals that do not seem to fit in any phyla were also present (figures 4.4P–T). But the phylum relations of some of them are now known. The frightful anomalocarids (figure 4.4T) were predators up to a meter long and the largest animals of the Cambrian, with giant, spiny appendages in front of their mouth. Though anomalocarids lack arthropods' distinctive walking legs, new fossils reveal that their eyes are the compound eyes of arthropods. Thus, they indeed are arthropods or the nearest relatives of arthropods. Vetulicolians (figure 4.4R) were large, eyeless swimmers with a giant "gill basket." They are puzzling because their gill basket resembles the pharynx of a tunicate, but their jointed posterior is like that of an arthropod. Evidence suggests that their posterior part contained a chordate notochord, so they may indeed be related to tunicates.[12]

The Cambrian Explosion

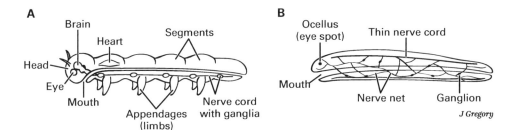

Figure 4.5
Bilaterian ancestor reconstructed, complex (A) versus simple (B) versions. The ancestor lived in the Ediacaran, perhaps 580 mya.

What do the relations among the bilaterian phyla tell us about the bilaterian ancestor? Just as the fossil record of Ediacaran worm trails suggests that the first bilaterians were worms, so does a study of the modern animal phyla (see figure 4.2). That is, many of the phyla that arose early on the tree (the more "basal" protostome and deuterostome branches in figure 4.2) are worms: for example, the arrow worms, phallus worms, and acorn worms.

But the attempts to understand this bilaterian ancestor in more detail have led to two conflicting reconstructions, the complex worm and the simple worm.[13] The complex version (figure 4.5A) has image-forming eyes in a distinct head, body segments, appendages (limbs), a heart, and blood vessels. It also has a large, three-part brain and a nerve cord. This complex reconstruction was inspired by the fact that the distantly related vertebrates and arthropods share all of the characteristics just mentioned, and they also share many of the gene pathways that signal these characteristics to form during embryonic development. By contrast, the simple model of the ancestor (figure 4.5B) has only a crude head with simple eyes (ocelli) that could sense light but could not form an image. It has no body segments, no appendages, and no heart or blood vessels. No brain is present, just a nerve net at the body surface.

Many biologists favor the simple bilaterian ancestor. This is because so many of the bilaterian phyla lack the complex traits of image-forming eyes, brains, segmentation, and limbs (see figure 4.2). That is, an unreasonably large number of existing phyla would have to have lost all these traits later, if their common ancestor had really had them. Figure 4.2 makes this point for the *brain* trait, following studies by R. Glenn Northcutt.[14] Yes, all bilaterians do have the genes that can signal the complex structures to develop. But this could just mean that in the ancestor these genes dictated simpler body structures such as cell types (light-detecting cells, rather than whole eyes) or that these genes ancestrally controlled the location of certain cells within the developing body, rather than dictating body segmentation. By this interpretation, several different bilaterian phyla separately modified the original gene pathways when independently evolving the more complex traits. Only this interpretation can explain why the nonbilaterian animals, such as sponges and cnidarian sea anemones, also have and express many of the same genes but utterly lack the complex bilaterian traits. This is why we accept the "simple bilaterian ancestor," with the simpler nervous system.

Could such a simple ancestor, with only a nerve net, perform all the food-finding and reproductive behaviors necessary for survival? Yes, because food was abundant and easy to find on the well-stocked buffet of the Ediacaran microbial mats. The ancestral worms needed only to follow chemical "reward gradients" of odors and taste molecules toward the food. For reproduction, they simply could have secreted and followed sex pheromones (attractant chemicals) to find mates on the mats. Reflexes and simple motor programs; no need for consciousness there!

The ancestral worms had evolved a few of the neurobiological features that we listed in table 2.1 as basic requirements of consciousness: a nervous system with interconnected nerve cells, plus reflexes. But the special features on the list, those involving more neural complexity, were not present in these simple worms.

Causes of the Explosion

We have presented the animals that burst forth in the Cambrian explosion, and have reconstructed their Ediacaran ancestor as a worm, so now we can ask what caused the worms to diversify so much and so fast. Some evidence says that oxygen, building up in the atmosphere for billions of years since the days of the first oxygen-producing (photosynthetic) bacteria, finally reached a high enough level to support large and active animals.[15] But the most important cause of the explosion seems to have been the evolution of the first predators. In simplistic terms, one worm found it could eat other worms of a different species. Then, in response to the attacks, the preyed-upon animals on the ocean floor evolved survival strategies. For example, some burrowed deeper to hide from the predators overhead (bottom of figure 4.3B). These burrowers churned the sediment, thereby bringing in nutrients and oxygen from the overlying water and mat. This fertilized the sediment, which then provided food for many new phyla of burrowers. And their fossilized trails and burrows show that their food-foraging behaviors had become more complex and varied. Nervous-system evolution was on the move![16]

Predation also structured the Cambrian radiation of animals that lived above and on the ocean floor, in a spectacular way (figure 4.3B). Cambrian predators, which we recognize from the fossils by their resemblances to modern predatory animals, similarly had spines, limbs, and claws for grasping and tearing prey. They included many of the arthropods (figure 4.4G, 4.4I, 4.4T, and 4.6A), which have manipulative limbs, and some spiny-topped worms (figure 4.4K, 4.4L). These predators put tremendous pressure on prey species to adapt or die. Prey evolved many defenses, including the first shells and hard parts of protective armor (mineralization); hiding among plants, behind rocks, or with camouflage; faster locomotion for escaping; improved distance senses for vision, hearing, and smell that detected the predators approaching; and growing too large for predators to kill. At the same time, natural selection favored *predators* with these very same characteristics, such as mineralized hard parts (here, for piercing or cutting prey, or for defense against other predators), hiding while stalking prey, improved distances senses that detected prey from afar, faster locomotion to chase prey, and growing larger to subdue ever bigger and more nutritious prey. These reciprocal interactions between predators and prey led to an escalatory arms race, to many different survival strategies, and to a fast increase in animal diversity, behaviors, and size.

One hypothesis may answer why the predatory arms race led to sophisticated senses and brains in some of the emerging groups of Cambrian animals. Roy Plotnick, Stephen Dornbos, and Junyuan Chen point out that the ocean bottom of the Cambrian, on and above which many new animals moved and shifted, was a more complex and varied environment than had been the monotonous and still mats of the Ediacaran (figure 4.3).[17] As such, the environment sent out many more signals (different tastes, both new and lingering odors, complex patterns of light and color, vibrations traveling through the water or sediment) that could help any animal who sensed and processed such stimuli. The animals used this information to orient themselves, react to danger, and find things in the landscape. In many cases, the more kinds of signals that could be sensed, the better. This drove the evolution of improved eyes, organs of smell and taste, touch and vibration receptors, and so on.[18] The simplest way to process sensory information from stimuli is into elementary reflexive reactions, but a better way is to assemble and organize the sensory information into maps in the nervous system and in a brain, mental maps that record and reconstruct the animal's environment (visual maps, smell maps, and maps of the outer body surface as it receives touch stimuli). Then, these maps can guide the animal as it moves and interacts with its environment. The sensory maps were under constant selective pressure to improve in resolution and to offer as complete a representation of the world as was necessary for survival. That is because the cost of overly crude or incomplete perception is very high: death from failing to see or smell a cleverly hidden predator, for example. With the improving senses, the whole nervous system rapidly evolved greater complexity.

During the Cambrian explosion, two of the most mobile clades of animals followed this path of ever improving sensory systems: the arthropods and the vertebrates. But most of the other emerging phyla did not. This is because sensory processing is costly in terms of the food energy required to fuel it. Also costly are the energy demands of moving around so much and being so active. Therefore, most of the other phyla adopted less expensive "shortcuts" to survival that in some cases even involved neural regression. So went the newly armored and burrowing groups that did not move much and could avoid predation without the full set of senses: the clamlike molluscs and brachiopods protected in their shells, and various blind, burrow-dwelling worms. In summary, high cost explains why so few bilaterian groups went down the road of extreme sensory and cognitive enhancement.[19]

Cambrian Explosion Explained

Sensory improvements helped to trigger the Cambrian explosion of animal diversity because they contributed to the rapid evolution of predation and a variety of prey defenses. The most active predators and prey relied more on sophisticated senses and complex, newly cognitive brains than on shelly armor, hiding and camouflage, or other passive ways of surviving. In the predatory arms race, scientists may have found a key cause of the Cambrian explosion and the solution to Darwin's dilemma.

Arthropods versus Vertebrates

In some ways, arthropods and vertebrates are very different. Arthropods have an outer skeleton (called the *exoskeleton*: their cuticle) and many jointed limbs, whereas fish are more streamlined for swimming, have an internal skeleton, and were originally limbless (figure 4.4D–I). Yet, during the Cambrian explosion, these two groups independently evolved high mobility, advanced sensory processing, and brains. Arthropods and vertebrates reflect the opposite sides of the Ediacaran-to-Cambrian predatory race, as predators versus prey, the warriors and the farmers, the lions and lambs.[20]

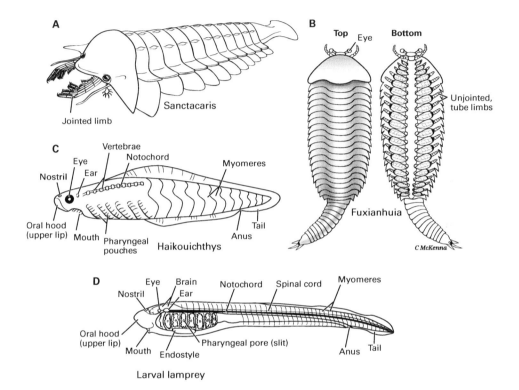

Figure 4.6
Comparison of Cambrian arthropods with vertebrates. A. *Sanctacaris*, an arthropod with prey-grasping appendages. Up to 9 cm long. B. *Fuxianhuia*, the most basal arthropod. From 4 to 7 cm long. C. *Haikouichthys*, a Cambrian fish shown in more detail than in figure 4.4D. About 3 cm long. D. Modern larva of a lamprey. Shown at 1 cm long, but grows to about 13 cm.

Arthropods occupied the highest levels of the Cambrian food chain, making up most of the predators (although not all these arthropods were predators).[21] Arthropods evolved from tube-limbed lobopodian worms (figure 4.4J), close relatives of today's velvet worms. This probably occurred late in the Ediacaran, and involved adapting the locomotory limbs to handle food, especially the limbs on the head nearest to the mouth. Figure 4.6A shows a good example of this adaptation in a predatory Cambrian arthropod, *Sanctacaris*, with prey-grasping appendages in front of its mouth. Its name means "saintly claws" and is a pun on Santa Claus. The uses of arthropod limbs in food manipulation included accurately reaching for, grasping, and shredding prey. Such actions require advanced spatial skills. They benefit from spatial vision, and most Cambrian arthropods had imaging-forming eyes, far more elaborate than the simple ocelli of some lobopodians. Figure 4.6B shows the most basal of all known arthropods, *Fuxianhuia*, which had good eyes and the arthropod brain but still had the primitive tube limbs, similar to those of their lobopodian ancestors.[22]

Jointed limbs gave arthropods a tremendous advantage. Neither the vertebrates nor any other limbless animals could at first match the food-handling prowess that spurred the arthropod radiation. Arthropods ruled the Cambrian seas as the most abundant and diverse of the animal phyla, including species with the largest individuals (see figure 4.3B).

Now, the vertebrates. The rare, Early Cambrian vertebrates were small fish, only about 3 cm long,[23] with no signs of having been predators. That is, they had no teeth, no jaws, and no limbs, as shown by the oldest fossil fish, *Haikouichthys* (figure 4.6C). Instead, they were probably filter feeders because that is how all their closest modern relatives feed. These relatives are the tunicates, the fishlike cephalochordates (see figure 4.2), and the young larvae of lampreys (figure 4.6D), with lampreys being the least specialized of the living jawless fish. A remarkable thing about today's larval lampreys, which live in stream beds and filter algae cells and detritus from the water, is that they feed on the densest, most nutrient-rich food suspensions of any filter feeder in the entire animal kingdom.[24] This suggests that their ancestors, the first vertebrates like *Haikouichthys*, fed on the rich microbial mats in the Cambrian. Actually, with the mats mostly gone by then from grazing by so many kinds of animals (figure 4.3B), it must mean that the Cambrian vertebrates found the richest surviving patches of mat, and traveled from patch to distant patch to feed.

The vertebrate swimming mechanism easily allows such travel. Fish have muscle segments called myomeres along their trunk (figure 4.6C, D). We can see this row of myomeres when we order trout at a restaurant. To swim, a fish contracts its myomeres one after another from front to back, bending the body into waves that propel the fish forward. It is an efficient mechanism for locomotion, working together with the streamlined, blade-shaped body of the fish. It requires an internal skeleton, the relatively light notochord, but no heavy exoskeleton as in the arthropods.[25]

Muscular myomeres, streamlining, and the fish way of swimming are fully developed in the related invertebrate amphioxus (figures 3.3E), which probably resembles the earliest prevertebrates (see chapter 3). Amphioxus can move like a bullet through the water but it lacks image-forming eyes and an organ for smell. This implies that fast, myomeral swimming evolved before the vertebrates did, and before the vertebrates evolved their keen distance senses. Perhaps the prevertebrates were so speedy that no predator could catch them even if the prevertebrates failed to sense the predator until it was nearby. And perhaps that explains why the first vertebrates did not need or have any protective armor on their body surface. But in the relentless arms race of the Early Cambrian, ever faster predators provided strong selective pressure for vertebrate senses to become keener, to sense danger from afar, for better vision, smell, and so on. Sharper distance senses also would have helped the early vertebrates find the richest parts of their patchy food mats.

The first fish (figure 4.6C) had advanced senses and sensory-processing abilities, which together with their powerful swimming made these fish the Cambrian's most effective farmers and filter feeders and also allowed them to escape predators. Surprisingly, we evolved from these hyperalert but gentle vegans of their day. We say "surprisingly," because these first vertebrates were quite small and apparently not very abundant or diverse. Yet, at the dawn of the vertebrates, their entire genome duplicated, to twice as many genes as in their invertebrate ancestor, and soon it duplicated *again*.[26] We know this because all the living vertebrates have so many more genes than does amphioxus or any other invertebrate. This "genomic quadrupling" provided genetic novelty for evolutionary innovation, which made early vertebrates by far the most complex animals on Earth, genetically, in their body structure, body functions, and in their nervous systems—and they remain so today.[27] Their more complex genomes meant vertebrates could have more complex neural hierarchies (discussed in chapter 2), with more levels. Cambrian vertebrates proved to be a group with great evolutionary potential, to which, as we will argue in the next chapters, consciousness contributed.

5 Consciousness Gets a Head Start: Vertebrate Brains, Vision, and the Cambrian Birth of the Mental Image

How ancient is consciousness? We will build the case that in vertebrates it evolved during the Cambrian explosion before 520 million years ago when critical advances in brain complexity made consciousness possible. If we are correct, this would make consciousness very old, much older than has typically been assumed. In this chapter and the next, we focus on the evolution of sensory consciousness of the external environment, that is, *exteroceptive consciousness*. In chapters 7 and 8, we discuss the evolution of *sentience*, the capacity for internal feelings and affects. But we will argue that the basic time frame for all forms of sensory consciousness is roughly the same.

Figure 5.1 is a phylogenetic tree that dates and shows the initial radiation of the vertebrates and other chordates. It marks the key advances in the evolution of the distance senses, the vertebrate head, and sensory consciousness, to be laid out in this chapter. Along with living animals, it shows the Cambrian fossil vertebrates and near-vertebrates that will help us to present our story (*Haikouella*, *Haikouichthys*, and *Metaspriggina*).

Vertebrate Brains

This chapter covers brain evolution because the Cambrian explosion that produced the vertebrates also produced the vertebrate brain.[1] Amphioxus has shown us that the ancestral chordate already possessed the three-part brain (see figure 3.5), with a forebrain, midbrain, and hindbrain; and this brain pattern can even be reconstructed in the bizarre tunicates (see figure 3.4). In vertebrates, the three brain parts are much enlarged, more complex, and better differentiated (figure 5.2), and they contain high levels of well-developed neural hierarchies.

Whereas figure 5.2 shows a generalized brain with the parts shared by all the vertebrates, figure 5.3 shows the brains of specific vertebrates, from fish to

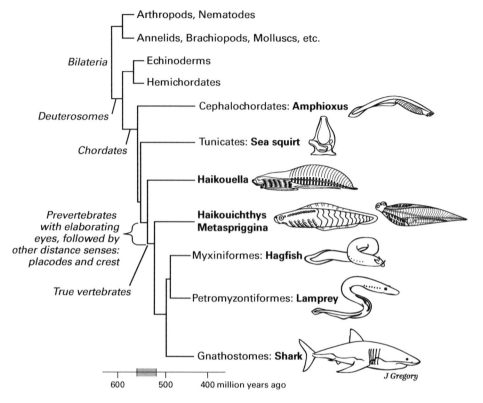

Figure 5.1
Relations and evolution of the chordate and vertebrate animals. Appearance of key sensory features is indicated at left. The animal sizes are not drawn to scale.

amphibian to reptile to bird to mammal. In addition, the human brain is shown in table 5.1. By labeling the brain structures that are held in common, figure 5.3 shows how these structures vary across the different vertebrates in relative and absolute size. These brain structures play a role through the rest of this book, so table 5.1 lays them out systematically and presents their functions. Then, table 5.2 shows which of the brain structures are also present in amphioxus and tunicates, and which ones are unique to the vertebrates. From this, the vertebrate-only structures will reveal the neural advances that occurred in the first vertebrates.

Here is a summary of tables 5.1 and 5.2, along with the brief survey of how brains vary across vertebrates. The *hindbrain* has a caudal part or medulla oblongata and a rostral part, the pons and cerebellum. The *midbrain* has a roof called the tectum and a tegmentum ventral to that. Together, the medulla, pons, and midbrain form the *brainstem*. The *forebrain* has a diencephalon

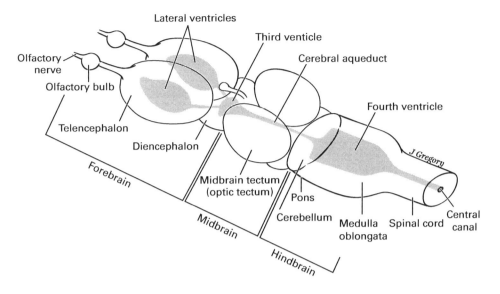

Figure 5.2
Main subdivisions and parts of the vertebrate brain (forebrain, midbrain, and hindbrain). Rostral is to the left. Modified from Butler and Hodos (2005). The hollow center is shaded.

(mostly thalamus and hypothalamus) and telencephalon (cerebrum, with two cerebral hemispheres). The brain is hollow with a connected system of fluid-filled *ventricles* in its center. Caudal to the brain is the spinal cord, whose hollow center is the *central canal*.

In vertebrate evolution, when the brain enlarged, this usually involved expansion of the cerebrum, especially of its dorsal part, the pallium or cerebral cortex, which is associated with higher mental functions. This expansion is best seen in the giant cerebrum of mammals and birds (figure 5.3F, G, and table 5.1), but also in some jawed fish and in reptiles (figure 5.3E). But not all enlargement involves the cerebrum. Any other part of the brain that performs a function at which a vertebrate species is expert can also be enlarged. For example, in fish with an extremely sharp sense of taste or of sensing electrical fields (electroreception),[2] the parts of the brainstem that start to process these sensory stimuli are large.[3]

The cerebral pallium functions in memory and learning, fear, processing smells, complex processing of most kinds of sensory stimuli, making decisions, influencing body movements, and cognition. Attaching to the front of the cerebrum is the olfactory bulb for the initial processing of smells. This

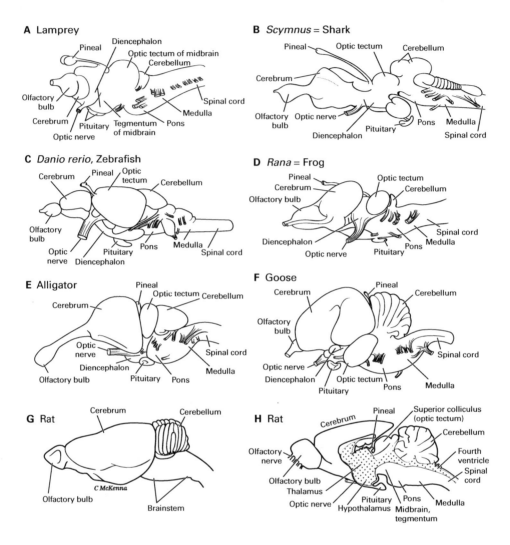

Figure 5.3

Brains of various groups of vertebrates in side view. Rostral is to the left. A. Lamprey. B. Shark, *Scymnus*. C. Zebrafish, *Danio*. D. Frog, *Rana*. E. Alligator. F. Goose, *Branta*. G. Rat, *Rattus*. H. Rat brain cut in half to show its interior; dotting shows the ventricles.

Consciousness Gets a Head Start

Table 5.1 Parts of the central nervous system of vertebrates, emphasizing the brain

I. Spinal cord
  Receives sensory inputs and sends motor outputs via the spinal nerves. Reflexes and simple movement commands involving the body below the head.
II. Brain, from caudal to rostral
  A. Hindbrain, or rhombencephalon
    1. Medulla oblongata, or myelencephalon
      - Attaches to last five of the twelve cranial nerves (8–12), so is involved with innervation of the head.
      - Contains part of the reticular formation (see below), with "centers" for controlling autonomic (inner-body) functions such as swallowing, heart rate, blood pressure, and breathing rate.
    2. Metencephalon, or pons and cerebellum
      a. Pons
        - Attaches to three of the twelve cranial nerves (5–7), so is involved with innervation of head.
        - Its reticular formation has centers for breathing, swallowing, and urinating.
        - Its "pontine nuclei" work with the cerebellum.
      b. Cerebellum
        - Calculates how to make body movements smooth and coordinated.
        - Helps with balance and posture.
  B. Midbrain, or mesencephalon
    1. Tegmentum
      - Attaches to one cranial nerve (3), which helps move the eyes to look at things.
      - Its reticular formation starts some coarse body movements: midbrain locomotor center.
      - Contains caudal parts of the striatum, which selects among basic motor programs (see below).
    2. Tectum ("roof"), or optic tectum (named superior colliculus in mammals)
      - Visual reflexes and extensive visual processing.
      - Also receives and processes other kinds of sensory information (touch, hearing, etc.).
      - Isomorphic sensory maps.
      - Attaches to one cranial nerve (4), which helps move the eyes to look at things.
    3. Periaqueductal gray, around the hollow center of the midbrain
      - Signals the bodily responses of panic.
  C. Forebrain, or prosencephalon
    1. Diencephalon
      a. Pretectum
        - Mediates responses of eye to changes in the intensity of light.
      b. Thalamus (egg-shaped in figure 5.4)
        - Relay center by which all other parts of the brain and spinal cord communicate with the pallium (see below), including with the cerebral cortex of mammals.

Table 5.1 (continued)

- Processes the information going through it.
- Role in alertness and wakefulness.
  - c. Epithalamus (most superior part of diencephalon)
    - i. Pineal organ
      - Has light receptors that tell direction of the light and time of day or year, but does not form images. Secretes the hormone *melatonin*, which prepares body for night time of the day-night cycle.
    - ii. Habenula (habenular nucleus)
      - Inhibits body movements, during sleep and to prevent behaviors that are inappropriate or dangerous.
  - d. Prethalamus, or subthalamus
    - A part of the striatum, which selects among basic motor programs (see below).
  - e. Hypothalamus (most inferior part of diencephalon)
    - Signals motor responses of affects or emotions.
    - Second cranial nerve attaches here: optic nerve of vision.
    - Retina of eye is an embryonic outgrowth of this part of brain.
    - Pituitary gland attaches to its floor, contains the pituitary's neurohypophysis.
    - Main brain region for monitoring and controlling the autonomic or visceral functions; controls pituitary's secretion of many hormones for such purposes.
    - Has centers for thirst, hunger, temperature regulation, digestion.
    - Controls some basic behaviors for survival, including sexual behaviors and biorhythms (generated by its suprachiasmatic nucleus).
2. Telencephalon, with two cerebral hemispheres (see also figure 5.4)
   a. Pallidum (ventrally located)
     i. Striatum, or corpus striatum; essentially the same as "basal ganglia."
       - Selects which motor program should be carried out, among the many different motor programs that are coded in the brainstem and spinal cord: such programs are for breathing, chewing, swallowing, locomotion; fast but inflexible decisions about what movement patterns to perform.[1]
     ii. Septal nuclei, septum pellucidum
       - Affects,[2] moods, anxiety
     iii. Amygdala, its pallidal or "subpallidal" part
       - Affect[2]-directed responses and actions. Fear or fearlike behaviors.
   b. Olfactory bulb and tract
     - First cranial nerves (olfactory) attach to bulb.
     - Initial smell processing occurs in the bulb.
   c. Pallium (dorsally located; see esp. figure 5.4)
     ii. Dorsal pallium
       - In mammals, includes most of the cerebral cortex.
       - Processes information from many senses, elaborate isomorphic maps in mammals.
       - Complex decisions.

Table 5.1 (continued)

- Influences actions and behavior, some motor control (of fine, voluntary movements in humans).
- Cognition
- Some affective[2] or emotional roles.

    iii. Lateral pallium
- Olfactory processing of odors, important part of olfactory pathway.
- In mammals, is called the olfactory cerebral cortex.

    iv. Medial pallium, hippocampus
- Constructs spatial memories of places and memories of objects and events.

    v. Ventral pallium, or pallial part of amygdala
- Affective[2] learning, especially of fears.

III. Other parts of nervous system, which involve multiple regions of the brain
    A. Brainstem
- This is the medulla oblongata, pons, and midbrain, all continuous with one another.

    B. Hollow center of spinal cord and brain (see figure 5.2)
        1. Central canal of spinal cord
        2. Ventricles of brain
            a. Fourth ventricle, in hindbrain
            b. Cerebral aqueduct, in midbrain
            c. Third ventricle, in diencephalon
            d. Lateral (first and second) ventricles, one in each cerebral hemisphere

    C. Autonomic nervous system (ANS)
- Motor neurons that issue from the brain and spinal cord, reach the visceral organs and control inner-body functions: digestion, heart rate, urination, etc.
- Technically defined as the part of the peripheral nervous system that activates (or inhibits) the glands and smooth muscles in the organs of the body; also, the muscle of the heart wall.
- Parasympathetic and sympathetic divisions. Parasympathetic signals organs for "resting and digesting," whereas sympathetic signals them for "fright, fight, or flight."
- Much of this ANS output is controlled or influenced by the brain and spinal cord.
- Contrasts with the *somatic motor system*, which signals *skeletal* muscles to contract and move the limbs or body.

    D. Reticular formation
- In the core of the brainstem: in medulla, pons, and midbrain.
- Contains the *reticular activating system*, for arousal, directing attention, and controlling the intensity of emotions (see figure 10.8).
- Signals and controls muscle contractions for basic or coarse body movements.
- Has centers that influence visceral functions via the ANS.

Table 5.1 (continued)

  E. Greater limbic system (figure 5.4)
   - For affects[2] (and for emotions in humans).
   - Motivates the basic behaviors for survival.
   - Maintains visceral functions and a constant state of the inner body (homeostasis).
   - In forebrain, includes amygdala, hippocampus, basal forebrain, septal nuclei, hypothalamus, and more.
   - In brainstem, includes periaqueductal gray, reticular formation, and more.

1. Striatum's motor functions are described by Grillner et al. (2005).
2. Affects are positive and negative feelings (see chapters 1, 7, and 8).

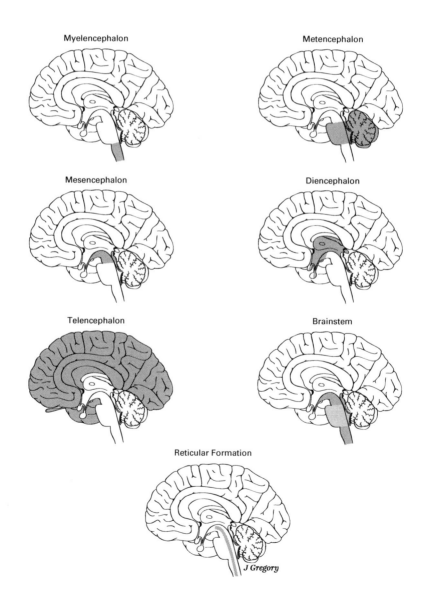

Table 5.2 Vertebrate CNS versus that of tunicates and amphioxus: a comparison of parts: (+) present, (–) absent

	Vertebrates	Amphioxus	Tunicates
1. Spinal cord	+	+	+
2. Hindbrain	+	+	+
A. myelencephalon	+	+	+
B. metencephalon	+	–	–
i. pons	+	–	–
ii. cerebellum	+	–	–
3. Midbrain	+	+	+
A. tegmentum	+	+	+
B. tectum	+	–	–
4. Forebrain	+	+	+
A. diencephalon	+	+	+
i. pretectum	+	–	–
ii. thalamus	+	?	?
iii. pineal	+	+	+
iv. prethalamus	+	–	–
v. hypothalamus	+	+	+
vi. neurohypophysis	+	+	+
B. telencephalon, with cerebral hemispheres	+	–	–
5. Hollow center of spinal cord and brain	+	+	+
6. Autonomic nervous system	+	–[1]	–[1]
7. Reticular formation	+	+	?
8. Greater limbic system	+	+[2]	?

The sources for amphioxus and tunicate structures are Burighel and Cloney (1997); Lacalli (2008); Mackie and Burighel (2005); Moret et al. (2005); Wicht and Lacalli (2005). See also chapter 3.

1. The visceral peripheral nervous systems of amphioxus and tunicates are different from the autonomic nervous system of vertebrates, with no contribution from the neural crest.
2. See the "neuropile" in amphioxus brain in figure 3.5B.

bulb is relatively largest in fish and amphibians and in some mammals like rats, who rely heavily on their sense of smell.

The optic tectum of the midbrain is a center for processing visual information throughout vertebrates, though less so in mammals. The tectum also processes the other kinds of sensory information (except smell). It is a relatively large part of the brain in many fish (figure 5.3A, C). These facts make the tectum and cerebral pallium the main brain regions for processing information from the senses of vision, touch, hearing, and smell.

The cerebellum (not to be confused with the cerebrum) is a hindbrain structure that varies in size across vertebrates. It senses the movements of one's own body and calculates how to make these movements smooth and coordinated, as well as helping with balance. It is largest in active and well-coordinated animals like birds and humans.

Mostly invisible from the outside is a deep or medial *limbic system* of the brain (the greater limbic system of Nieuwenhuys).[4] All subdivisions of the brain contribute to the limbic system (figure 5.4). That is, it has parts in the brainstem, diencephalon, and telencephalon. It is involved in running core-body and basic survival functions, motivations, and affects. It includes the brain's "primitive core region" described in chapter 3.

From the limbic core through the rest of the brain, we see much similarity of brain structure across all the vertebrates. The most obvious exception to

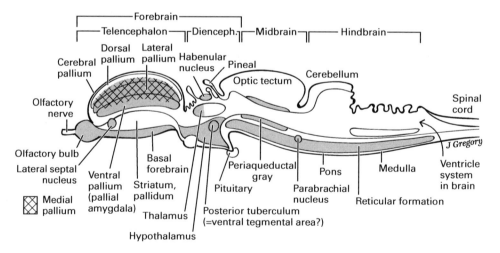

Figure 5.4
The limbic system, seen as the shaded structures in this mid-sagittal view of a generalized vertebrate brain. This figure also gives the best view of the parts of the telencephalon: the pallium, pallidum, and so on.

this is the greater complexity of the cerebral pallium in mammals, birds, and reptiles.

Also note that vertebrates have critical brain structures that are lacking in amphioxus, our proxy for the earliest prevertebrates (table 5.2). This was a theme of chapter 3, but it becomes even clearer now that we have presented the vertebrate brain in more depth. Judging from the list in table 5.2, the most important brain structures that evolved anew in vertebrates are the cerebrum, a well-developed thalamus, the optic tectum, and the metencephalon (cerebellum and pons). All these new brain regions have major sensory-processing roles.

Vertebrate Senses

We will now reconstruct the senses of the prevertebrate ancestor, starting with light detection. Based on amphioxus from chapter 3 and our reconstruction of the original bilaterian worms from figure 4.5B, the early prevertebrates must have had light-sensitive cells or *photoreceptors*, some of which were partly shielded by a pigment screen, allowing them to tell the direction of light sources but not to form a visual image[5] (see figure 5.5A, Stage II). For their other senses, early prevertebrates would have had *mechanoreceptors* in their skin, detecting various kinds of touch stimuli. We say this because amphioxus has a wide variety of such receptors.[6] These prevertebrates also must have had *nociceptors*, for responding to noxious or harmful stimuli, as almost all animals do, and *chemoreceptors* for senses akin to smell and taste that detected odor and taste molecules dissolved in the water. The claim for chemoreceptors seems a risky deduction because scientists have not yet been able to find any definitive chemoreceptor organs in amphioxus and tunicates.[7] But these chordates do have some of the genes associated with chemoreception,[8] and chemoreceptors are thought to be a primitive character of all bilaterian animals. All the kinds of receptors—photoreceptors, mechanoreceptors, and chemoreceptors—would have been located in the body-surface layer of the prevertebrates, their skin, although some of the photoreceptors were in the brain and spinal cord.

Even though the vertebrates' ancestor had these basic sensory systems that allowed some simple processing of environmental stimuli, we reconstruct these systems as entirely reflexive, as in amphioxus and tunicates. For conscious processing to arise, a more powerful and enlarged brain would have to evolve, with complex sensory hierarchies. The logical question, then, is how and why did this explosion in the sophistication of sensory processing occur?

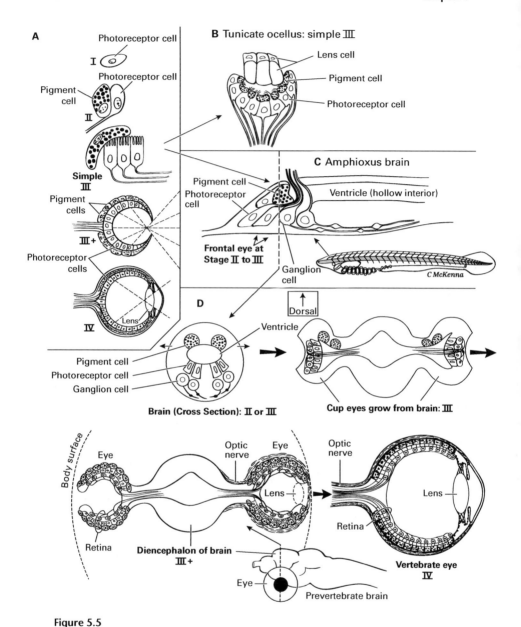

Figure 5.5

Stepwise evolution of the vertebrate eye. Reconstructed using amphioxus and tunicates, and from Dan-Eric Nilsson's stages of eye evolution that were based on the range of eye types across the invertebrates. A. Nilsson's four stages I–IV. B. Tiny ocellus "eye" of a tunicate larva (homologous to the pineal, "third-eye" of vertebrates). C. The small frontal eye of the amphioxus larva, in a side view of the brain. D. Proposed stages in vertebrate eye evolution starting from an amphioxus-like frontal eye. These cross sections were inspired by the art in articles by Trevor Lamb.

Vision First?

In the previous chapter, we deduced that the different distance senses of vertebrates evolved more or less together as Cambrian prevertebrates scrambled to detect and avoid the many new predators, most of which were arthropods. But in this nearly simultaneous appearance, did one of the senses actually evolve first, getting a head start, with the others soon following along?

One hypothesis is that light detection was the first of the distance senses to evolve sensory images. This hypothesis was proposed by Andrew Parker and built upon by Michael Trestman,[9] both of whom focused on arthropods, although the claim applies to vertebrates as well. Parker called it the Light-Switch hypothesis and said that the appearance of true, image-forming eyes in arthropods allowed them to become the first predators, which then led to the Cambrian explosion itself. Trestman refined this to say that the new eyes did not merely form images, but formed *spatial* images that could identify discrete objects in the visual field and could reveal the positions, distances, and movements of these objects, as well as which objects lay in front of others. This spatial vision, he said, allowed prey to be spotted, tracked, struck, and manipulated with ease, and then placed in the mouth. Although the *arthropods* evolved such eyes first and threw the light switch, the vision of their *prevertebrate* prey quickly improved, via natural selection, to see the predators coming. The reasoning behind the vision-first hypothesis is that of all the senses, focused vision yields the most information about the environment, so it is most likely to have enabled the complex predatory activities and the evasive behavioral defenses of the Cambrian arms race. Once begun by vision, this race then led to the elaboration of other senses—smell, touch, and hearing—all of which improved survival in both predators (arthropods) and prey (prevertebrates).

By what anatomic steps did the image-forming eyes evolve in vertebrates (figure 5.5)? The fossil record is of no help until the last stages, so we must reconstruct the sequence using modern amphioxus and tunicates, plus some general trends in eye evolution that can be determined from other modern invertebrates. That is, dozens of groups of invertebrates have gone through the initial stages and evolved simple eyes independently, especially a variety of worms and snails, and the stages are mostly the same in all these animal groups.[10] From such comparative-animal evidence, Dan-Eric Nilsson has divided eye evolution into four functional and structural stages, each of which could easily and logically lead to the next (I–IV in figure 5.5A).[11] The four stages go from isolated photoreceptor cells (I), to an added pigment cell (II), to cupping (III and III+), to camera-style eye with high visual resolution (IV). Now for a more precise description of the stages.

Stage I is basic, nondirectional photoreception in which individual photoreceptor cells, unshaded by any pigment, just monitor the intensity of light. An animal in this stage can tell if it is night or day, or if it is in a dark place.

Stage II is directional photoreception. Here, the photoreceptor cells are partly shielded by pigment cells so they receive and sense light from some directions but not from others. This directionality lets the animal move away from or toward light, and lets it detect shadows of moving objects such as approaching predators. The ocelli ("little eyes") of the blind, ancestral bilaterian shown in figure 4.5B would have been Stage II directional photoreceptors. The frontal "eye" of larval amphioxus, which is inside the brain and homologous to the eyes of vertebrates, is at this Stage II, or else barely into the next Stage III. We show this frontal eye of amphioxus in figure 5.5C (also in figure 3.5).

Stage III is low-resolution vision. A simple cupping of the photoreceptor and/or pigment cells is all that is needed to go from Stage II to Stage III (figure 5.5A, B, D). With the partial screen of light-blocking pigment, each photoreceptor cell only receives and responds to light that enters the cup from the unique *direction* that that particular photoreceptor faces. By assembling the inputs from all its photoreceptor cells, the eye forms an image—a crude one, but one that's sufficient to track moving objects or to avoid bumping into things while swimming and navigating. Some animals with Stage III eyes have evolved a lens (figure 5.5B), not effective enough to focus an image but sufficient to bend and concentrate enough light onto the photoreceptor cells so that the eye can operate in dim light.

Figure 5.5D models how the prevertebrates went through this Stage III and beyond, with the eyecups enlarging as lateral outgrowths of the brain, and becoming the paired eyes. Note that in a proposed "III-plus" stage, the cupping became so deep that the eye acted like a pinhole camera and produced higher-resolution vision than did the shallower cup of the initial Stage III. Here our ancestors really began to see *images*—not perfectly, but well enough for earliest Cambrian times. Presumably their neo-arthropod predators, with which the prevertebrates coevolved, could not yet see any better. These two clades of visual unsophisticates fit an old saying, translated from its original Latin: "In the land of the blind, the one-eyed man is king." And vision was rapidly improving.

Stage IV is high-resolution vision. Detailed vision required little more than changes in the lens so that this lens sharply focused images on the retina, and along with this a larger number of retinal photoreceptor cells. This is shown as the final step in figure 5.5D, labeled "Vertebrate eye."

The new high-resolution images in the eye, if not to be useless, must have been accompanied by a great increase in visual processing in forebrain and midbrain. The detailed, isomorphic visual representations that resulted from such processing were the brain's first conscious mental images, and they allowed the first vertebrates to perform a wide range of vision-guided behaviors. In this way, vision signaled the dawn of consciousness.

The retina of the vertebrate eye comes from the diencephalon (figure 5.5D), which means it is a part of the brain and belongs to the central nervous system. Therefore, even though the eye lies out near the body surface, it is not part of the peripheral nervous system (PNS), as one would expect. By contrast, all the other senses are received in the PNS.

Smell First?

We have taken the view that vision was the first of the distance senses to evolve in vertebrates and arthropods, but some of our colleagues, Roy Plotnick, Stephen Dornbos, and Junyuan Chen, disagree. They say that smell came first.[12] They argue that at the start of the Cambrian, the ancestral worms lived under microbial mats on the ocean floor (as discussed in chapter 4), where odor is important but light cannot penetrate. They said that vision evolved later, when the animals started to swim above the ocean floor. But we counter that many ancestral worms must have grazed *on top* of the mats, where lots of light did reach. Plotnick and coworkers also argue that vision is not necessary for predators that live on the ocean bottom, pointing to modern starfish and snails. Our counterargument is that many visual predators do live in this environment, and they take advantage of the complex visual landscape of the ocean floor. Visual bottom dwellers among today's arthropods and fish include lobsters, horseshoe crabs, flounders and stingrays. Finally, Plotnick and coworkers argue that the earliest fossil animals with image-forming eyes (such as trilobites) did not appear until 520 mya, which is 20 million years after the Cambrian began, so eyes must have evolved late. This argument seems unwarranted, however, because no body fossils of any animals at all are available from about 540 to 520 mya—an unfortunate gap in the rock record from which only small scales, shells, and fragments of the early animals are known[13]—meaning neither eyes *nor smell organs* could be known from this gap, whether they existed then or not.

We will reemphasize why we reason that vision was the key initial sense: because light provides so much more information. "Smell images" of space do exist[14] and can be constructed by the vertebrate brain from odor plumes and concentration gradients of odor molecules emerging from smelly objects in the water, but these smell images cannot reveal the sharp borders of objects or the precise distances that light images can provide. And visual images are so much better for following objects that move fast, such as an attacking predator. Granted, vision does not work well in murky waters, but such murk is often caused by turbidity currents that also disperse and confuse odor signals, likewise interfering with the sense of smell.

However, smell is special in one important way. Unlike sights, sounds, or touch stimuli, smells linger after the source of the stimulus goes away. Odors remain, of body wastes, body secretions, and other spoor that is left behind by predators, mates, prey, and other animals. Smells tell of the past. Because smell has this unique time dimension, as soon as olfaction appeared natural selection linked it to memory: from this smell, I *remember* and recognize what odor-emitting object used to be here and still may be nearby, so I can know to avoid or approach it. The intimacy of this relationship explains why memory and smell structures evolved so near one another in the incipient telencephalon of the ancient vertebrate brain.

Now back to the argument that vision evolved before smell had elaborated beyond basic, primitive, chemoreception. In addition to the theoretical reasons, the physical evidence also favors this view. First, a method of identifying cell types by the genes they express has shown that the frontal eye of the amphioxus larva really is homologous to the vertebrate eye, in details ranging from its retinal cells to its pigment cells. Thus amphioxus has an "eye" while lacking any known organ of smell or hearing. Second, a study by Martin Šestak and his colleagues found that a surge of genes involved with vision evolved earlier on the tree of chordate animals than did the genes associated with the other senses such as smell, hearing and balance. Third, the candidate, fossil prevertebrates from Cambrian rocks show more evidence of eyes than of nasal (smell) organs.[15]

We will return to these fossils and to vision, but first we must explore the *nonvisual* distance senses of the ancient prevertebrates, the senses that evolved after vision. These other senses did not merely tag along as afterthoughts behind the blossoming of vision. Instead, their elaboration was part of a

Consciousness Gets a Head Start

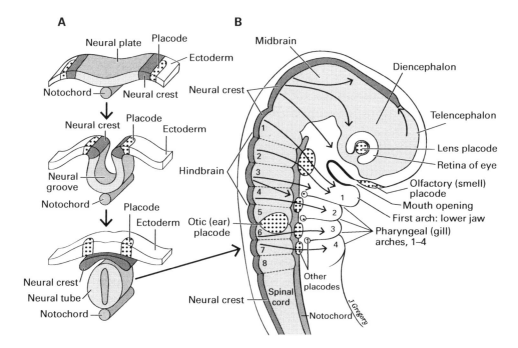

Figure 5.6
The embryonic tissues that are unique to vertebrates: neural crest and placodes. A. Time sequence of their formation and development in the back (dorsal side) of the embryo, arising from ectoderm near the neural tube. B. A vertebrate embryo, side view of its head and neck regions, showing the placodes and neural crest. The long arrows in the body show that the neural crest migrates widely.

reorganization of the ancestral body and nervous system that contributed as much to the evolution of brain and consciousness in vertebrates as did the new visual images.

The Neural Crest and Placodes

We have established that the focusing lens of the eye is essential for high-resolution vision and the evolution of the visual image. But where did this lens come from? Nor have we explained the body source of the neurons of the vastly improved nonvisual senses of vertebrates. Unexpectedly, these things (and more) came from brand new embryonic cells that evolved near the dawn of the vertebrates, specifically from neural crest and placodes (figure 5.6).[16]

Neural crest and placodes do not exist in any animals except the vertebrates,[17] and they develop into so many important adult structures that they are widely recognized to be key contributors to the evolutionary success of the vertebrates. They especially contribute to the head and neck regions, and it is thought that their evolutionary origin led to the great complexity of the vertebrate head. They form in the early embryo, on the dorsal head and back, around the periphery of the *neural plate*, which is a long strip of surface tissue (ectoderm) extending from head to tail (figure 5.6A). The neural plate soon pushes down and folds into a neural tube, to become the spinal cord and brain. When this folding begins, ridges form at the margins, with the top of the ridge becoming the neural crest and the lateral walls giving rise to the placodes. As shown in figure 5.6A, the crest and placodes are paired on the right and left sides of the body.

We will discuss the neural crest first. Its cells develop into both nervous and nonnervous structures. They give rise to all the sensory neurons in the trunk and many in the head, especially neurons for the somatic senses of touch, pain, temperature, and so on. In addition, many crest cells migrate away from their original location in the back, to become widespread pigment cells called *melanocytes*, motor neurons of the gut, some of the skull, and more.

The placodes, whose official name is *ectodermal* or *neurogenic placodes*, develop in the head and neck region, where they mostly form sensory structures. The placodes are indicated by stippling in figure 5.6B. Most of them stay at the body surface, and overall they do not migrate as deeply or widely as does the neural crest. The olfactory placode gives rise to the receptor cells for smell in the nose, the lens placode gives rise to the lens of the eye, and the otic placodes give rise to the cells in the ear for the senses of hearing and balance. Other placodes contribute, along with neural crest, to the somatic sensory neurons of the head. They also form visceral sensory neurons to digestive organs including those to the taste buds, and the sensory cells of the lateral line (a line of mechanoreceptors in fish and water-dwelling amphibians, in the skin of their head and the lateral side of the body, for detecting vibrations in the water).[18]

Nearly all of the nonvisual sensory PNS of vertebrates comes from the placodes and neural crest. Amphioxus and tunicates, by contrast, have a sensory PNS but no crest or placodes; at best they only have rudimentary precursors of the crest and placodes that contribute little to their PNS.[19] This difference implies that there was an almost complete replacement of the sensory PNS around the time of the dawn of the vertebrates. Our story of vertebrate sensory evolution has emphasized how giant eyes ballooned from the brain to gather vast quantities of visual information, but in some ways that was not as impressive as the near-total reboot of all the other senses, mechanoreceptive and chemoreceptive, that came with the neural crest and placodes.

When did the neural crest and placodes evolve, in relation to the stages of eye evolution mapped out in figure 5.5D? We deduce that they originated no later than between the low- and high-resolution stages (III and IV) in the eye sequence. Our reasoning is that a lens had to be present for that transition to sharp vision to occur, so the lens placode had to exist by then to form the lens.

The sensory revolution near the dawn of vertebrates affected the brain (figure 5.7). The new waves of sensory information arriving from the placodal and crest senses of olfaction, hearing, an increased sense of touch, and so on

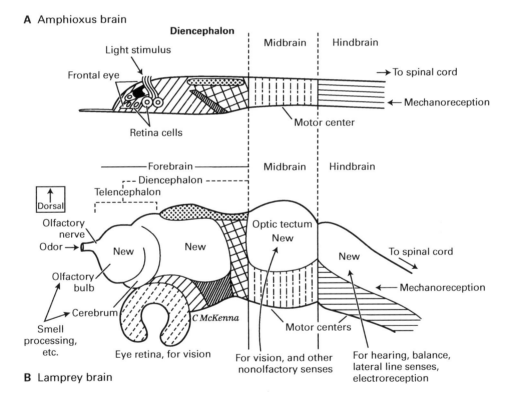

Figure 5.7
The brain of amphioxus (as a proxy for the ancestral, prevertebrate brain) compared to the brain of a lamprey (a proxy for the brain of the first true vertebrate). Identical hatching patterns mark the corresponding regions in the two brains. Notice the large "New" regions of the vertebrate brain. Largely after the studies of Thurston Lacalli.

joined the flood of visual information from the image-forming eyes, demanding elaborate multisensory processing by the brain. As proposed by Lacalli, the resultant expansion of brain regions explains the differences between the vertebrate and amphioxus brains that we laid out in chapter 3 (figure 3.5).[20] New, mainly dorsal regions of the brain appeared and enlarged. These are labeled "New" in figure 5.7B. From the dorsal and rostral part of the forebrain grew the new telencephalon, with the cerebrum. The vertebrate telencephalon receives the input of smell, so its origin was related to the appearance of this olfactory sense. The optic tectum of the midbrain enlarged to process visual information (along with other senses), and the dorsum of the hindbrain enlarged also, to process information from the senses of hearing and balance, and from the lateral line. Judging from all living vertebrates, each of the new and improved senses arrived in a neatly mapped, isomorphic arrangement. Thus, the retina, neural crest, and placodes supplied sensory input to the expanding dorsal parts of the brain, in which evolved the most elaborate sensory maps and images.

Candidates for the First Conscious Organisms on Earth

We have proposed that the senses of vertebrates evolved in two overlapping waves: image-forming eyes grew out from the brain, and then the other senses elaborated from the placodes and neural crest. Our colleague Ann B. Butler actually proposed this first, in the year 2000, in her stepwise hypothesis of vertebrate origins.[21] To illustrate her hypothesis, she reconstructed an intermediate stage between an amphioxus-like ancestor and the first true vertebrate, which she called the "cephalate" animal (figure 5.8A). Her hypothetical cephalate had evolved paired eyes and an enlarged brain with its visual system based on the diencephalon and midbrain. But it still lacked the other sensory systems of vertebrates, also lacked neural crest and placodes, and had no telencephalon. Its visual pathways, she proposed, served as a *circuitry template* for the new or improved sensory systems of smell, taste, touch, and hearing. That is, the sensory paths from the newly evolved crest and placodes

Figure 5.8
Prevertebrates and early vertebrates. A. The hypothetical "cephalate" animal, conceived by Ann Butler as a sort of intermediate between amphioxus and true vertebrates. B. *Haikouella*, a fishlike fossil animal. C. *Haikouichthys*, an Early Cambrian fossil fish. D. *Metaspriggina*, a Middle Cambrian relative of *Haikouichthys*.

Consciousness Gets a Head Start

A Cephalate

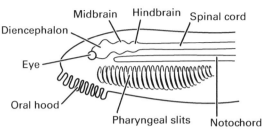

B Haikouella

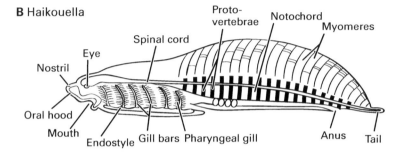

C Haikouichthys

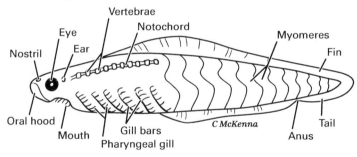

D Metaspriggina

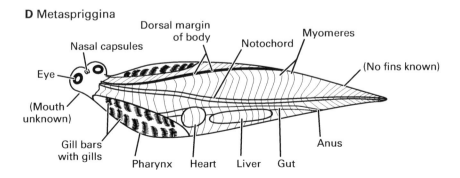

adopted vision's neuronal patterns in the brain. The sensory hierarchies of neurons that formed for the nonvisual senses mimicked the preexisting visual hierarchy.

About the time Butler presented her cephalate model, some fossil animals by which to test it were being discovered. As pictured in figure 5.8B, these Cambrian fossils are the soft-bodied, blade-shaped, inch-long yunnanozoans from the Chengjiang Shale of the Yunnan province in southern China.[22] This shale is 520 million years old and contains the oldest known whole-body fossils of bilaterian animals (discussed in chapter 4). The best-preserved and most informative yunnanozoan fossils are of a species named *Haikouella lanceolatum* (figure 5.8B). The specimens of *Haikouella* were interpreted by Junyuan Chen, coauthor Mallatt and their colleagues as fish-like and the closest relative of vertebrates (figure 5.1).[23] By this assessment, it had a notochord, paired eyes, and a large brain with diencephalon and hindbrain parts, with all these structures in the same positions as in living fish such as lampreys. But *Haikouella* differed from vertebrates in having no skull or ear capsules, and only hints of a telencephalon and nostrils (for smell) were evident.

With its paired eyes and well-developed diencephalon but no ear and an uncertain "nose," *Haikouella* seems to fit well with the vision-first idea and with the proposed cephalate stage in which the crest and placodal senses had not yet evolved. Yet, it had fishlike gill bars, which develop from neural crest, and a dotlike lens in the center of each eye, which develops from the lens placode. This suggests that *Haikouella* was closer to the vertebrates than was the cephalate, as the latter was originally conceived. *Haikouella*'s eyes were small, just a fifth of a millimeter in diameter. This is much larger than the tiny frontal "eye" of amphioxus, whose diameter is a hundredth of a millimeter, but still was probably too small to have formed a well-focused image.[24] Other, clearly visual Cambrian animals, mostly arthropods, had eyes closer to a millimeter or larger.[25] Still, the fact that *Haikouella* had paired eyes at all, lateral to the diencephalon rather than within it, gives some support to the vision-first hypothesis.

Haikouella and the other yunnanozoans are controversial, and some paleontologists think their "eyes" are just artifacts of fossilization that never existed in life, and that the yunnanozoans are not even chordates.[26] If that should turn out to be true, it leaves another group of inch-long fossil animals from the Chengjiang Shale, *Haikouichthys* and its close relatives,[27] as most likely to reveal the early evolution of the vertebrate nervous system (figure 5.8C). Unfortunately, *Haikouichthys* has a name that sounds like *Haikouella*, which naturally causes many to confuse these two different animals with each other. The best way to avoid confusion is to remember that "-ichthys" means "fish" (a vertebrate), and "-ella" is the dimunitive ending meaning "little-bit," or "not fully a vertebrate." Since it was discovered and described as a vertebrate by Degan Shu and his coworkers in the late 1990s, *Haikouichthys* has generated surprisingly little controversy among scientists. It is almost universally agreed to be a true vertebrate, a jawless fish, because its fossils show vertebrae, obvious eyes with a diameter of 0.6 millimeter, and ear and nasal (olfactory) capsules. Its fossils also show the characteristic gill bars, gill pouches, body fins, and curved myomeres of vertebrate fish. Without hesitation, we used *Haikouichthys* as a representative Cambrian vertebrate in chapter 4 (figures 4.4D and 4.3B). With eyes that probably formed visual images, which presumably went to its brain for processing, *Haikouichthys* had visual consciousness in our view. Recall that the 520 million-year-old Chengjiang Shale yields the oldest good record of bilaterian animals, meaning that consciousness has existed since that time.

More recently, paleontologists Jean-Bernard Caron and Simon Conway Morris identified another *Haikouichthys*-like fish, *Metaspriggina*, from the slightly younger Burgess Shale of British Columbia in Canada (505 mya).[28] *Metaspriggina* is shown in figure 5.8D. We drew it as if twisted, with its head up but its trunk shown in typical, lateral view. The head of *Haikouichthys* looks almost exactly like this in top view,[29] supporting a close relation between these two Cambrian fish.

Although *Haikouichthys* and *Metaspriggina* tell much about the first true vertebrates and even suggest when consciousness arose, they reveal less about how the senses and brain evolved along the road to vertebrates. First, their brains are not preserved—although their eyes and noses imply that they had a vertebrate brain. Second, though *Haikouichthys* and *Metaspriggina* lived among the most ancient bilaterians for which we have body fossils, each was already *too good* of a vertebrate, a full vertebrate in every way that can be learned from their fossils, so they cannot reveal the *pre*vertebrate stages of sensory evolution. To uncover those stages, we must rely on the theory-driven logic from earlier in this chapter, namely the logic that favors the vision-first view, and on the yunnanozoans, which show the brain expansion.

To summarize the chapter so far, somewhere between 560 and 520 mya, the unpaired frontal "eye" of a creature like amphioxus, which recorded the direction of light and moving shadows, evolved into the vertebrate "camera eyes" that recorded images in high resolution (figure 5.5D), a process that was completed by 520 mya. The isomorphic visual images were processed by the expanding brain into mental images, which we propose marks the arrival of consciousness. Other distance senses followed and entered conscious perception. As an attempt at more precise dating (figure 5.1), we deduce that the larger-eyed *Haikouichthys* and *Metaspriggina* were conscious but the smaller-eyed yunnanozoans may not have been. The first vertebrates had the first consciousness.

Common Pattern of Pathways and Images, for All the Senses

Butler proposed that as good vision evolved in the prevertebrates and the brain took on visual processing, the original reflex paths gained more levels of neurons in a well-organized, isomorphic, and hierarchical pattern. Soon thereafter, the nonvisual senses formed chains of processing neurons that copied the visual pattern. Indeed, all the different sensory systems of vertebrates are similar to each other in this way. The similarities are shown in figures 5.9 through 5.12, which portray the visual, touch (somatosensory), auditory, and olfactory pathways, respectively.[30] Then, table 5.3 shows how each pathway can fit into the same, five-part scheme, with the circled numbers in the pathways in the figures corresponding to the column numbers in the table (1–5). The sensory paths are illustrated for mammals, as is standard procedure, although in mammals the topmost parts are more complex than in fish, amphibians, or reptiles. The paths of fish and amphibians are included in table 5.3 and shown in figure 6.4.

The pathways and the figures are intricate and can seem overwhelming to those learning them, reminding us of a frequent complaint of our young neurobiology students who ask, "Why is the vertebrate brain, which is the most complicated thing in the whole universe, so detailed and challenging to learn?"—a question that answers itself. But we will keep things as simple as possible. We will consider just the general, shared features of the different sensory pathways and how they fit together in the basic parts of the brain.

The shared features are as follows. All the sensory pathways start with sensory information that stimulates receptors (1), which signal sensory neurons (2). The sensory neurons then send the signal to first-order multipolar neurons (3) in the central nervous system. Then in most cases these neurons project to further neurons (4) in the midbrain's "tectum" (= superior colliculus) or in the thalamus; and the thalamus projects to the higher processing center (5), the cerebrum, which is the "cortex" in each figure. For the

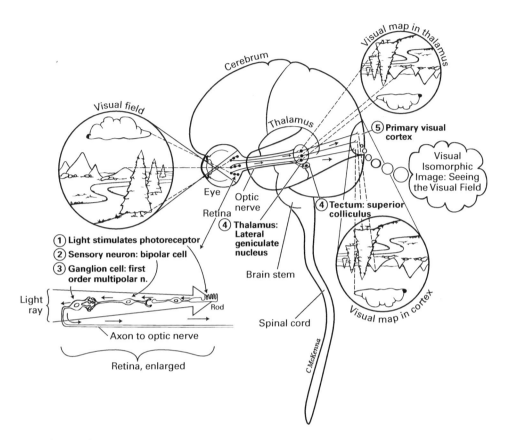

Figure 5.9
Vision: visual pathway in the human. Path goes from where light stimulates a photoreceptor (left), to bipolar sensory neuron, to ganglion cell, to thalamus (or to superior colliculus = optic tectum), to primary visual cortex of the cerebrum. Throughout the pathway, the nervous structures are arranged according to a map of the visual field, as projected onto the retina, an isomorphic feature called retinotopy. Numbers correspond to the column numbers in table 5.3. The visual image might emerge higher in the cerebral cortex than is shown here, and the same may be true for the other senses shown in figure 5.10 to 5.12 (Panagiotaropoulos et al., 2012; Pollen, 2011).

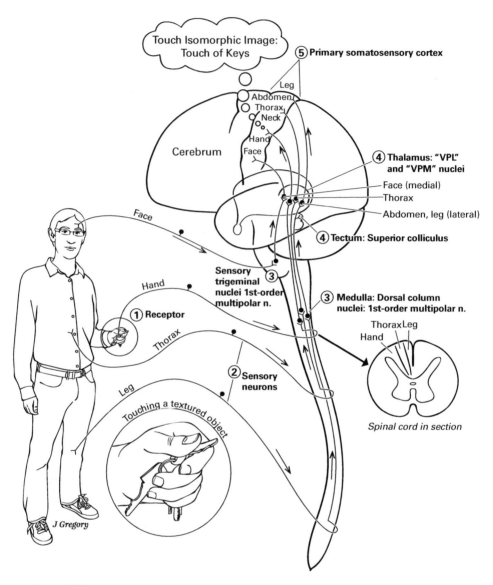

Figure 5.10
Touch: somatosensory-touch pathway in the human. Path goes from a textured object stimulating mechanoreceptors, through sensory neuron, to dorsal column nuclei in medulla oblongata of hindbrain, to thalamus (or to superior colliculus = optic tectum), to primary somatosensory cortex of the cerebrum. Throughout the pathway, the neural structures are isomorphically arranged according to a map of the body, a feature called somatotopy.

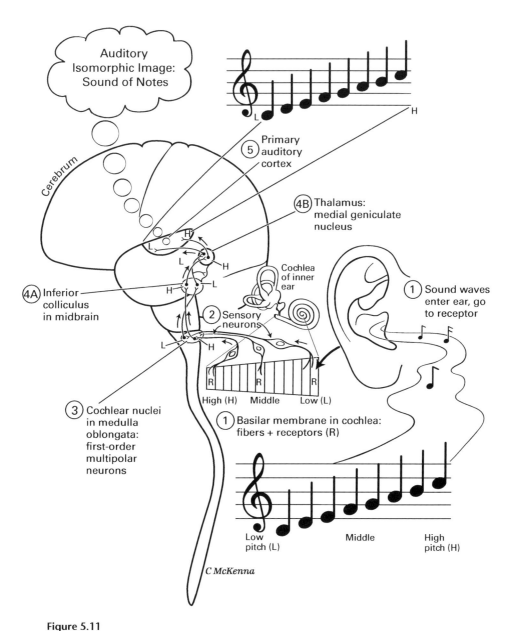

Figure 5.11
Hearing: auditory pathway in the human. Path goes from sound waves stimulating receptor cells (R) in the cochlea of the inner ear, to sensory neurons, to cochlear nuclei, to inferior colliculus, to thalamus, to primary auditory cortex of the cerebrum. Throughout the pathway, the neural structures are arranged according to a map of the basilar receptor-membrane in the cochlea, which in turn is mapped by tone (high, middle, and low pitch), an isomorphic feature called *tonotopy*.

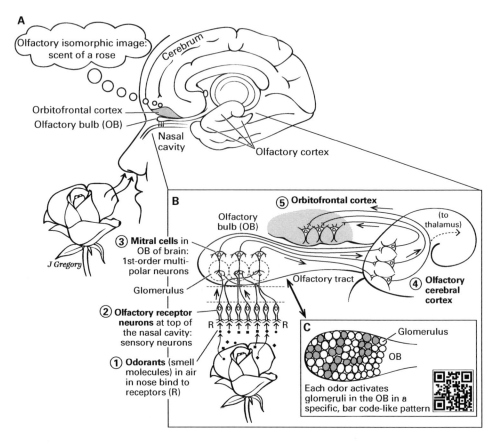

Figure 5.12

Smell: olfactory pathway in the human. A. Overview showing the medial part of the cerebrum in relation to a nose inhaling the scent of a rose. B. Path goes from smell molecules stimulating smell receptors (R), through sensory neuron, to mitral cells in the olfactory bulb (OB), to the olfactory cerebral cortex, to the orbitofrontal cortex. C. Inset showing how an odor activates a specific set of glomeruli in the olfactory bulb, as a scent's "bar code." In this pathway, the neural structures are arranged according to maps that indicate specific odors.

Table 5.3 Summary of the major sensory pathways of vertebrates

1 Sense, receptor	2 Sensory neuron	3 First-order multipolar	4 Next order	5 Next order[1]: cerebrum
Vision Photoreceptor cells (figure 5.9)	Retina's bipolar cells	Retina's ganglion cells	Thalamus; Optic tectum[2]: *visual maps and images*	Primary visual cortex: *visual maps and images*
Somatic senses Mechanoreceptors of somatosensory neurons[3] (see "R" in figure 5.10)	Somatosensory neurons (sensory ganglia)	Dorsal column nuclei of hindbrain or dorsal horn of spinal cord (trunk); sensory trigeminal nerve nuclei (face)	Thalamus; Tectum[2]: *somatosensory maps and images*	Primary somatosensory cortex: *somatosensory maps and images*
Hearing Mechanoreceptors: hair cells (see "R" in figure 5.11)	Sensory neurons (spiral ganglion in cochlea of inner ear)	Cochlear nuclei in medulla of hindbrain	Inferior colliculus of midbrain and thalamus; Torus semicircularis to tectum[2]: *auditory maps and images in tectum*	Primary auditory cortex: *auditory maps and images*
Equilibrium Mechanoreceptors: hair cells	Sensory neurons (vestibular ganglia in inner ear)	Vestibular nuclei in hindbrain	Thalamus; Torus semicircularis to tectum[2]: *tectal maps and images of body position and motion*	Primary vestibular cortex: *maps and images of body position and motion*

Table 5.3 (continued)

Taste Chemoreceptors: taste cells (see figure 7.1A)	Taste sensory neurons (cranial ganglia)	Gustatory nucleus of solitary tract in hindbrain	Thalamus; Tectum[2]: *taste maps and images*	Cortex of insula: *taste maps and images*
Smell Chemoreceptors of olfactory sensory neurons (figure 5.12)	Olfactory sensory neurons in nose	Olfactory bulb = glomerulus: mitral cells	Olfactory cortex	Orbitofrontal cortex: *olfactory maps and images*

1. "Next order: cerebrum" applies only to the mammals (for all senses but smell), and, in mammals, to still higher-order processing of all senses, leading up to the frontal lobes of the cerebral cortex. It is likely that further processing in these higher areas and a merging of all classes of sensory information there are required for complete, multisensory conscious images in mammals: Boly et al. (2013); Kandel et al. (2012); Maier et al. (2008); Pollen (2011).
2. "Tectum" here means in fish and amphibians. In chapter 6 we argue that in these vertebrates, the images for these senses are in the optic tectum with help from the thalamus, but no higher.
3. For the general inner-body senses, pain, and affective senses, see chapters 7 and 8, especially figure 7.1.

exceptions, the auditory path has an extra neuron (thus, "4A" and "4B" in figure 5.11), and the main olfactory path does not go to the thalamus or tectum (figure 5.12).[31] Without exception, however, all the sensory paths show isomorphic organization through their successive levels, which is important in mapped sensory consciousness. This isomorphism is emphasized in all four figures, 5.9 through 5.12.

How Many Levels in a Conscious Sensory Hierarchy?

Table 5.3 is good for showing the shared features of the pathways for the different senses and how they all have the same plan, but it is not ideal for showing how many *neuronal* levels are in each pathway. The latter is of interest because we seek the minimum number of levels a sensory hierarchy can have to produce consciousness. We realize that number of levels can only be a crude indicator of consciousness because the amount of processing within and between the levels in a neural hierarchy is as important as the total number of levels. But level number is far easier to measure and count than is within-hierarchy communication, so we will adopt it as a "marker," to seek the requirements for consciousness.

Upon first consideration, the minimum number of neuronal levels in a conscious sensory hierarchy seems to be four. In humans, the only animals known with certainty to be conscious, this value of four is obtained by counting neurons in the somatosensory pathway of figure 5.10 (where we counted the somatosensory cortex as one of the neuron levels), and in the smell pathway of figure 5.12 (where we counted the orbitofrontal cortex as a level). Five neuronal levels, by contrast, characterize the human visual pathway in figure 5.9 (if the photoreceptors are counted as neurons and the primary visual cortex is also counted as a level) and also the hearing pathway of figure 5.11 (again, where the primary auditory cortex is included as a level). We must stress that the actual number of neuronal levels for consciousness may be higher than this; there is much debate over whether consciousness really emerges in the primary sensory areas, as traditionally thought, or else higher in the cortex.[32] More, higher-order processing areas may be required, leading all the way up to the prefrontal cortex.

If conscious sensory images can emerge in the optic tectum of fish and amphibians, as we claim in table 5.3 and we investigate in chapter 6, then the minimum number of neuronal levels in a sensory hierarchy is only three. Such a three-neuron chain makes up the *somatosensory* pathway in these particular vertebrates, from (1) the sensory neurons to (2) relay neurons in the spinal cord or brainstem to (3) the optic tectum (see figure 6.4, the pathway named "s"). However, *four* levels characterize the all-important visual and hearing pathways to the tectum of the fish and amphibians, and the smell pathway to the dorsal pallium (see pathways "v" and "oL" in figure 6.4). Again, the actual number of levels may be higher because extensive sensory processing occurs *within* the tectum.[33]

We conclude that sensory hierarchies can be conscious if they have four or more levels of neurons projecting to (and including) the highest processing area, or in some cases, just three levels.

To summarize this chapter, the evolution of image-forming eyes in vertebrates led to the first complex sensory hierarchy, to brain expansion, "mental images," and the "dawn of consciousness." The known, fishlike Cambrian fossils are consistent with this view. Eye evolution and the visual hierarchy were closely followed by the appearance of additional game-changing structures, the placodes and neural crest, so that similarly constructed hierarchies of smell, touch, hearing, and so on arose and enabled these early brains to create integrated "multisensory images," thereby enriching sensory consciousness. Isomorphic organization characterized the successive levels of these hierarchies, contributing a mapped order to the resultant sensory images. Upon counting from these hierarchies in various living vertebrates, we deduced that the minimum number of neuronal levels required for consciousness is four in mammals (figures 5.9 through 5.12), and three or four in fish (figure 6.4).

Our concept of neurobiological naturalism postulated a set of special neurobiological features required for sensory consciousness (table 2.1). In modeling how the simple reflex pathways of an amphioxus-like ancestor expanded into complex sensory hierarchies with isomorphic organization as the vertebrates evolved, we have shown how the first vertebrates could have met these requirements and been conscious. But we have not yet considered any specific *living* vertebrate that might give clues to the origin of consciousness. For that purpose, the next chapter will explore the nervous system of the lamprey, which is in the lineage of living vertebrates that arose first. We will show how the optic tectum of its brain could hold the key to sensory consciousness.

6 Two-Step Evolution of Sensory Consciousness in Vertebrates

As we discussed in chapter 1, sensory or phenomenal consciousness entails the kind of consciousness that is subjectively experienced as images of the outside world. So far we have considered how vertebrate consciousness evolved from an amphioxus-like, invertebrate stage. Now we will investigate this problem from the perspective of the living vertebrates and everything that is known of their brains. We will present evidence that today's jawless fish, represented by lampreys, are conscious, which would mean that every other vertebrate is also conscious. Then we will argue that long after consciousness originated in the first fish in the Cambrian (before 520 mya), it progressed in a second great step in the first mammals (around 220 mya) and the first birds (around 165 mya). Throughout the chapter, we will show how our interpretations differ from the favored view that says consciousness needs an enlarged dorsal pallium or cerebral cortex and thus exists only in humans or in all mammals and birds. Much of this chapter is adapted from our two recent publications on this same topic.[1]

Introduction to the Vertebrates

First it is necessary to describe the kinds of vertebrates, both living and extinct, so that we can investigate consciousness through their long evolutionary history.[2] Figure 6.1 pictures the major vertebrate clades and their phylogenetic relations. The animals pictured at left along the tree's backbone are reconstructions of the ancestors of some of the clades. Recall the earliest known vertebrates, *Haikouichthys* and *Metaspriggina* from the Cambrian ocean (see figures 5.8C, D), which were jawless and probably filter feeders. Additional lines of jawless fish enjoyed modest success from the Late Cambrian until about 360 mya,[3] still feeding on small or suspended food particles. A few of these fossil jawless fish are shown in figure 6.2. Some invaded fresh waters

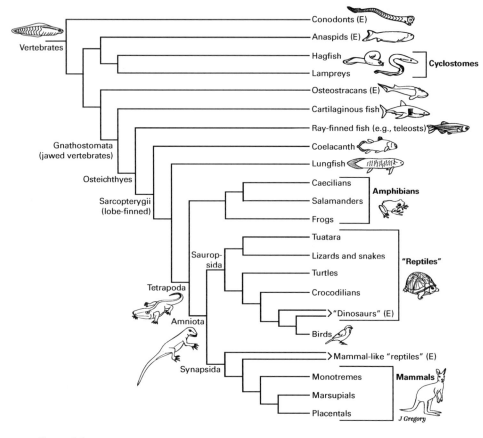

Figure 6.1
The clades of vertebrates and their phylogenetic relationships, according to current understanding. A few of the fossil groups are included and labeled "E" for extinct.

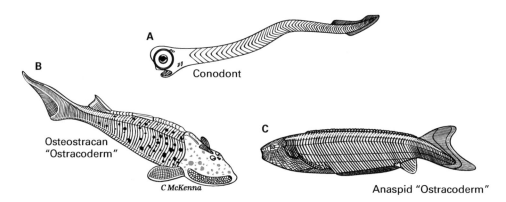

Figure 6.2
Fossil jawless vertebrates from the late Cambrian and after. A. Conodont. B. Osteostracan fish. C. Anaspid fish.

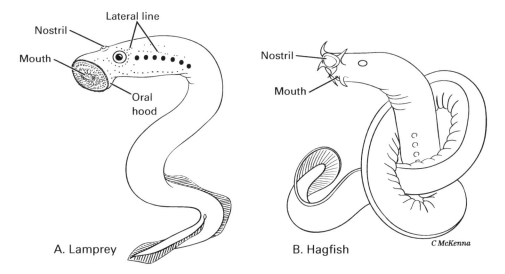

Figure 6.3
Cyclostomes: Modern lamprey (A) and hagfish (B).

(around 440 mya), and one line gave rise to the eel-shaped lampreys and hagfish (figure 6.3). These two cyclostomes ("round-mouths") are the only jawless vertebrates that survive today.[4] Both are predators and carnivores, using a pistonlike "tongue" tipped with horny blades to slice up flesh and prey animals, and to bring this food into the mouth. The adult lamprey is a free-swimming parasite that uses its round oral hood as a spiked suction cup to stick to the side of a larger fish, then rasps away the fish's skin and removes body fluids and muscle pieces. Vicious vampires, adult lampreys feed very differently than they did as young and gentle filter-feeding larvae (see figure 4.6D). Hagfish live in burrows in the muddy floor of the deep sea and emerge to scavenge dead fish and whale carcasses that have fallen from the water column above. They also prey in mud on live worms and burrowing bony fish.

Lampreys and other fish have all the senses with which people are familiar. Most fish have good vision, smell, taste, and a sense of touch, inherited from the first vertebrates. Fish also have the senses of hearing and of equilibrium (balance), monitored by receptors called hair cells in their inner ear. They also have some less familiar senses, in the *lateral line* that monitors vibrations and electrical fields in the water (chapter 5). The lateral line of a lamprey is shown in figure 6.3A.[5]

One line of ancient jawless fish gave rise to the jawed vertebrates, or gnathostomes (figure 6.1). This occurred in the ocean about 460 mya, and jaws were a major advance leading to a big evolutionary radiation. Ninety-nine percent of all vertebrate species alive today are gnathostomes. Jaws are the most anterior of a fish's hinged gill bars, enlarged around the mouth. They allowed the first jawed fish to feed on animals by clasping prey that had been sucked into the mouth. Now, large and highly nutritious food items could be obtained for the first time in vertebrate history. Soon, teeth evolved on the jaws, allowing the fish to hold the prey more firmly. Finally, the vertebrates, which lacked the grasping and food-manipulating appendages of arthropods, had found a way to become top predators.[6]

An early split of the gnathostomes was into (1) cartilaginous fish, represented today by the sharks, skates, rays, and ratfish, and (2) bony fish ("Osteichthyes": figure 6.1). Bony fish in turn split into the ray-finned and lobe-finned clades. The ray-finned *teleosts* are the most abundant and familiar, including zebrafish, trout, perch, and eels, among others. The lobe-finned Sarcopterygii, represented today by only lungfish and coelacanths, gave rise to the land vertebrates, or tetrapods, about 360 mya. The tetrapods are the amphibians and the amniotes, and the amniotes are the reptiles, mammals, and birds. The amniotes are considered in the second half of the chapter, but we should note one thing here. Informally, all vertebrates that are not amniotes are called *an*amniotes (literally "not-amniotes"). That is, all fish and amphibians are anamniotes.[7]

Evolutionary Step 1: Consciousness Appears

Lampreys
Because the jawless clade to which lampreys belong arose before the gnathostomes did (see figure 6.1), investigating whether lampreys are consciousness can tell us whether the ancient ancestor of all today's vertebrates was conscious. Before we discuss whether lampreys are conscious, however, remember that they are not the only jawless vertebrates. The others are the hagfish (figure 6.3B). However, the blind hagfish are thought to have become specialized for life in and on the mud of the dark, deep seafloor, with some senses having regressed (vision, electroreception) and others having become much sharper (smell, touch, and taste).[8] Having experienced so much secondary modification, hagfish seem poor models of the ancestral vertebrates.

On the other hand, the free-swimming lampreys seem to have few specializations except their sucker mouth and rasping tongue. At least, their neural features are more similar to those of the jawed vertebrates. As mentioned, lampreys have all of the vertebrate distance senses and an obvious forebrain, midbrain, and hindbrain. They also have the neural crest and placode derivatives, and peripheral nerves. The sensory pathways to their brain are typical and as complete as in the other clades of anamniotes. Their major sensory pathways, shown in figure 6.4, can be compared to those of mammals (see figures 5.9–5.12 and table 5.3).[9]

The nervous system of lampreys, however, is not as highly developed as that of the jawed vertebrates.[10] It lacks impulse-speeding myelin on its axons, has a simpler visceral nervous system, and the lamprey cerebrum is the smallest in any vertebrate (figures 6.4A and 5.3).

Lampreys and the other anamniotes have some of the brain structures that are linked to consciousness in humans, structures we will use to test whether these anamniotes have consciousness. Scientific knowledge of the *functions* of the brain structures in anamniotes, however, is much less complete, especially for the telencephalon, the brain region that is most involved in consciousness in humans. Scientists know the special value of lampreys for evolutionary studies, so the lamprey brain is comparatively well-studied.[11] Regrettably, the functions of two key parts of the lamprey forebrain are still poorly known, namely the pallium and thalamus.

In spite of spotty information, enough is known about the brains of all vertebrates to let researchers explore when consciousness first appeared. Helping this search is the vertebrate-informed theory of the neuronal arrangements that produce conscious mental images.

Theory of Isomorphic Neural Hierarchies

Much of the theory of isomorphic neural hierarchies was introduced in earlier chapters, but we will elaborate upon it here (figure 6.5). The idea is that chains and nets of neurons run in the vertebrate nervous system, sending sensory signals from one processing center to another in a hierarchical series, throughout which this information is organized isomorphically (topographically).[12] Along the pathway, the centers receive information from other parts of the brain ("OI" in figure 6.5: outside input), and they process the information in increasingly complex ways. The hierarchy's levels communicate back and forth with one another, and also communicate with other parts of the nervous system. Near the top, some of the outside input can come from other

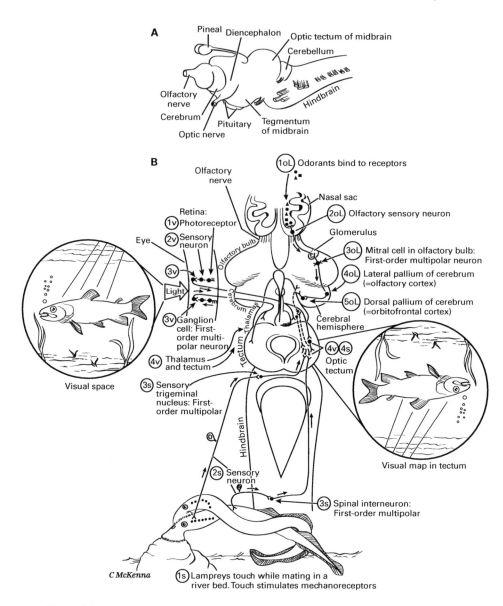

Figure 6.4

Lamprey brain and sensory pathways. A. Brain in side view. B. Brain and spinal cord in dorsal view (center) showing the sensory pathways for vision, touch/somatosensations, and olfaction. The parts of the paths are labeled from 1 to 5, with 1oL meaning first stage in the olfactory path, 3v meaning third stage in the visual path, 2s meaning second stage in the somatosensory path, and so on. Numbers correspond to the column numbers in table 5.3.

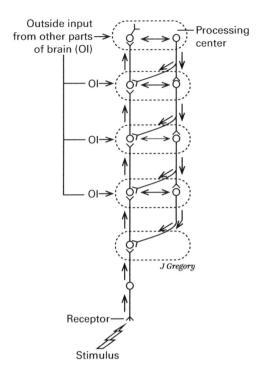

Figure 6.5
Simple diagram of a sensory neural hierarchy. Neurons carry sensory information through increasingly complex processing centers.

sensory hierarchies, as the visual hierarchy receives input regarding sounds, for example. In this way, the isomorphic mapping becomes more complex, more multisensory, and more abstract toward the top of the processing hierarchy. Then, the most highly processed one-sense or multisensory representations are experienced as conscious *mental images*.

The firing of neurons in coordinated, oscillating patterns also seems necessary for sensory consciousness, as suggested by many studies.[13] These back-and-forth oscillations between brain regions are mostly gamma-frequency "brainwaves." They are thought to encode sensory information and then to integrate—or bind into one perception—all those neurons in a sensory hierarchy that are responding to the same sensed object or scene. Not confined to one-sense hierarchies (for example, not for vision alone), the synchronized oscillations would also bind the different types of sensation into one percept: joining the sound, texture, and visual appearance of an object being sensed. This would produce the unified, multisensory, image of consciousness, its "unitary scene."[14] Besides unifying the conscious experience and providing awareness of sensed objects, the oscillations are also thought to focus one's attention on specific stimuli, in *selective attention*.

These three themes—topography, hierarchy, and oscillatory binding—are most often applied to the cerebral cortex of mammals, where electrical recordings show that they involve the thalamus communicating with the cortex. But such evidence would not preclude fish and amphibians, who have no cortex, from generating consciousness in some lower part of their brain. That is, sensory consciousness may have moved to higher brain centers when the mammalian cortex evolved. Whether fish and amphibians are conscious should not be decided from the human state but from evidence from these animals themselves (for more discussion of whether consciousness needs a cerebral cortex, see pp. 209–211). In the anamniotes, most kinds of sensory information merge in the optic tectum of the midbrain. We will now consider this structure as a possible site for consciousness.

Optic Tectum

Recall from chapter 5 that the optic tectum forms most of the roof of the vertebrate midbrain (figure 5.2).[15] It is relatively small in mammals where it is called superior colliculus. It is proportionally largest in the anamniotes, where it is sometimes larger than the cerebrum, but it is also large in birds and reptiles in spite of their larger cerebrums (figures 6.4, 6.6, and 5.3). As the main vision-processing center in fish and amphibians, their tectum receives more visual input from the retina than does any other part of their brain. A fish whose tectum has been destroyed acts as if it is blind.[16]

In all vertebrates, the tectum shows extreme isomorphic mapping, not just of the visual input but also from other senses: touch, hearing, taste, mechanical from the lateral line, and electroreception—but not smell. As we recently stated, this multisensory mapping "is reflected in the tectum's neuronal layering or lamination [figure 6.6B], where the different layers receive different classes of sensory input, and the mapped inputs from the different senses are in topographic register with one another."[17] This patterning yields a unified and coherent map of sensory space. The tectum also connects to and interacts with many other parts of the brain. The tectum of lampreys (figure 6.4) is not especially simple or primitive but is typical for a vertebrate.[18]

The tectum has a number of functions, most of which are conserved across the vertebrates: (a) in its sensory function, it processes the sensory input it receives; (b) in its motor function, it signals behavioral actions; and (c) in its attentive function, it selectively directs attention.

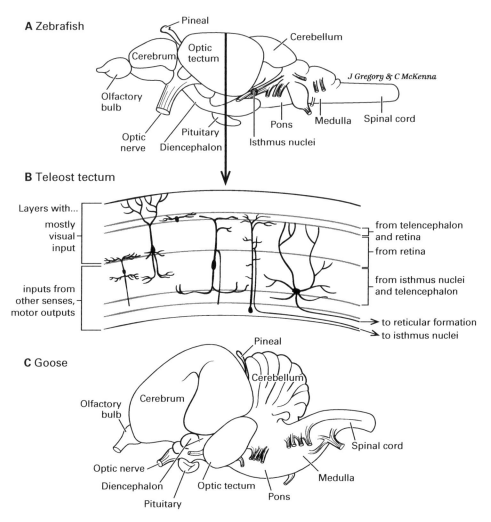

Figure 6.6
Optic tectum, showing its size in a teleost zebrafish (A) and a bird (C), with its layered neuronal structure in a teleost (B). Part B was redrawn from Northmore (2011).

For its sensory roles, the tectum helps to recognize and distinguish between the objects being seen by their size, shape, newness, movement, orientation, and rate of change. It also determines the salience (importance and distinctiveness) of viewed objects. For example, it assigns high salience to objects that are fast moving, new, large, or threateningly looming.[19]

For its motor functions, the tectum signals the brain's reticular formation (see table 5.1).[20] Through these signals it directs the eyes to move toward—and the body to orient toward—viewed objects, and to track such objects as they move. It also signals the animal to approach or avoid these objects. Furthermore, the tectum guides the body's vision-driven movements through space, as when seeking food or dodging obstacles. The movements it directs require a lot of spatial resolution.

For its role in attention, the tectum teams up with nearby centers in the brainstem named the *isthmus nuclei* (figure 6.6A).[21] Together, these structures allow selective attention to sensed stimuli. That means they find out which object in the mapped, multisensory space is the most salient and then direct attention to that object.

The tectum is for fast rather than deep analysis.[22] It makes the first evaluation of sensory stimuli, or of a situation, needing immediate attention or a quick decision and a direct reaction. The fish neurobiologist David Northmore has described this as fast reactions to moment-to-moment changes in the environment. Next, the pallium of the cerebrum may carry out a further, slower, and more thorough evaluation (more on the pallium below).

Why does the tectum receive input from so many different senses, when it is foremost a *visual* part of the brain?[23] The obvious answer is that adding the other senses provides a more complete world map than vision alone can produce. But the additional senses also check for consistency. They check whether all the senses consistently support the same reconstruction of the world. And the other senses can correct any areas of weak or unclear visual input in the visual map. The tectum's closely stacked and topographically aligned neuronal maps for the different senses keep the connections between neurons as short as possible, and therefore the maps are highly efficient. As Jon Kaas explained, the close alignments "effectively group neurons that most commonly interact, thus decreasing requirements for long, slow, and metabolically costly connections."[24] These are the advantages of joining the different senses into a single isomorphic image in the tectum. The advantages reveal why, when vision evolved in the first vertebrates (chapter 5), isomorphic input from most other senses soon projected to the visual tectum. This convergence marked the origin of the key feature of mental unity of the exteroceptive senses (table 2.1).

Among the functions of the anamniote tectum are kinds of sensory discrimination that seem advanced enough to call conscious. As Ursula Dicke and Gerhard Roth put it:

> In amphibians as in all other anamniote vertebrates ... the tectum is the major brain center for integrating visual *perception* and visuomotor functions. In the amphibian tectum, localization and *recognition* of objects and depth perception takes place. Three separate retino-tectal subsystems for object recognition exist, which process information about (i) size and shape, (ii) velocity and movement pattern and (iii) changes in ambient illumination. These kinds of information are processed at the level of different retinal ganglion cells and tectal neurons in close interaction with neurons in other visual centers.[25]

In this passage, we've italicized "perception" and "recognition" as implying consciousness, which we also infer from the work of Mario Wullimann and Philippe Vernier, who concluded that the fish tectum is for "object identification and location."[26] Other investigators will disagree with us and claim that these tectal functions are only unconscious reflexes (although complex reflexes of the highest order). But to us, *real consciousness* is indicated by the fish and amphibian tectum making a multisensory map of the world, then attending to the most important objects in this map, and then signaling active behaviors that suggest the animals are *accessing* and making use of the map.[27]

Researchers have recorded gamma waves oscillating between the tectum and the isthmus nuclei.[28] Recall that the tectal-isthmus system acts in selective attention to salient stimuli, and the scientists who recorded these tectal oscillations remarked how similar they are to the gamma oscillations for spatial attention and consciousness in the mammalian cerebral cortex. This could be further evidence of tectal consciousness. However, we must warn that the best evidence for the oscillations comes from birds, whose tectum is especially elaborate and large, and who are amniotes rather than anamniotes. But some tectal gamma oscillations have been recorded in frogs and teleost fish;[29] and all vertebrates have the required neuronal connections between the optic tectum and the isthmus nuclei. This includes lampreys.[30] Further electrical recordings should be done on anamniote animals, to determine whether they really do have these tectal-isthmus "oscillations of consciousness."

Telencephalon and Pallium

Does the cerebral-cortex homologue of fish and amphibians, which is the dorsal pallium of their telencephalon (figure 6.7A), consciously perceive every sense as does the human cerebral cortex? The anamniote telencephalon was once thought to be just a "smell brain" and not involved with the other senses,[31] but it is now known to receive visual, hearing, touch, and other kinds of nonsmell sensory information.[32] Anamniote telencephalons are also known to have the same nonsensory parts as the amniotes, such as the hippocampus (medial pallium) for constructing memories and the striatum (basal ganglia) for choosing which behavioral motor-programs to carry out (table 5.1).

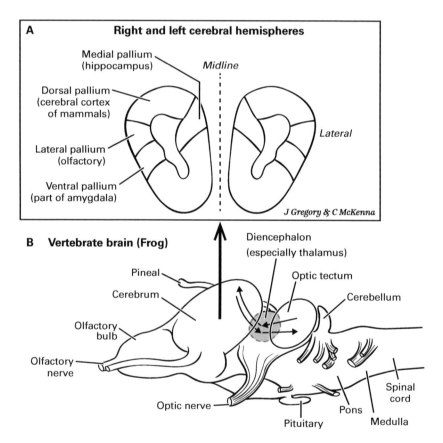

Figure 6.7
Pallium of the cerebrum of the vertebrate brain. A. Cross section through the right and left cerebral hemispheres in an embryo, showing the four basic parts of the pallium. B. Orientation diagram of an adult vertebrate brain showing the plane of section through the cerebrum in part A.

But it is still likely that smell processing was the main function of the telencephalon's pallium in the first vertebrates. This is because the pallium of today's most basally arising lines of fish—lampreys, hagfish, and the cartilaginous sharks and their relatives—receives mostly olfactory projections, with smell processing dominating it far more than in the other vertebrates.[33]

Yet it remains undeniable that the anamniote pallium receives all modes of sensory information, not just smell. So we still have to ask if the pallia of fish and amphibians map the nonsmell senses into conscious images, as does the cerebral cortex of mammals. Evidently not, because the projections of these senses to the pallium are not separated by the sensory classes (modes), nor do they reach the pallium in any isomorphic arrangement. Walter Wilczynski said it this way:

> There is no evidence for separate representations for each sensory system, no indication of topographically preserved projection ... to any telencephalic area ... In essence, there is no evidence for the distinct, unimodal, mapped sensory representations that are so prominent in the mammalian cortex. A possible exception may be the core olfactory-recipient regions of the lateral pallium.[34]

Wilczynski was writing about amphibians, but his statements also apply to the pallia of the fish that have been studied.[35] In conclusion, in fish and amphibians, all the isomorphic sensory representations except for smell are in their optic tectum and not the pallium.

Where are the smell representations in fish and amphibian brains? All vertebrates including lampreys have the same olfactory pathway, targeting the telencephalon (figure 6.4, "oL").[36] Therefore, if anamniotes have any olfactory consciousness, it must arise in their dorsal pallium (as in amniotes), even though for their other senses the consciousness arises in the optic tectum (unlike in mammals).

The cerebral pallium and the optic tectum communicate back and forth through the anamniote diencephalon, which lies between them. This interaction is shown by the four arrows in figure 6.7B. The connections between the tectum and the diencephalon are especially well documented and abundant.[37] Perhaps the tectal-pallial crosstalk somehow integrates and unites the two kinds of conscious images in anamniotes—the olfactory with the nonolfactory sensory kinds.

If the pallium of anamniotes is not involved with mapped consciousness based on multiple senses, and is therefore so unlike our own cortex, then what functions can this pallium possibly perform, other than smell perception? New evidence from lampreys suggests it influences motor actions,[38] as it does in the amniotes, but we want to approach this question from another direction. The reason why pallial function in anamniotes is such a tough problem is that a fish whose pallium has been removed or destroyed can see, catch prey, and seems to act normally. However, such a fish cannot learn from its experiences or from the consequences of its actions. Nor is it able to learn the locations of objects in space. This is a memory problem, and the medial and dorsal pallia of vertebrates are known to store memories.[39]

Dorsal and medial pallia are also the main regions of the vertebrate brain where smell input from the rostral telencephalon meets the input from every other sense that is relayed up from the thalamus and optic tectum (figure 6.4B).[40] And all this sensory information has been highly processed by the time it enters the dorsal-medial pallium. Therefore, because it always receives more fully processed information than the tectum, the dorsal-medial pallium could profitably override the motor commands ordered by the tectum, which were based on faster and cruder sensory assessments and thus are more likely to be in error. The dorsal-medial pallium could be a "wiser" motor-controlling center than the tectum.

We deduce that the dorsal-medial pallium, even though small and not conscious in anamniotes, has certain advantages, of (1) memory storage and (2) a location where it receives *every* class of sensory information in highly processed form. These advantages, we will propose, made it the main site of brain expansion in those vertebrates that evolved greater mental capabilities (see "Evolutionary Step 2").

Memory

As especially emphasized by Gerald Edelman, a giant in the field of consciousness studies, memory plays an important role in sensory consciousness. In conscious perception, images are continually called up from memory, then modified by the new, incoming sensory information into an updated image of the world as it is being experienced. Thus, there is no need to build completely new mental images from instant to instant, just a need to adjust and update the existing image. And even more basically, almost nothing sensed can be recognized without prior learning and training, and that depends on memory.[41]

The medial pallium (hippocampus) constructs the memories.[42] These are not just the *episodic* memories of objects and experiences but also memories of mapped space, so the animal can use recall to navigate accurately through its environment. To build these memories, the hippocampus receives all kinds of sensory information. It also has many back-and-forth connections to all the main regions of the forebrain. A hippocampus has been found in every clade of jawed vertebrates,[43] which is almost enough to indicate that memories have served consciousness since the very first vertebrates. However, a hippocampus is not well documented in lampreys, where the only evidence for it is some hippocampus-defining genes expressed in the developing brain (specifically, the *Lhx 1/5* and *2/9* genes).[44]

Recall the strong evolutionary link between memory and smell: animals can track prey or avoid predators by remembering the source of lingering odors (chapter 5). As a way of maximizing the adaptive efficiency of this smell–memory relationship, the medial pallium lies close to the region of olfactory input (lateral pallium: figure 6.7A), as does the amygdala, the brain region for emotional memory (see figure 5.4 and chapter 8); and the olfactory pathway projects extensively to both the hippocampus and amygdala. In other words, an especially large input of odor information goes to the memory brain nearby. One would expect this proximity, as it often is said that smells evoke the deepest memories. By our reasoning, these smell-evoked memories enter consciousness in all the vertebrates.[45]

Arousal

Arousal relates to attention because an aroused, alert animal is more attentive to sensory stimuli in general. Arousal is also a part of consciousness—of the *level* of consciousness, which is defined as the degree or intensity of consciousness, of arousal, or of wakefulness. All vertebrates have the same neural structures for arousal, namely the reticular formation's activating system (RAS) and parts of the basal forebrain (BF), including "septal nuclei."[46] These structures are pictured in figure 5.4. In mammals, they also help the cerebral cortex to direct its attention to specific, salient stimuli,[47] but apparently not in fish and amphibians, where the isthmus-tectal mechanism is responsible for such selective attention.

The arousal component is well developed in fish and amphibians. As in the amniotes, their RAS and BF project and signal widely throughout the brain, and these two regions likewise release chemicals that are associated with general arousal, alertness, and vigilance. Important examples of these neurochemicals are norepinephrine and acetylcholine.[48]

We have made the case that anamniotes have all the ingredients of sensory consciousness. These are topographic maps for mental images (mostly in their tectum but in the pallium for smell); a place to form, store, and retrieve memories (in their pallium); and brain mechanisms for selective attention and arousal.

Is Amphioxus Conscious?

With our new, vertebrate-based insights into the ingredients of consciousness, we can ask again if amphioxus, the invertebrate whose brain most likely resembles that of prevertebrates, shows any marks of sensory consciousness. Recall that amphioxus has photoreceptors but no image-forming eyes, and it has touch receptors and presumed chemoreceptors on its body surface.[49] In chapter 3, we reported that Lacalli, who studies the larval amphioxus brain, thinks it is too simple for consciousness.

Indeed, amphioxus lacks the markers for consciousness. First, remember that sensory neurons are the crucial first level in the conscious sensory hierarchies of vertebrates. Yet, as emphasized in chapter 5, the sensory neurons of amphioxus do not develop from neural crest or placodes, as do those of vertebrates, so the sensory neurons are not even homologous to one another in the two animal groups. Second, most sensory pathways to the brain of larval amphioxus have only one or two neurons leading to just one processing center of neuropile (figure 3.5). This seems too few for consciousness.

Third is a problem of topography. It is not known for sure if the sensory pathways of amphioxus are organized topographically, but Nicholas Holland and Jr-Kai Yu[50] found hints of a curious topography of the touch-sensory axons in the larval spinal cord. The axons from both the dorsal and ventral body were grouped together, separate from the axons from the midbody. This contrasts with the mostly head-to-tail topographic arrangement in the CNS of vertebrates. It is unlike the vertebrate version of body mapping.

Amphioxus has a reticular formation (the "primary motor center": figure 3.5), so it may have a basic arousal system, if only to prime it to swim away from danger. But tellingly, its forebrain lacks both arousal neurochemicals, norepinephrine and acetylcholine.[51] So amphioxus may not have the same arousal mechanism as vertebrates. In conclusion, four lines of evidence indicate that larval amphioxus is not conscious.

Adult amphioxus is also unlikely to be conscious. The adult brain is more complex than in the larva (figure 3.5A), but it lacks a distinct tectum, its frontal "eye" is simple and cannot form images, and this diencephalic eye lies in the very front of the brain, so a telencephalon is probably absent. No tectum, no camera eye, no telencephalon, so we deduce no consciousness.

Genes and Consciousness

In our first big study of conscious origins, we explored sensory consciousness from the angle of the genes involved. That is, we searched the published literature and found many genes that operate throughout all vertebrates and signal the embryonic development of structures related to consciousness, from the three-part brain, to the neural crest and placodes, to the eyes. However, we could not find any shared genes that signal isomorphic organization across all the sensory systems, as our theory says must exist. Recently, we found reports of such genes, the *Eph/ephrin* group. Researchers have documented the isomorphic-signaling function of these genes in the visual, auditory, and somatosensory-touch systems of vertebrates. Does amphioxus have genes of consciousness? No; *Eph/ephrin* genes do not seem to participate in the development of its central nervous system, nor do any of the other genes that have been linked to consciousness in vertebrates.[52]

Summary of Step 1, Conscious Origins and Evolutionary Adaptation

We can now update our story from chapter 5 of how consciousness evolved, by adding more vertebrate-brain anatomy to it. As the ancestors of vertebrates evolved image-forming eyes, this led to visual images of the world in their new optic tectum. These mental images emerged from an isomorphically organized hierarchy of neurons and visual-processing centers that extended from the retina to the nearby tectum. The tectal images contained the first subjective qualities of consciousness, the visual qualia, which soon were improved by isomorphic inputs from other senses. These were the new distance senses of hearing, balance, and lateral line, and the rebooted senses of touch and taste. At the same time, the olfactory hierarchy was evolving and smell images arose in the pallium of the new telencephalon.

All these consciously perceived sensory images were adaptive and favored by natural selection. The mental reconstructions were broad and accurate enough that they allowed reliable behavioral interactions with the *real* environment. By referencing these images in their minds, the first fish could now target their actions precisely to any location in the increasingly complex, fast-moving, and dangerous world of Cambrian seas, to find food, escape from attack, and mate. We also propose that the brain manipulated its conscious images to give simple *predictions* of the immediate future, allowing faster behavioral responses, a survival advantage for the early fish in the fast-moving Cambrian environment. Many researchers now interpret consciousness as a "prediction device," and this is our take on the prediction theme.[53] In addition, the conscious mental images were highly "evolvable" from the very start. This means that their *detail* could easily be increased over time for any or all senses, simply by adding more neurons to the existing hierarchies, especially to the processing centers.[54]

Also vital to consciousness were brain systems for selective attention, such as the isthmus-tectal complex. As the earliest vertebrates began building mapped neural simulations of the world, they had to be gaining the ability to attend to the important objects in that world map. And the RAS system for overall arousal adjusted the level of consciousness and of alertness to all new stimuli. Every living vertebrate has these elements of consciousness, but their invertebrate relative, amphioxus, has few or none.

Evolutionary Step 2: Consciousness Jumps Ahead in Mammals and Birds

Why Mammals and Birds?
If consciousness started with the eyes and optic tectum of the first fish before 520 mya, then it took a second big step several hundred million years later, independently in two clades of land vertebrates. These vertebrates were the first mammals and birds, whose ancestors split from the early reptile-like amniotes about 350 mya into the synapsids (mammal-like reptiles) and the sauropsids (the ancestors of today's reptiles and birds) (figures 6.1, 6.8, 6.9A).[55] At the origin of both mammals and birds, things changed so that the dorsal pallium became the predominant brain region for sensory processing and the end site of sensory consciousness.

At this stage, the cerebrum surged in size, first in the earliest mammals around 220 mya (figure 6.8E) and then later in the first birds (figure 6.8G), who evolved from dinosaurs around 165 mya.[56] Although the brains of today's mammals and birds are on average ten times larger than those of today's reptiles when adjusted for body weight, fossilized skulls show that the brains of the first mammals and birds were "only" three to five times larger than those of their "reptile" ancestors.[57] Still, these size increases of 300 to 500 percent are very impressive considering they happened over relatively brief time intervals of just tens of millions of years.

In both mammals and birds, the size increases mainly involved the sensory-processing pallium. This fact has led many researchers, who believe consciousness requires a large pallium (or cerebral cortex), to say that consciousness *first* evolved in these clades, and that only mammals and birds are conscious.[58] These researchers include Ann Butler, Rodney Cotterill, Melanie Boly, Bernard Baars, David Edelman, and Anil Seth. The mammal-bird hypothesis has more supporters than our newer hypothesis that says all the vertebrates are conscious.

Perhaps the brain advances in the first mammals and birds were just part of a gradual climb that started 350 mya when the first amniotes arose

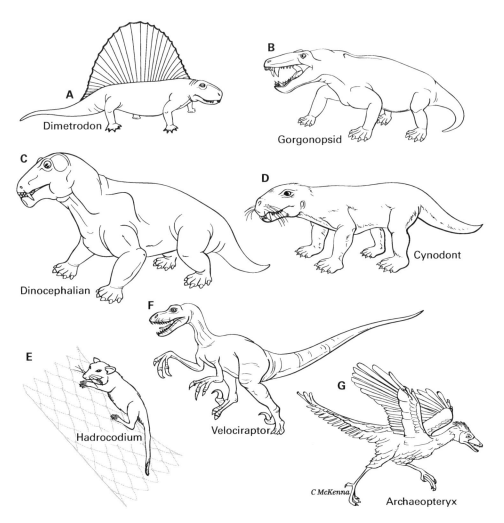

Figure 6.8
A sampling of extinct synapsid and sauropsid amniotes, from 280 to 70 mya: mammal-like reptiles, a mammal, a dinosaur, and a bird. The mammal-like reptiles, as arranged from A to D, are progressively more mammal-like. A. Sail-backed *Dimetrodon*. B. Gorgonopsid ("gorgon-faced"). C. Dinocephalian ("terrible-headed"). D. Cynodont ("dog-tooth"). E. Early mammal *Hadrocodium*. Parts F and G are sauropsids: F. A dinosaur related to birds, *Velociraptor*. G. Early bird *Archaeopteryx*.

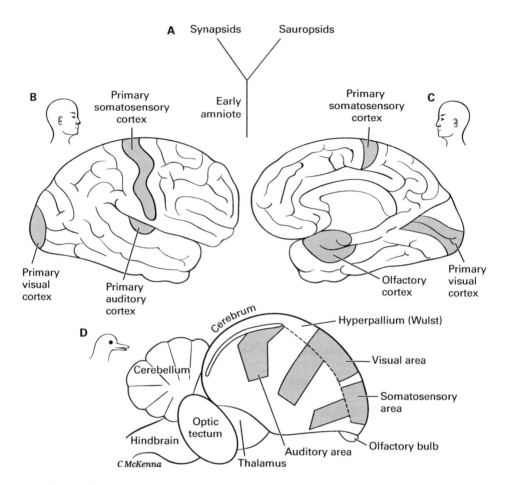

Figure 6.9
Comparing the brains of different amniote vertebrates. A. Simplified phylogenetic tree showing relation of synapsids (mammal-like reptiles and then mammals) and sauropsids (the other reptiles and then birds). B, C. Primary sensory areas of the cerebral cortex of a mammal (human) in lateral (B) and medial (C) views of a cerebral hemisphere. D. The sensory areas in the bird pallium. E. Six layers of the cerebral cortex in mammals, I–VI, in a rat brain cut in half through the midline. F. Corresponding regions I–VI in dorsal pallium of a bird. G. Dorsal pallium of a modern reptile. D–G are based on the work of Erich Jarvis and his collaborators.

Two-Step Evolution of Sensory Consciousness in Vertebrates 121

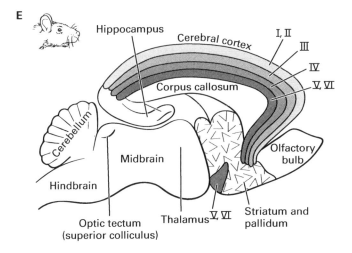

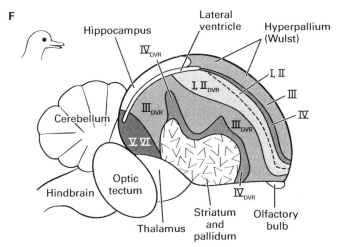

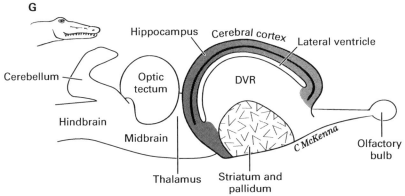

Figure 6.9 (continued)

by splitting from a line of amphibians.[59] Gradual advance seems unlikely, however. Although direct fossil evidence is sparse, the living reptiles have brains less than twice as big as those of living amphibians.[60] Not much size increase there. No, the right place to seek the huge advances in brain size and consciousness is in the first mammals and birds.

Mammals and birds made the shift to pallium-centered consciousness millions of years apart and with different dominant senses, smell versus vision. Given these differences, it is hard to find any commonalities that could reveal why consciousness surged in both lineages at roughly the same time in Earth's history.

Maybe there is a shortcut to the answer. Some investigators have hypothesized a link between consciousness and the fact that mammals and birds are both homeothermic (endothermic or "warm-blooded"), with a high, constant body temperature.[61] With this, mammals and birds have a faster metabolism and are more active than the ectothermic ("cold-blooded") reptiles that represent the ancestral condition. Apparently, the hypothesis proposes that the increase in metabolic rate allowed neurons to send signals faster and process more information to meet the high computational demands of increased consciousness.[62] While this hypothesis may have value, it faces a problem: scientists do not know just when the animals of the evolving mammal and bird lineages became more active and homeothermic. Instead of appearing in the first true birds and mammals, homeothermy could have evolved slowly and gradually, or in earlier synapsids that lived long before mammals (figure 6.8B through 6.8D), or in the prebird dinosaurs named "coelurosaurs" (figure 6.8F).[63] Because the dates of brain expansion and homeothermy may not match, we cannot use homeothermy to explain the common advance in consciousness. The homeothermy shortcut fails, and we must tackle the problem head-on by examining the mammals and birds individually.

The mammalian jump seems easier to explain. During the transition from mammal-like reptiles to mammals about 220 mya, the increase in pallial/cortical size accompanied a shift from vision to smell (and touch) as the main sense. This was documented by the biologists and paleontologists, Timothy Rowe, Francisco Aboitiz, Margaret Hall, and others.[64] They interpreted the change as an adaptation to night living on the ground, where little can be seen in the dark but the many smells are very informative, as are the touch stimuli monitored by surface hairs all over the body (bend some hairs on your forearm and you will feel that they are good sensors of light touch). The smell-associated olfactory bulbs and lateral pallium enlarged in the first mammals, as shown by fossil brain casts. So did the touch (somatosensory) region of the cerebral cortex, to which, we presume, the touch senses now projected topographically. The decline in vision in these first mammals can explain the relatively small optic tectum of all mammals.

We deduce that when vision and the mammalian tectum declined, the cerebral cortex became the end-site of consciousness for the senses that formerly had reached consciousness in the tectum: touch, hearing, vision, and balance. Logically, this seems to be a shift from a declining part of the brain to an enlarging one, a shift that enriched the increasingly useful smell images in the pallium with more information from the other senses. The clear logic is why we consider the mammalian shift "not so puzzling."

The brains of the living mammals uphold this explanation. Today's mammals are the egg-laying monotremes such as the duck-billed platypus and the echidna ("spiny anteater"); the pouched marsupials, including opossums and kangaroos; and placentals, which are all the others: apes, rodents, wolves, horses, aardvarks, porpoises, and so on. In all three of these clades, the cortex is large and contains an isomorphically organized primary-sensory area for each conscious sense (figure 6.9B) and nearby sensory-association areas.[65] Additionally, in the basal members of each clade, the olfactory cortical regions are large. Given this shared plan, the cerebral cortex should be the end-site of sensory consciousness in every one of today's living mammals. Fossils date the last common ancestor of monotremes, marsupials, and placentals to between 190 and 220 mya.[66] This is near the 220 mya age of the first fossil mammals and their explosively enlarging cerebrums, supporting that as the date when mammalian consciousness became fully cerebral.

In birds, the shift to pallial consciousness is harder to understand. This is because their sensory specializations differ from those in mammals—almost the opposite. As birds evolved from a line of dinosaurs, their already fine vision grew even sharper, and their sense of smell stayed the same.[67] This challenges our own hypothesis with a question: Why did birds move their visual consciousness to the largely olfactory pallium when the optic tectum was already performing this visual role and their sense of smell was not improving?

The reason why the dorsal pallium enlarged rapidly in both vision-driven birds and smell-driven mammals must lie in some built-in advantage of the pallium over the tectum. As pointed out earlier in the chapter, in every vertebrate the dorsal pallium differs from the tectum in receiving information from all the senses including smell; and the dorsal pallium more thoroughly processes higher-order sensory inputs, which it compares to memories called up by the nearby hippocampus. Referencing memory like this lets the pallium choose from a wide variety of more accurate options for directing behavior, to make better decisions based on past learning.

Expanding and elaborating the memory-storing dorsal pallium[68] meant more memories could be stored: more smell memories in mammals and more visual memories in birds.[69] With more memory, the conscious experiences of these animals became less fleeting and "in the moment," as they began to contain more of the recalled past. Through increased memory, many more of the objects being sensed could be *recognized* and then attended to.[70] In fact, the increase in memory supplied the mammals and (presumably) birds with more of an advanced kind of selective attention, in which the pallium directs attention to those stimuli whose importance has been *learned*.[71]

While memory took mammals and birds to the next level of primary consciousness (Step 2), it also set the stage for some higher aspects of consciousness and for increased cognition. That is, combined with pallium's increased capacity for information processing, the memory enhancement allowed more *interpretation* of the primary images and experiences, and building up more knowledge about the world.[72]

But these increases in pallial processing and memory, along with the advance in primary consciousness, required additional neural calculations, which took more time and demanded extra energy. Something so expensive is not helpful to all animal species in all habitats. It could only have evolved in special habitats under special conditions.

With these considerations, the question becomes whether the first birds and mammals lived in special environments that favored the evolution of advanced pallial sensory processing and more memory. Both groups are believed to have inhabited forests, or for the first mammals, other places that had protective cover such as among rocks, in burrows, or under the cover of ground plants. We imagine the nocturnal mammals down on the forest floor scurrying among the litter, where there were many things to smell and to touch, especially with their sensitive vibrissae (whiskers). We picture the first birds as active in the daytime up in the trees, an intricately three-dimensional world that they had to master with keen vision in order to fly from tree to tree while avoiding collisions and falls. The animals of both lineages had become *miniaturized*, with small bodies and both were probably hunted by larger, predatory dinosaurs. The best deductions and descriptions of these special environments have been given by the paleontologists Michael Lee, Michael Benton, Zofia Kielan-Jawarowska, and their colleagues.[73]

The above interpretation may need to be adjusted a bit for the birds. Amy Balanoff, Michael Lee, and their colleagues found that enlarged brains characterized an extensive radiation of bird-related dinosaurs, whose bodies got progressively smaller, shrinking from the size of a one-and-a-half-year-old human child down to the size of a crow in the first true birds (from 10 kg down to 0.8 kg).[74] That is, the brain enlargement did not all happen in the first true bird. Still, the full radiation was very fast (within a brief 5 million years, according to Lee and colleagues), and all its members were feathered and had the "ability to fly in some form" (according to Balanoff). Thus, no matter whether consciousness advanced early or late in this radiation that led to birds, it still fits the flight-driven explanation we gave in the previous paragraph.

Even though they were relatively weak and small, both mammals and birds had safety and protection. When not hidden in trees, birds flew or glided away from danger, fast and far. The nocturnal mammals were cloaked by darkness and by the ground cover under which they lived or hid. Relatively safe from predation, the birds and mammals had enough time to weigh options and decide among a variety of choices based on the many sensations they constantly picked up.[75] Although their complex sensory processing took a lot of time, their sometimes slower decisions about danger were no handicap in the "safe neighborhoods" where they lived. Additionally, the first birds and mammals ate insects—abundant, high-energy meals that can fuel the energy demands of increasingly complex brains, including the high computational needs of advancing consciousness.

Therefore, both birds and mammals originated in the unusual kinds of environments that favor advanced, memory-enhanced consciousness and pallial expansion. We see this as a case of *convergent evolution* (see box 4.1), in which unrelated clades independently evolve the same traits when exposed to similar habitats, predators, and competitors.[76]

Alternatively, the same evolutionary sequence we just presented could explain how birds and mammals each evolved consciousness anew from utterly unconscious reptile ancestors, in the rival, pallial hypothesis that says the optic tectum was never a site of consciousness. That hypothesis, however, says that consciousness originated twice in the vertebrates (in mammals, then in birds), so it may be more unwieldy than our hypothesis of a single origin, in the first fish. But our hypothesis has a possible weak point of its own: it implies that the bird brain has two distinct, dominant parts for conscious visual images: the tectum and the pallium. It raises the awkward question of whether two highly detailed and distantly generated mental images, tectal and pallial, could be effectively united into one.

Pallial Diversity, Another Puzzle

Whether consciousness evolved anew in the first birds and mammals or just advanced in these clades, another mystery arises. Structurally, the pallia of mammals and birds look very different (figure 6.9). Yes, the enlarged pallia of both clades have the isomorphically organized sensory regions for vision, hearing, touch, and smell (figure 6.9B, 6.9C, 6.9D),[77] and both these pallia also perform high-level cognitive functions.[78] But the neurons in the cortex of mammals are arranged differently from those in the corresponding dorsal pallium of birds, and neuroscientists struggled for many years to find any similarities for comparison. The mammalian cortex has six layers of neurons, or laminae, over its full caplike extent (I–VI in figure 6.9E). In contrast, the dorsal pallium of birds has a large, unlayered part called the dorsal ventricular ridge (DVR) plus an outer, cortexlike "hyperpallium" with several thick layers that do not much resemble the mammalian cortical laminae (figure 6.9F).[79]

Recently, the mystery of these differences was solved by studying the axon connections, neuron circuits, and patterns of gene expression in the pallia of mammals and birds. Major contributions to this breakthrough came from Harvey Karten, Jennifer Dugas-Ford, Erich Jarvis, Ann Butler, and their colleagues.[80] The results are surprising. The bird equivalents of the six mammalian laminae spread and scatter widely through the dorsal pallium, as cell clusters (nuclei) or thick bands instead of stacked laminae (compare figures 6.9E and 6.9F). And, the *visual* area of the bird pallium lies much farther forward than in the mammalian cortex (compare figures 6.9B and 6.9D).

Yet the new studies found that many of the functional groups of neurons and microcircuits are the same in birds and mammals, as are the long axonal connections to and from the primary sensory regions. Thus, the pallia of birds and mammals must share many ways of generating conscious sensory images. Nonetheless, the fact remains that in other ways the pallial anatomy of the two clades has evolved quite differently.

It is hard to tell when and how these differences between the bird and mammal pallia evolved, or if either the bird state or mammal state is primitive. This is because the pallium of their common ancestor, the early and extinct reptile-like amniote, is so difficult to know. The dorsal pallium of the living reptiles (figure 6.9G) has a large DVR as in birds and a cerebral cortex as in mammals, although the reptilian cortex is much simpler with just three laminae.[81] Using the living reptiles is unreliable, however, because all of them are sauropsids (bird relatives: figures 6.1, 6.9A), meaning we cannot suppose a DVR existed in the premammal synapsids or the early amniotes. Judging from the closest living relatives of amniotes, which are the amphibians and lungfish,[82] the first amniote had a simple cortex with just one layer of neurons. But whether it yet had the multiple, isomorphic sensory regions that we associate with consciousness is unknown. Discovering that these regions exist in the living reptiles would help, but the functions and neuronal homologies of the reptilian pallium are not well enough known.[83] Further investigation of the understudied reptiles would help greatly to unravel the evolutionary history of consciousness.

In spite of the many unknowns, the confirmed differences between birds and mammals show that the neural structures for pallial consciousness differ in different vertebrates. But the pallial diversity across vertebrates is even greater than this, as shown by some fish. The dorsal pallium has enlarged independently in hagfish, in certain sharks, in skates, rays, and some clades of teleost fish—not in exactly the same ways, but always by adding novel and specialized brain nuclei or stacks of laminae.[84] These different elaborations of the pallium, though not yet studied at the circuit level, always correlate with some especially acute sense(s) and the advanced sensory processing that characterizes these specific fish groups. And all the jawed fish on the list originated in the Mesozoic,[85] the same era when the pallia and sensory consciousness were expanding in birds and mammals. Did some of these clades use their increasing awareness to compete against each other in a race for survival, making the Mesozoic an age of widely advancing consciousness? This could have been an echo of the earlier Cambrian arms race to consciousness (see chapter 4).

To summarize the chapter, the type of sensory consciousness that involves building and experiencing mapped images of the external world underwent two great advances during vertebrate history. In Step 1, when vertebrates originated before 520 mya, consciousness first appeared. It arose from well-mapped sensory representations in the upper levels of hierarchical chains of neurons. The first end-sites of consciousness were (1) the optic tectum, for vision and the other nonsmell senses; and (2) the telencephalon for olfactory consciousness. Also essential were systems for arousal (in the reticular formation and basal forebrain) and for selective attention (such as the isthmus-tectum system).

Lampreys, representing the earliest-arising vertebrates, have all these features. Hagfish, the lampreys' bizarrely specialized relatives, should also be conscious because their brain is three times bigger[86] and it bulges with large sensory processing areas for smell, touch, and taste.[87]

From the beginning, consciousness raised the behavior of vertebrates beyond reflexes and simple patterned actions. It let vertebrates map out and then assess their environments so they could accurately interact with the world. Isomorphic neuronal organization is the key to this mapping, and *Eph/ephrin* genes help signal the isomorphic connections to form.[88]

Step 2 occurred in the Mesozoic, when mammals and birds evolved. The dorsal pallium of both expanded with new, isomorphically organized sensory regions and became the brain's only end-site of sensory consciousness. No longer was it for olfactory consciousness alone. The expansions happened by convergent evolution, first in mammals and then in birds. In both clades, this change stemmed from the dorsal pallium's favorable position in the brain, located where all classes of sensory input meet and also near the memory-recall center (the hippocampus). The leap in memory-enhanced consciousness built on the dorsal pallium's original role of analyzing multisensory input slowly to make complex decisions from prior learning. This was an advantage, we propose, in the stimulating but relatively safe habitats of the first birds and mammals.

We find that isomorphic sensory consciousness is more *widespread* than many researchers have thought, because it exists in all the vertebrates. It is also *older*, being over 520 my old. It is also more *diverse*, being tectal and pallial, and stemming from different pallial arrangements in mammals versus birds. With this, we begin to see the diversity feature of consciousness, postulated back in chapter 2 (table 2.1). Next, for comparison with isomorphic sensory consciousness, we turn to the affective consciousness of good and bad "feelings" to see when that aspect evolved and which vertebrates have it.

7 Searching for Sentience: Feelings

What Is "Sentience"?

While the neurobiology of primary consciousness or the nature of "something it is like to be" is often treated as a single question, our view is that this is an oversimplification of a far more complex problem. In the previous two chapters, we focused on the origin of exteroceptive consciousness—that is, sensory consciousness created by isomorphically (topographically) mapped neural representations of the world that are experienced subjectively as referred mental *images*. But there are other types of qualia that are not referred to the world as externalized images but rather are subjectively experienced as internal *feelings* or body *states*. The name *sentience* is most often applied to these types of experiences.

The term "sentience" means "capable of feeling" from the Latin word *sentiens* meaning "feeling." While the term in some contexts refers broadly to all aspects of sensory consciousness[1]—both exteroceptive and interoceptive (internal) awareness—in the current philosophy of ethics and especially in the animal-rights literature, "sentience" refers more specifically to an animal's capacity to experience pleasure or pain. These sensory experiences are created when an organism becomes consciously aware of its own internal bodily and affective feelings. For present purposes, we will adopt the following definition: for an animal to be "sentient" it must be *capable of experiencing an affective state*.

Affective experiences have a *valence*—the aversiveness (negativity) or the attractiveness (positivity) of the stimulus or event. The most basal affective experiences are the capacity to have sensory experiences of a negative (noxious) or positive (pleasurable) nature: feeling good or bad.[2]

We, like others, define "affect" as a conscious experience.[3] However, some authorities such as the renowned neuroscientist Joseph LeDoux claim that affects can be unconscious.[4] "Affect" is sometimes defined as experiencing an *emotion*, but different authors apply widely different definitions to "emotion." These definitions range from the affective feeling itself, to a cognitive reaction to a value-filled event, to experiencing a physiological imbalance in the body, to the physiological and motor *responses* to that imbalance, to name just a few.[5] For these reasons, we will avoid "emotion" as much as possible and use the terms "affect," "affective state," or "affective feeling" when discussing sentient feelings. In the rare cases where we do use "emotions," it is to mean affective feelings, mainly those of humans.

Three Types of Sensory Consciousness: Exteroceptive, Interoceptive, and Affective

One of the most important differences between exteroceptive sensory consciousness and sentient awareness is that exteroceptive consciousness does not *in itself* assign valance to its images. Thus, while a sight of a lion or the smell of a rose may eventually—but not necessarily—evoke fear or bliss, these exteroceptive isomorphic images evoke the feelings only after additional processing by different brain systems that provide the affective and inner-body aspects. Thus, exteroceptive and affective consciousness are two types—and we must bring in *interoceptive* consciousness as well (table 7.1).[6]

Exteroception

Since we already described exteroceptive qualia in the previous chapters, here we simply summarize their features. Exteroceptive qualia (part 1 of table 7.1) are projicient and are experienced as externalized mental images. They are subjectively "local," meaning that points on their image correspond to point loci in the environment (for vision, touch) or to points on a chemical map (of smells or tastes). They are highly isomorphic, but their qualia primarily lack affective properties, and only invoke these properties when relayed to affective parts of the brain.[7]

Interoception and Pain

Interoceptive sensory consciousness (part 2 of table 7.1) is intermediate between the other two types of sensory consciousness and has a dual aspect, with features of both the exteroceptive and affective systems. Its interoceptors

Table 7.1 Three different aspects of sensory consciousness (qualia) and "something it is like to be"[1]

1. *Exteroceptive*: From distance receptors: vision, smell, hearing, taste, discriminative touch, sharp pain in skin.
   A. "Isomorphic sensory consciousness."
   B. Projicient (refers to external environment), and is about external mapped objects.[2]
   C. Experienced as local "mental images."
   D. High sensory isomorphism and low intrinsic affect/valence.
   E. Anatomic pathways (see chapter 5, figures 5.9–5.12): visual, discriminative touch, and auditory paths are exteroceptive alone, but pain and chemosenses (smell, taste) are both exteroceptive and interoceptive. Paths reach cerebrum, optic tectum.
2. *Interoceptive awareness and pain*: skin pain, visceral pain, mechanical senses in body (e.g., stretch or irritation in organ walls), chemoreception (e.g., taste, smell).
   A. "Interoceptive sensory consciousness."
   B. Has a "dual aspect": is both projicient and visceral (refers both to the inner body and the external environment). The input for homeostatic responses.
   C. May be experienced as both local "mental images" and global inner-body/visceral and affective states.
   D. Moderate level of sensory isomorphism (combines features of isomorphic system and nonisomorphic affective system); sensations suffused with affect.
   E. Primary anatomic pathways and structures: those along the spinothalamic and trigeminothalamic paths plus insula and anterior cingulate cortex (see figure 7.1).
3. *Affective/limbic*: pleasure/displeasure, all the felt emotions.
   A. "Affective awareness."
   B. Nonprojicient, referred to "self."
   C. Experienced as global affective states (feelings, emotions); assigns values to internal and external sensory processes.
   D. Low or no sensory isomorphism and high affect/valence.
   E. Anatomic structures are mostly "limbic" (figures 7.1, 7.2): e.g., reticular formation, periaqueductal gray, mesolimbic reward system, basal forebrain, amygdala, insula, anterior cingulate cortex, parts of prefrontal cortex.

1. Adapted from table 4 in Feinberg and Mallatt (2016a).
2. Tsakiris (2013).

sense physiological and mechanical changes in the body via interoceptive neuronal endings that are distributed widely. They measure deep and shallow pain, temperature, chemical states, and sensations in internal organs such as the digestive organs, lungs, bladder, heart, and blood vessels. These internal sensations range from simply recording stretch or irritation in the wall of a distended organ, to the tickle feeling that precedes a sneeze, to the complex sensations of hunger, thirst, nausea, "air hunger" from oxygen deprivation, and fatigue. These varied signals provide the input by which the body adapts itself to change and adjusts its internal environment to its optimal state, the process of maintaining homeostasis (*homeo* = constant; *stasis* = standing, state).

The interoceptive information follows isomorphically organized paths to the brain, but these paths vary in their degree of isomorphism. As a rule, they are more diffuse and crudely organized than is exteroceptive isomorphism,[8] except sharp pain in the skin is highly isomorphic (somatotopic).[9] Conscious interoceptive sensations may be referred to specific parts of the body (localized to the stomach, lungs, muscles, etc.), but their qualia are often experienced as more body-wide or global, as affective and motivational—for instance, in states of fatigue, starvation, oxygen deprivation, nausea, and thirst.

Traditionally, tactile and pain sensations from the skin were not included with the internally generated sensations in the category of "interception." More recently, however, the neurobiologist A. D. (Bud) Craig pointed out that the skin "exteroceptors" that are responsible for pain (nociception), temperature, and itch are necessary for homeostasis and therefore should be considered interoceptive and homeostatic.[10] All these noxious experiences from the skin have strong affective aspects and they generate drives and emotion. Yet they also display unequivocally exteroceptive features. For instance, a pinprick creates an exquisitely precise projicient image of the pin's tip, as well as the noxious affective qualities of the unpleasant sharp sensation. Pain's dual, extero-intero nature is also seen in its subtypes: whereas sharp pricking pain is precisely localized on the skin (isomorphic, like exteroception), slow, dull, and burning pain is less localized, covering larger areas (more crudely isomorphic, like interoception).[11]

Additionally, taste and smell—the chemosenses—display both exteroceptive and interoceptive aspects. They are exteroceptive in that their stimuli come from the outside world and their pathways are isomorphically organized (chapter 5). But they are interoceptive in that their receptors are located inside the body and they have strongly affective aspects: tastes and smells are readily judged to be good or bad, and they rapidly evoke motivation.

Therefore, not only is interoceptive consciousness more directly affective than exteroceptive consciousness (e.g., vision), it also blends with certain exteroceptive senses to give affective feeling (in skin nociception, taste, smell). Interoceptive is "intermediate" and dual in so many ways.

Interoceptive pathways in mammals (based on humans)

Figure 7.1 and table 7.2 show the interceptive and pain pathways of humans.[12] Taste, as another partially interoceptive sense, also takes this interoceptive route (figure 7.1A).[13] The pathways form hierarchies of neurons, with their sensory information being processed in various brain nuclei along the way. The sensory neurons of the interoceptive pathways have thin axons named C and Aδ fibers (table 7.2). For nociception, the C fibers signal the slow, dull pain that is associated with prolonged suffering, whereas Aδ fibers signal the fast, sharp pain of the initial wound or injury. The figure and table are provided, not to overwhelm the reader with detail, but so you can compare them to the *extero*ceptive pathways that were presented back in table 5.3 and figures 5.9 to 5.12. Here we will just present the basics.

The sensory neurons from the body below the head signal the ascending *spinothalamic tract*, whereas the sensory neurons in the head nerves (cranial nerves) signal either the ascending *solitary tract* or the *trigeminothalamic tract* (figure 7.1). All these tracts relay through the thalamus to reach the cerebral cortex for consciousness (as is the case for the exteroceptive pathways: table 5.3), but along the way they send many more branches to the subcortical, affective parts of the limbic brain. That is, compared to exteroceptive information, much more interoceptive information goes directly to limbic centers such as the hypothalamus, the parabrachial nucleus at the pons-midbrain junction, limbic reticular formation, ventral tegmental area of midbrain, and amygdala (figure 7.1). This inflow provides the strong physical link between interoception and affective feelings.

Figure 7.1 also shows how the interoceptive pathways combine the features of isomorphism (as in the exteroceptive pathways) and nonisomorphism (as in the affective networks). That is, as indicated by shading, the paths start with isomorphic organization (dark gray), but this fades and is lost higher up, in the subcortical limbic structures (light gray).[14]

The part of the interoceptive pathway that reaches the cerebral cortex (figure 7.1) retains its isomorphism. This part reaches the primary viscerosensory

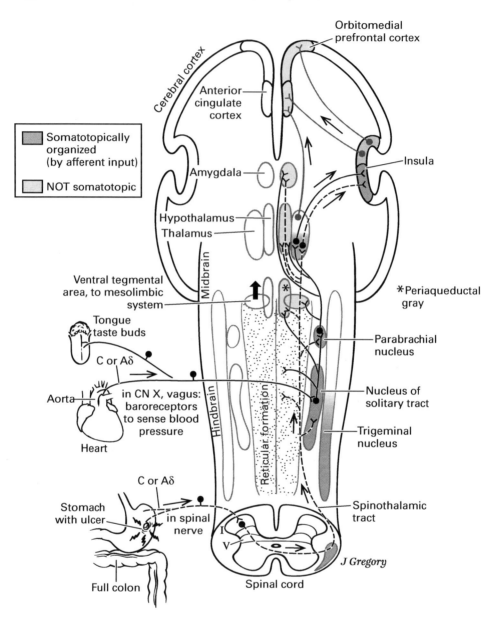

Figure 7.1

Interoceptive and pain pathways to the brain of humans. A. The paths for all the interoceptive senses, with the spinal part below and the cranial part above. B. The pain pathway, which is interoceptive but has some special features. Shading indicates degree of isomorphic organization: dark means some, light means none. For more information, see table 7.2. CN: cranial nerve. In part B, Aδ means fast pain, C means slow pain.

Searching for Sentience

B   Pain

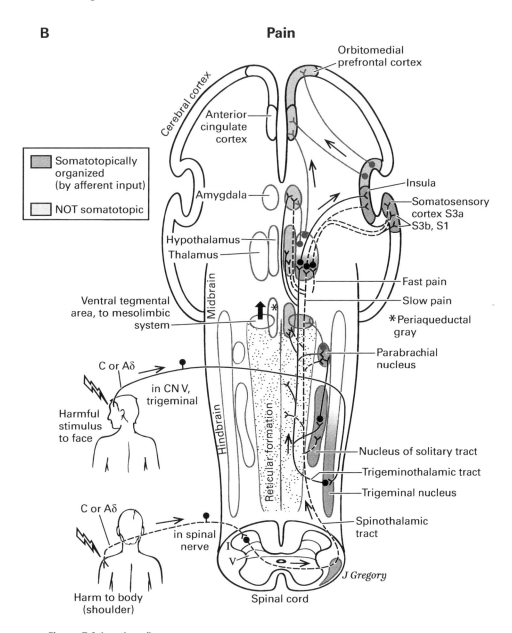

Figure 7.1 (continued)

Table 7.2 The interoceptive and nociceptive pathways in mammals

A. *Interoceptive affective pathways (including pain) (figure 7.1A).*

1. Spinal interoceptive pathways: from spinal nerves of neck, trunk, limbs	2. Cranial interoceptive pathway: from various cranial nerves: trigeminal, facial, glossopharyngeal, vagus
a. Sensory neurons relay to Lamina I of spinal cord, to HPC (heat, cold, pinch) neurons and visceroceptive neurons.	a. Sensory neurons relay to neurons in the nucleus of the solitary tract.
b. Ascending tracts: all in the lateral spinothalamic tract of figure 7.1A: "paleospinothalamic,"[1] spinoreticular, spinomesencephalic, spinoparabrachial, spinohypothalamic tracts.	b. Various tracts, ascending through brain.
c. Proceeds, mostly multisynaptically, to nucleus of solitary tract, parabrachial nucleus, medullary reticular formation, periaqueductal gray, tectum, VMpo of thalamus, hypothalamus, amygdala.	c. Proceeds from nucleus of solitary tract to parabrachial nucleus, reticular formation, periaqueductal gray, VMpo of thalamus, hypothalamus, amygdala.
d. C and Aδ fibers.	d. C and Aδ fibers.
e. Slow and somatotopic (= viscerotopic, organotopic).	e. Slow and somatotopic (= viscerotopic, organotopic).
f. Mostly information on tissue damage and organ fullness (of stomach or intestine, for example).	f. Chemoreceptors, mechanoreceptors, baroreceptors (sense blood pressure), hunger, thirst, nausea, respiratory sensations, taste.
g. Final target: insula, anterior cingulate cortex, somatosensory cortex S3a, 3b, S1 (pain), orbitomedial prefrontal cortex; or, in fish and amphibians, to various parts of the pallium and telencephalon.[2]	g. Final target: insula, anterior cingulate cortex, orbitomedial prefrontal cortex; or in fish and amphibians, to telencephalic targets.[3]

B. *Nociceptive pathways for pain (figure 7.1B).*

1. Spinal pain path: from neck, trunk, limbs.		2. Cranial-trigeminal pain path: from head.
a. Affective pain path (medial: closest to old idea of paleospinothalamic tract[1]).	b. Discriminative pain path (lateral: closest to old idea of neospinothalamic tract[1]).	c. Affective and discriminative pain paths from head.
i. Sensory neurons relay to Lamina I of spinal cord, HPC (heat, cold, pinch) neurons.	i. Sensory neurons relay to Lamina I, NS (nociceptive-specific) neurons and Lamina V WDR (wide-dynamic-range) neurons.	i. Sensory neurons relay to the Lamina I—equivalent in spinal trigeminal nucleus.

Table 7.2 (continued)

a. Affective pain path (cont.)	b. Discriminative pain path (cont.)	c. Head pain path (cont.)
ii. Ascending tracts: all in the lateral spinothalamic tract of figure 7.1B: "paleospinothalamic,"[1] spinoreticular, spinomesencephalic, spinoparabrachial, spinohypothalamic tracts.	ii. In spinothalamic tract.	ii. Trigeminothalamic tract.
iii. C fibers.	iii. Aδ fibers.	iii. C and Aδ fibers.
iv. Slower and somatotopic.	iv. Faster, highly somatotopic (Lamina I) or not somatotopic (Lamina V).	iv. Slow and fast components, somatotopic.
v. Second (burning) pain, heat, pinch, cold.	v. First (sharp) pain (Lamina I), and many other sensory modalities (Lamina V).	v. First and second pain, heat, pinch, cold.
vi. Final target: insula, anterior cingulate cortex, orbitomedial prefrontal cortex, S3a of somatosensory cortex; or, in fish and amphibians, to various parts of pallium and telencephalon.[2]	vi. Final target: somatosensory cortex (S3b and S1).	vi. Final target: insula, anterior cingulate cortex, orbitomedial prefrontal cortex; or, in fish and amphibians, to various parts of pallium and telencephalon.[2]

The references for interoceptive pathways are: Craig (2003b); Critchley and Harrison (2013); Nieuwenhuys (1996), p. 570; Saper (2002). The references on nociceptive pathways in mammals are: Almeida, Roizenblatt, and Tufik (2004); Craig (2003a, 2003b); Saper (2002); Todd (2010); Vierck et al. (2013); Yang et al. (2011). Unique views are presented by Sewards and Sewards (2002). Nociceptive references for other vertebrates, which are not as well known, are: Butler and Hodos (2005), pp. 552–553; and Stevens (2004).

Note: Key facts are that the spinoreticular pathway, trigeminal pathway, nucleus of the solitary tract, and parabrachial nucleus, at least, occur across all the vertebrates (see the text).

1. References for the old paleospinothalamic and neospinothalamic views: Butler and Hodos (2005), pp. 552–553; http://neuroscience.uth.tmc.edu/s2/chapter07.html; Almeida, Roizenblatt, and Tufik (2004); Kandel et al. (2012).
2. Demski (2013); Dunlop and Laming (2005); Rose et al. (2014), p. 111.
3. Rink and Wullimann (1998).

cortex in the insula (the insula being a lobe of the cerebrum: figure 7.2B). For pain, the path also extends to the somatotopically organized primary sensory cortex (figure 7.1B). Up beyond the insula and primary sensory cortex, the interoceptive isomorphism is lost.[15]

Interoceptive pathways in other vertebrates

In the nonmammalian vertebrates, the interoceptive pathways are not as well studied, especially not their spinal parts that carry the signals up from the body below the head. Based on what is known, however, these other vertebrates share enough features with mammals to reveal that they are basically the same. Here are the known similarities: Although a mammal-like spinothalamic tract is not universally present, all vertebrates have the "spinoreticular" part of the interoceptive pathway (left column in table 7.2).[16] The taste path has been worked out in all vertebrates and has the same fundamental plan throughout.[17] The trigeminal pathway of all vertebrates has the spinal trigeminal nucleus, which is associated with nociceptive fibers in mammals.[18] And all vertebrate groups have a nucleus of the solitary tract, parabrachial nucleus, and reticular formation of the brainstem.[19] In fish and amphibians, the poorly studied highest levels of their interoceptive pathways are probably simpler than in mammals, given the less elaborate cerebrum of their brains.

Affective/Limbic System

The third aspect of sensory consciousness is the affective (part 3 in table 7.1), and it involves the so-called limbic system of the brain.[20] We introduced the limbic system in chapter 5 and saw some of its parts in figure 7.1; all its parts are labeled in depth in figure 7.2.

Affective consciousness is global and involves the entire "self." It assigns the basic affective states (of good and bad feelings) and is also responsible for the complex human emotions of sadness, joy, shame, despair, fear, and so on. The affective-limbic aspect relates more directly to internal motivations, drives, and behavioral responses than do the other two aspects of sensory consciousness.[21] Positive affects (liking, pleasure) motivate us to approach a rewarding stimulus, and negative affects (dislike, displeasure, discomfort) motivate avoidance or escape from a noxious or threatening stimulus. In contrast to the exteroceptive system, the affective system is not isomorphic and not in direct contact with the external environment.

While the affective system is more closely related to the "internal milieu,"[22] let us not forget that exteroceptive aspects of sensory awareness (the beauty

Searching for Sentience

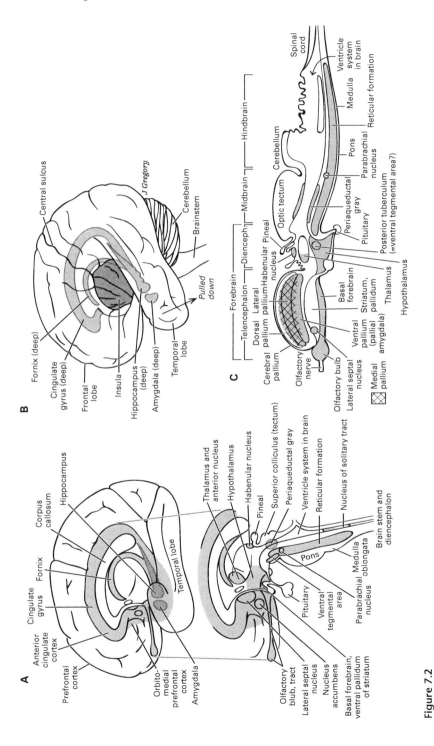

Figure 7.2

Limbic system in the vertebrate brain (the parts shaded gray). A. Medial view, vertebrate, in mid-sagittal cut. Rostral (forehead) is to the left. In A, the brainstem and the diencephalon have been drawn down and away from the telencephalon, so they do not block the view of the telencephalon's limbic structures.

of a rainbow or a concerto) obtain affective consciousness via their interactions with this affective system. Nor should we forget that affective consciousness is even more closely related to the *interoceptive* system.

Some experts, starting with the famous nineteenth-century thinkers William James and Carl Lange, say that this interoceptive-affective relation is extremely close.[23] According to this "James–Lange" or "Jamesian" account (View 1), *all* valent stimuli first incite physiological changes in the body, which are then sensed by interoceptors and sent to the brain to create affects. As an interesting corollary, this would mean that *extero*ceptive stimuli must be double-translated, first by causing a physiological response and then through interoception, in order to induce affects. Other investigators, by contrast, argue that interoceptive and exteroceptive stimuli incite their affects in the limbic system more directly, without first having to induce physiological changes (View 2).[24] For example, they claim that the interoceptive inputs go directly up the interoceptive path in figure 7.1, targeting the limbic centers along the way. There is evidence for both View 1 and View 2. In support of the James–Lange View 1, experimentally increasing interoception by stimulating physiological changes (higher blood pressure) can increase a person's affective responses to many different experiences. But in support of View 2, people react emotionally to external and internal stimuli even after suffering a severed spinal cord, a sliced vagus nerve, or bilateral removal of the insula, all of which would greatly reduce or stop interoceptive signals from reaching their brain.[25] Thus we combine the two views, and conclude that interoceptive and exteroceptive inputs reach the affective brain directly but also indirectly through sensed physiological changes.[26]

Past Theories of the Origin and Basis of Interoceptive and Affective Consciousness

Interoceptive Theories
Interoceptive awareness and "the sentient self" (Craig)
Bud Craig proposes an interoceptive awareness model of the origins of the "sentient self."[27] In his model, the dorsal posterior part of the insula serves as a primary target for all interoceptive and homeostatic information (including pain, itch, thermal sensation, etc.) from the entire body (figure 7.1). This information is somatotopically represented in the dorsal posterior insula and then is progressively re-represented further downstream in the mid and anterior insula, culminating in a "cinemascopic representation" of the sentient

self, or what he also calls the "material me."[28] Craig makes the further claim that the high concentrations of "von Economo neurons" that are peculiar to the anterior insula and anterior cingulate cortices of the largest-brained mammals are critical to these functions, and therefore only humans, chimps, gorillas, whales, porpoises, and elephants have sentience, self-awareness, and presumably, conscious pain. Finally, he suggests that the conscious awareness of these affective feelings, as they are continually reconstructed from moment to moment, is responsible for human emotion. Craig regards this view as consistent with the Jamesian model of awareness and consciousness and with the views of Antonio Damasio in which "feelings" emerge from sensory (and homeostatic) signals from the body.[29]

Craig's "insula" theory has the appealingly simple features of localizing pain and sentience to a relatively limited area of the brain and their origin to a relatively limited time period (those mammals with von Economo neurons appeared within the past 35 million years). However, the insula may exist in all the amniote vertebrates,[30] which would instead imply that sentience originated back at 350 mya.[31]

Those who challenge Craig's theory point to case studies of men whose right and left insula were destroyed by herpes simplex virus, but whose emotions tested as normal.[32] This, along with findings from studies that electrically stimulated and recorded from various parts of the human brain, suggests that while the insula is involved in pain and homeostatic processing, it need not account for sentience.[33]

Charles Vierck and his colleagues on pain

Like Craig, Charles Vierck and his coworkers studied the nociceptive pathways from the skin to the cerebral cortex in a way that may yield clues to the evolutionary origins of sensory consciousness.[34] From investigations of monkeys and humans, they discovered that the C pathway for slow, burning pain not only projects to the cerebral cortex's insula but also to the primary somatosensory cortex, or SI (its S3a part: see figure 7.1B). In the sensory cortex, this C nociception is purportedly integrated with the nociceptive input from the Aδ pathway (fast, sharp pain) to produce the fully conscious pain experience.[35] The proposal of Vierck and coworkers that SI, rather than the insula, is the site of pain perception distinguishes their theory from Craig's.

Vierck's team did not propose a date for the origin of sentience. But according to their account of pain perception, since all mammals have a SI area of cerebral cortex, and mammals evolved about 220 mya (chapter 6), sentience could date to that time. On the other hand, they only studied monkeys and humans, whose last common ancestor lived about 35 mya,[36] so that might be a better date. Thus, let us say between 220 and 35 mya.

Affective Theories
Primordial emotions (Derek Denton)

Derek Denton, an Australian physiologist, proposes that basic "primordial emotions," the subjective aspects of instinctual drives controlling vegetative functions of the body, were the first forms of consciousness of any sort.[37] He proposes that consciousness first appeared in animals when "homeostatic functions," including thirst, hunger for food, air hunger, and hunger for specific minerals, evolved from complex reflexes into conscious feelings. In a nice use of language, he says that when unsatisfied these basic needs are "imperious sensations" that lead to "compelling intentions" toward the goal of "gratification," a strong type of primal affect. The neural centers for these bodily needs are in the vertebrate brainstem and hypothalamus, so these brain regions are where the neural circuits for these drives and emotions evolved. Then later, as additional behavioral programs and motivations evolved, new emotional circuits appeared and contributed to the progressively rostral development of the telencephalon, specifically to its insula and limbic areas. According to Denton's theory, the more complex emotions evolved after the primordial ones. Also evolved secondarily were the exteroceptive aspect of consciousness and the ability of exteroceptive stimuli to induce affects.

Although Denton involves both brainstem and forebrain structures in the creation of conscious emotions, he does not clearly commit to a particular time when they first appeared. On the one hand, he presents evidence that decorticate mammals (lacking a cerebral cortex, through injury or a birth defect) show emotional behavior, and he attributes these emotions to brainstem regions that exist in all vertebrates and even in cephalochordates.[38] That would mean that emotions appeared at the dawn of the chordates, as early as 560 mya. On the other hand, he says the "early substrate" for primordial emotions includes the full-blown cerebrum and neocortex that is only present in mammals.[39] That would mean that consciousness originated only with the first mammals, 220 mya. But then again, he dates it to the origin of true thirst, in the first amniotes inhabiting dry land,[40] which would be 350 mya.

Affective consciousness and primal emotional feelings (Jaak Panksepp, Mark Solms, Douglas Watt, Bjorn Merker)

Jaak Panksepp is a psychologist and neuroscientist whose main interest is in the neurobiology of affect, especially in mammals. His theory has some similarities to Denton's in that he sees emotion as an evolutionary extension of homeostasis, but the emotions he considers are more varied and less physiological than those of Denton, who emphasizes thirst.[41] In Panksepp's account, certain nonhuman amniotes experience a set of "primal affects" defined as valenced phenomenal experiences (qualia) that come in desirable (positive) and undesirable (negative) varieties. He says these are the most basic kinds of emotions ("primary process emotions"). These affects result from "intrinsic brain value systems" that produce sensory, homeostatic, and emotional feelings that support instinctual survival demands. The ancient brain structures necessary for these basic emotional feelings are caudal (brainstem) and medial (mainly the limbic forebrain) (shaded in figure 7.2A). These structures were responsible for the first evolution of subjective experience. He argues that affect appeared before exteroception—that an "affectively responsive organism" appeared earlier in evolution than the emergence of "sophisticated distance receptors and their neocortical analyzers." The reasoning behind this argument is that early animals could not survive unless they first and foremost felt the need to maintain homeostasis and were motivated, by affects, to forage through the environment to meet these needs. Only then did they evolve the ability to sense exquisitely the great variety of stimuli in their external world.[42] Douglas Watt, an advocate of Panksepp's views, expressed similar reasoning.[43]

Panksepp proposes seven primal or primary emotions ("reptilian emotions") that are subserved by an "affective core self." These natural kinds begin with SEEKING to meet homeostatic needs (for food, water, maintaining the right body temperature, etc.), RAGE and FEAR (to avoid bodily harm), and LUST (to promote reproduction and survival of the species). Later, social emotions made their appearance "presumably in species existing before the divergence of birds and mammals." These are "uniquely social-emotional systems" such as CARE, GRIEF, and PLAY. Although not Panksepp's central concern, it appears he believes that these emotions were conscious when they made their evolutionary debut.

Panksepp has long toiled to refute behaviorist psychologists who say that only humans can be called conscious. In general, he assigns consciousness to mammals (220 million years old) or all amniotes (350 mya), yet most of the subcortical limbic structures to which he attributes basic emotions exist in all the vertebrates, implying a date of 560–520 mya. Indeed, in 2015 after many years, Panksepp went on record as supporting this ancient date.[44] Andrew Packard and Jonathan Delafield-Butt also understood the implications and claimed this Cambrian date of origin.[45]

Mark Solms, a South African psychologist, teamed up with Panksepp to probe the origin of affective consciousness.[46] The dominant view in the field is that affect involves only the cerebral cortex,[47] but Solms assembled the evidence for the opposite view, that affect comes from subcortical parts of the limbic brain instead. Panksepp had previously shown that using electrodes or drugs to stimulate these deep brain regions causes emotional or emotion-like reactions in nonhuman mammals (distinct facial expressions in rats, for example).[48] To this, Solms and Panksepp add that removing a rat's cerebral cortex makes it behave more "emotionally," not less.[49] Nor do humans with lesions to their cerebral cortex suffer much loss of their primal affects.[50]

Solms, like Denton and coworkers,[51] drew attention to the observations of the brain scientists Bjorn Merker and Antonio Damasio[52] on human children who were born with a condition called hydranencephaly. In this congenital disorder, the cerebral cortex is almost completely destroyed by a blood-vessel blockage before birth although the subcortical brainstem, diencephalon, and basal telencephalon are spared and functional. These children display a range of behaviors that suggest affects. As noted by Merker:

> They express pleasure by smiling and laughter, and aversion by "fussing," arching of the back and crying (in many gradations), their faces being animated by these emotional states. A familiar adult can employ this responsiveness to build up play sequences predictably progressing from smiling, through giggling, to laughter and great excitement on the part of the child.[53]

Though these children are cognitively and movement-impaired, and with limited exteroceptive consciousness because they lack the cortical sensory-processing areas, they can learn operantly (that is, learn from *rewarded* experience, e.g., to kick at or activate special toys), and they appear to display pleasure. Thus, their behavior indicates that they experience affects, and this implies that nonhuman animals with the same subcortical brain regions can experience the same affects.

According to Panksepp and Solms, the cortex does not generate raw emotions but only fine-tunes and cognitively inhibits or amplifies the primal emotions that originate in the subcortical limbic networks.[54]

Pleasure as the common currency that guides behavior (Michel Cabanac)

Michel Cabanac's theory starts by positing that consciousness originates with sensations and that the sensations have four dimensions: quality (such as blue color), intensity, duration, and affect.[55] Of these dimensions, affect is the most significant because it assigns a degree of pleasure to each stimulus, so the different stimuli that an animal constantly receives can be ranked in the order of their importance. That is, affect ranks by *salience* (chapter 6), the degree to which one stimulus stands out from the others. Those stimuli that produce the most pleasure have the greatest survival value and take precedence in the behavioral responses they will motivate or elicit. This ranking simplifies the choice of which behavioral response to make, so that the brain does not have to code and retain millions of specific preprogrammed reflexes, for each of the many stimuli that are encountered. In essence, affect promotes survival by increasing neural efficiency.[56]

In his original 1996 paper, Cabanac said that his theory addresses only why, and not when, affect evolved.[57] His logic, however, could imply that affect evolved over 560 mya, as far back as the early bilaterian worms that were able to make numerous "choices" among several different feeding, mating, or escape strategies, and even choose among different directions in which to move their bodies. And despite his disclaimer, Cabanac's 1996 paper had a table that actually assigned consciousness to the first vertebrates (560–520 mya).

Later, in 2009, Cabanac and his colleagues tackled the dating question directly.[58] The approach they took was to examine the various vertebrate groups for behaviors that they felt were likely to reflect pleasure and displeasure, and therefore the possession of affective states. These behaviors include trade-off decisions that balance an animal's exposures to good versus bad stimuli; play behavior that could indicate joy; and sleeping behavior that generates a certain class of brain waves. From this analysis, they concluded that affective consciousness arose in the first, reptile-like amniotes, which would be 350 mya.

Survival circuits and "global organismic state" (Joseph LeDoux)

LeDoux studies classical fear conditioning in mammals, especially rodents. In particular, he has experimented on learned responses to a "conditioned stimulus" that has been paired with a noxious or harmful stimulus, such as a sound before a shock. Across mammals, he found that fear conditioning involves neural signals from the outside world being processed by the limbic system, particularly by the amygdala and its connections. This then triggers the animal's defensive behaviors, such as fleeing or freezing. He believes, as do Denton, Panksepp, and Solms, that these circuits are very ancient and probably function in similar fashion across all vertebrates. In contrast to those authors, however, LeDoux is largely agnostic about whether nonhuman mammals have "conscious feelings or emotion," despite their conserved limbic structures. In fact, he seems skeptical that a rodent brain is complex enough for that. For LeDoux, conscious feelings require cognitive self-observation and cognitive appraisal of the organismal states, abilities that only humans have.[59] In saying that other mammals need not have true fear, he writes:

> Fear conditioning thus became a process that is carried out by cells, synapses, and molecules in specific circuits of the nervous system. As such, fear conditioning is explainable solely in terms of associations created and stored via cellular, synaptic, and molecular plasticity mechanisms in amygdala circuits. When the CS [conditioned stimulus] later occurs, it activates the association and leads to the expression of species-typical defensive responses that prepare the organism to cope with the danger signaled by the CS. There is no need for conscious feelings of fear to intervene. The circuit function is the intervening variable. Yet, I and others muddied the waters [in the past] by continuing to call the circuits involved in detecting and responding to threats the fear system.[60]

Accordingly, LeDoux suggests that we should not refer to "fear" or "emotions" at all except in humans; and in nonhuman animals, we should call these "defensive survival circuits" that also include the circuits for acquiring nutrients, fluid balance, regulating body temperature, reproduction, and motivations. Note that this list includes functions that Denton and Panksepp say are driven by true, conscious emotions, but LeDoux disagrees. He proposes that these survival circuits, conserved throughout vertebrates, activate "global (body-wide) organismic states" that maximize the organism's well-being and self-preservation but need not be consciously perceived. He points out that survival circuits even occur in the brains of many invertebrates (which he assumes to be nonconscious) and that non-neural defense mechanisms are present in single-cell organisms, as "defense against harm is a fundamental requirement of life."[61]

Therefore, perhaps counterintuitively, while LeDoux eschews talk of "emotions" and affects in any animals except humans, he goes far back in evolutionary history for at least the precursors of survival circuits and global organismic states that resulted ultimately in human emotions. But the fully cognitive emotions of his theory date only to the first modern humans, who evolved 200,000 years ago.[62]

Conclusions from Past Theories

Table 7.3 summarizes the key claims of six theories of the nature and origin of affective consciousness. Both similarities and differences are evident among the claims, but overall the differences stand out. The extreme differences among the theories are difficult to resolve, leaving three key questions unanswered about the nature and origin of sentience.

The first two questions are where and when affective consciousness evolved (columns 1 and 2). As seen in the table, some of the theories place the primary affects in the subcortical limbic system, while others place them in the cortex but disagree over which cortical regions are involved (insula? SI? prefrontal

Table 7.3 Summary of theories of the origin of affective consciousness

	1 Brain site	2 Date (mya)	3 Affective or exteroceptive evolved first?
A. Interoceptive-centered theories			
1. Craig[1]	Insula (ACC, PFC)	350 or 35	—
2. Vierck[1]	SI cortex	220 or 35	—
B. Affective-centered theories			
3. Denton	Subcortical limbic areas and cortex	~560, 350, or 220	Affective
4. Panksepp/Solms	Subcortical limbic areas	350 or 220 or 560–520	Affective
5. Cabanac	Cortex	before 560, or 560–520, or 350	Affective?
6. LeDoux	PFC, PaC	0.2	Both
Our own conclusions (chapter 8)	Subcortical limbic areas	560–520	Both

1. Craig's and Vierck's theories are human-centered and do not explicitly cover evolution. Thus, we inferred the dates when these theories imply that interoceptive consciousness evolved.

Abbreviations in column 1: ACC, anterior cingulate cortex; PaC, parietal cortex; PFC, prefrontal cortex; SI, primary somatosensory cortex.

Note: For the dates, 560–520 mya means first vertebrates, 350 means first amniotes, 220 means first mammals, 35 means common ancestor of humans and monkeys, and 0.2 means 200,000 years ago and first modern humans *Homo sapiens*.

Note: Precise references for each theory are given in the endnotes to the text.

cortex?), and some theories put the affects in both the cortex and subcortically. All this leads to an enormous range of estimated dates of origin—from over 560 mya to just 200,000 years ago. This interval spans all of vertebrate, or even bilaterian, history and is so unconstrained that it is meaningless to our quest.

The third question raised by past studies is whether affective or exteroceptive consciousness evolved first (column 3 in table 7.3). The studies in the table center on and favor the affective-first idea, but as noted earlier, many other theories, not included in the table, center instead on the exteroceptive aspect of consciousness and they might say the exteroceptive aspect evolved first.[63] Thus, the table is biased on this point, and the literature has not yet been unified in a way that could solve the problem of the first kind of consciousness.

In this chapter, we introduced the affective and interoceptive (sentient) aspects of consciousness and presented past theories for when these aspects evolved in the vertebrates. The different theories come to wildly differing conclusions regarding the date of origin and neural substrate of affect and sentience. Thus, this chapter framed the questions but did not solve them. In the next chapter, we answer these questions.

8 Finding Sentience

This chapter weighs the available experimental evidence in order to find which vertebrates have affective and interoceptive consciousness, and when these phenomena evolved. We do this first, by comparing the reported behaviors of the different groups of vertebrates, and second, by comparing the affective-limbic brain structures across the vertebrates. This second, comparative-neuroanatomic approach will be familiar to the reader by now, but we have not yet shown how behavioral experiments are used to test whether animals have sensory consciousness. To do this, we begin by exploring the behavioral studies of animal pain. Pain is a good place to start because it is one of the most thoroughly studied (and controversial) aspects of animal consciousness. Pain is also of interest because it produces the strongest and most recognizable negative affects in humans and is often treated as the prototypical sentient state.[1]

Pain in Animals? The Behavioral Evidence

To explore which animals experience pain, we first must define the term. The most widely accepted definition of pain is an unpleasant sensory and emotional experience that is associated with actual or possible tissue damage, or is described in terms of such damage; it is always subjective.[2] We will focus on the prolonged, burning pain that originates from the C fibers, because this relates directly to whether animals can suffer and therefore to the most humane ways for people to treat animals.

One of the most comprehensive, collaborative, and relatively unbiased efforts to examine whether animals suffer pain came from a committee of the National Research Council. Their report, entitled *Recognition and Alleviation of Pain in Laboratory Animals*,[3] was produced in 2009 by experts from the fields of neuroscience, animal behavior, pharmacology, and related specialties.

Central to the committee's report is the distinction between *nociception* and the *conscious experience of pain*. The concept of nociception was first described for the skin by Charles Scott Sherrington in 1906.[4] Nociception is the response of specialized neurons (nociceptors) to a "noxious stimulus," defined as a sensory stimulus that damages or threatens to damage tissues, such as a potentially harmful mechanical, thermal, or chemical stimulus. Cuts, scrapes, burns, crushing, and electrical shocks come to mind. Nociception leads to reflexive, or even complex, behavioral responses by the body, but not necessarily to conscious pain. That is, there are many circumstances in which nociceptive responses occur yet conscious pain and suffering are deemed impossible. For instance, nociceptive responses in organisms that lack a central nervous system are presumably only reflexive or reactive. In this category are the avoidance responses in single-celled animals with no nervous system at all[5] or in animals with simple nerve networks such as jellyfish and comb jellies.[6]

In previous chapters, we established that the brain's cerebrum is a center of many aspects of consciousness in mammals; and pain is such an aspect.[7] Given this, another behavioral example of nociception without conscious pain is the nociceptive responses of mammals who have had their cerebrum and diencephalon disconnected from the lower brainstem and spinal cord where the nociceptive processing is occurring. Such "decerebrate" rats and dogs still respond to noxious stimuli with an activation of their sympathetic nervous system that increases their heart rate, elevates their blood pressure, and dilates their pupils.[8] Decerebrate cats can also learn to keep a limb withdrawn to avoid electrical shocks, as can spinally transected rats, in which *all* of the brain has been cut away from the spinal cord.[9] Decerebrate rats still show a remarkable variety of defensive behaviors, such as reacting to the insertion of a feeding tube in their mouth by struggling, pushing it away with their paws, and vocalizing.[10] These rats, when receiving an injection, will bite at the syringe and at the experimenter's hand, and will lick the injection site.[11] After their foot has been injected with the noxious chemical, formalin (which is formaldehyde plus a kind of alcohol), decerebrate rats still shake and groom the injured foot, and hold it above the ground, just as normal rats do.[12] In summary, noxious stimuli can induce many painlike behaviors in mammals without participation of the cerebral cortex and thus without any actual pain.

In contrast to behavioral responses to noxious stimuli that remain after decerebration, some behaviors are lost, suggesting that these are better indicators of conscious pain. One class of behaviors that is largely lost is the capacity to be trained in an operant conditioning paradigm. *Operant conditioning*, or *instrumental learning*, is learning to associate a behavior with a consequence of that behavior, such as rats learning to press a bar to obtain a food pellet or cattle learning to avoid an electrified fence; it is learning from experience.[13] *Classical conditioning*, or *Pavlovian conditioning*, by contrast, is learning in which a conditioned stimulus (CS) such as the sound of a bell is associated with an unconditioned stimulus (UCS) such as the sight of food, until the animal responds to the CS alone (by salivating, for example). Operant behaviors, such as learned strategies for avoiding or escaping nociceptive stimuli, are much more likely to be eliminated by cortical lesions than are classical conditioning paradigms. Thus, decorticate rats show all the typical nociceptive responses, plus classical conditioning, yet they cannot learn or remember how to avoid the noxious stimulus[14] Operant conditioning is considered the more complex and advanced,[15] and researchers have demonstrated it only in some of the animal phyla. Yet almost every bilaterian phylum can learn classically, even those with the simplest nervous systems.[16]

Besides operant learning, another potential behavioral indicator of pain is based on the assumption that true, conscious pain persists longer after an injury event than does simple nociception. Thus, the committee of the National Research Council says that to conclude an animal is experiencing pain, its behavioral response must last long enough to rule out nonconscious reaction.

The committee also considered self-delivery of analgesics (pain relievers) to be a valid indicator of conscious pain. In one such demonstration, rats whose spinal nerves were ligated and compressed pushed a lever to receive the analgesic clonidine, whereas unligated, control rats did not.[17] Furthermore, oral self-administration of pain-relieving NSAIDs (non-steroidal anti-inflammatory drugs) was observed in lame rats and chickens that had arthritis, but not in their healthy counterparts.[18]

It is perhaps significant that such behavioral indications of pain are confined to mammals and birds. For fish, it may be a different story, and the question of fish pain is heavily debated.[19] We will address this later, after developing even better tools for recognizing sentience in animals and after completing our analysis of which vertebrates have affective consciousness.

Behavioral Criteria for Affective Consciousness

Informed by the behavioral criteria for recognizing pain in animals, we now can apply, generalize, and add to these criteria in order to identify other instances of affective states, positive as well as negative. For this analysis, we chose behaviors that best reflect positive and negative affects, as indicated by an animal approaching or avoiding beneficial or harmful stimuli, respectively. Yet, as we learned when considering pain, the chosen behaviors had to go beyond the simplicity or rigidity of mere reflexes and patterned motor programs. First, table 8.1 shows the behavioral criteria we *excluded* on these grounds.[20] Simple approach/avoidance is excluded because, as mentioned earlier, it occurs in one-celled organisms that do not even have a nervous system. Classical conditioning is excluded because mere reflexes can be so conditioned. The other behaviors in the table were excluded because they could be performed, for example, by decerebrate rats or rats with a disconnected spinal cord.

Next, table 8.2 presents the criteria that we ruled in as probable indicators of positive and negative affective consciousness.[21] These criteria were strictly chosen. That is, to increase the likelihood they are correct, they reflect *extended* or *multiple* levels of affect. The criterion of *global operant responses* is included because it reflects memory of the affective state in addition to the valence-related trait of approach or avoidance. *Behavioral trade-offs* indicate that two different valences are recognized and weighed against one another. *Frustration*

Table 8.1 Behaviors that *do not* indicate pain/pleasure (or negative/positive affect)

- Simple approach/avoidance
- "Spinal" learned responses
- Responses from classical conditioning
- Reflexive responses in decerebrate mammals (for example, rodents pushing away objects and syringes from the mouth, or lick or guard injured site, vocalize or jump)
- Innate, possibly reflexive, behaviors including those from basic motor programs (chewing for feeding, freezing, fleeing, etc.)

References: Committee on Recognition and Alleviation of Pain in Laboratory Animals, National Research Council (2009); Mogil (2009); see also Flood, Overmier, and Savage (1976). For more on spinal-cord learning see Grau et al. (2006) and Allen et al. (2009, pp. 137–138); such spinal learning includes a simple and incomplete version of operantly learned leg flexion. For classical conditioning being an inadequate marker, see Rolls (2014). For decerebrate mammals, see Matthies and Franklin (1992); Cloninger and Gilligan (1987); Rose and Flynn (1993); Mogil (2009); Woods (1964).

Table 8.2 Criteria for operant learned behaviors that *probably* indicate pain/pleasure (or negative/positive affect)

- Learning a global, nonreflexive *operant* response based upon a valenced result
- Behavioral trade-offs, value-based cost/benefit decisions
- Frustration behavior
- Successive negative contrast: degraded behaviors after a learned reward unexpectedly stops
- Self-delivery of analgesics, or of rewards
- Approaches reinforcing drugs/conditioned place preference

References: Committee on Recognition and Alleviation of Pain in Laboratory Animals, National Research Council (2009); Papini (2002) on contrast; Mogil (2009) on trained operant measures, trade-offs, and self-delivery of analgesics; Søvik and Barron (2013) on drug approach and self-delivery; Balasko and Cabanac (1998), Cabanac, Cabanac, and Parent (2009), and Appel and Elwood (2009) on trade-offs. See Rose and Woodbury (2008) on self-delivery of analgesics.

Note: "Play" might be another good criterion. It has also been suggested to indicate positive affect, and it characterizes all the groups of amniote vertebrates and possibly amphibians and bony fish: Cabanac, Cabanac, and Parent (2009); Burghardt (2013).

and negative contrast show prolonged duration of the negative affect. The final two criteria, *self-delivery of analgesics or reward* and *approaching drugs/conditioned place preference* extend beyond basic *approaching* to the higher level of *seeking* and *pursuit* of positively valenced stimuli. The behavioral criteria of tables 8.1 and 8.2 fall largely in line with those proposed by recognized experts, including the National Research Council's committee. However, we did not use the often-proposed criterion for pain of behavioral disruptions,[22] such as animals changing their activity levels, hiding, disrupted feeding, disrupted sleep, disrupted social interactions, inattention, anxiety, and so on. We rejected these behaviors because they are not specific to pain or affect, and they often cannot be distinguished from the general degrading effects of injury, stress, or shock.[23] Additionally, this criterion was developed for mammals and is especially difficult to apply or quantify in other vertebrates and invertebrates.

Nor did we use another criterion that has been suggested to indicate affect, which is *persistence* in pursuit of a rewarding goal.[24] Although such persistence could indeed reflect extreme desire and thus an affect-driven perseverance, this is difficult to distinguish from a strong neuronal excitement that activates the brain's pattern generators over and over. In other words, persistence could reflect aroused but unconscious habits.

Behavioral Evidence for Positive and Negative Affect across Species

Now that we have laid out the behavioral criteria for recognizing affective consciousness, we can use these criteria to examine which groups of chordates have these behaviors (table 8.3).[25] We used cephalochordates to represent the initial, prevertebrate state, but the apparently rudimentary behavior of amphioxus is so poorly studied that, to help us approximate the vertebrate ancestor, we brought in some unrelated (protostome) invertebrates that seem to perform at similar levels to amphioxus: nematode roundworms, platyhelminth flatworms, and the gastropod snails and slugs (see figure 4.2).

As shown in table 8.3, the results are fairly clear. Every vertebrate group shows most of the behaviors indicative of affective consciousness. The cephalochordates and the worms do not. The gastropods that have been studied, which show some amount of cephalization (head or brain specialization) within the mollusc lineage,[26] exhibit two or three of the five or six behaviors. Among the vertebrates, only mammals and birds show behavioral contrast (degraded behaviors that last for some time after frustration, behaviors such as agitation, aggression, leaving, or stress).[27] Behavioral contrast could indicate advanced affective feelings in the higher vertebrates, namely an anticipation of the reward and then a remembered, lasting depression or anger after the reward is not delivered. *But the most important finding of table 8.3 is the big increase in affective behaviors between the cephalochordate and vertebrate stages.*

Structural Criteria: The Affective-Limbic Structures

Having found that all the groups of vertebrates fit behavioral criteria for affective consciousness, we turn to the neuroanatomic criteria for affects. These are the structures that are associated with affective consciousness in humans, and their distribution across vertebrates is shown in tables 8.4 and 8.5. They are mostly limbic structures of the brain, which we introduced in previous chapters (figure 7.2C, table 5.1). Other investigators have linked limbic structures to affects in amniotes and anamniotes,[28] and we sought to investigate these links in more depth. To apply the anatomic criteria, we made the basic assumption that similar (homologous) nervous structures in humans and other vertebrates indicate similar affective functions, especially whenever the homologous structures are interconnected into similar networks in all the animals.

Protostome invertebrates are included in the tables as a reference category even though they lack all the structures of the vertebrate brain. If protostome nervous systems have any structures for affective consciousness, then they must have evolved them independently.

Tables 8.4 and 8.5 subdivide the structures. Table 8.4 is a general set of neural structures associated with affects, and table 8.5 lists the brain's mesolimbic reward system (MRS). The MRS is the division of the limbic system said to evaluate the salience (importance) of stimuli and assign positive or negative value.[29] The affect-related functions of the structures are listed in the first column of each table, these functions having been determined for mammals but at least partly demonstrated in other vertebrates as well.[30]

Again, the overall results are clear. The neural structures indicating affect are mostly absent in cephalochordates but mostly present throughout the vertebrates. This is because the structures are mostly limbic and, as mentioned, every group of vertebrates has a well-developed limbic system in the brainstem and basal forebrain. Of all the limbic structures, only a few are in the enlarged dorsal pallium and cerebral cortex of mammals and birds. This finding is consistent with the proposal of Solms, Panksepp, Damasio, Denton, and others, that core affects are experienced in the subcortical parts of the vertebrate brain (chapter 7, table 7.3).[31]

Besides showing that all the vertebrates have affective neurostructures, tables 8.4 and 8.5 also show evidence of a progressive evolution within the vertebrate clade. Counting from the tables reveals that lampreys have 13 of the 18 neurostructures, cartilage fish have 14, bony fish have 16, amphibians have 17, and the amniotes have all 18. But only mammals, birds, and possibly reptiles have the complex dorsal pallium/cerebral cortex (trait 9 in table 8.5) that acts cognitively to modulate the affects and feelings, which originate subcortically.[32] The affective system, although present from the start of vertebrate history, became more elaborate in the amniotes.[33]

Dating Complex Behaviors That Reflect Affective Consciousness: The Adaptive Behavioral Network

This book centers on sensory consciousness in vertebrates, not on the motor responses that conscious states induce. However, affects are thought to drive all the motor behaviors for survival, as directed by the brain's "social behavior network" or SBN.[34] This name is misleading for our purposes because the network is not only involved in social interactions but drives *all* the vital behaviors, ranging from reproduction to aggression against rivals and predators, to escaping danger, to feeding behaviors. For this reason, we rename the SBN the *adaptive behavior network* (ABN). This ABN (table 8.6) gets its input from the

Table 8.3 Distribution of the behavioral evidence for positive and negative affect across animals

	Protosomes			Chordates
	Nematodes	Platyhelminths	Gastropods	Cephalochordates
• Operantly learned responses to punishments or rewards	C. elegans?[1-3]	—	sea hare[4-8] pond snail[8]	—[9]
• Behavioral trade-off	—	—	predatory sea slug[21-23]	—
• Frustration/ Contrast	—/—	—/—	—/—	—/—
• Self-delivery of analgesics or rewards	—	—	land snail *Helix*?[40]	—
• Approaches reinforcing drugs/ conditioned place preference (CPP)	C. elegans?[50]	flatworm[51-54]	—	—

NOTE: Among fish groups, few of the behavioral tests have been performed in lampreys or chondrichthyan fish, so this is mostly teleosts. Effectively no learning has been demonstrated in the seldom-studied cephalochordates but their behaviors seem limited and largely reflexive (see chapter 3 and Casimir, 2009). The other invertebrate chordates, tunicates, have a highly regressed nervous system (chapter 3) and no behaviors beyond simple sensitization and habituation (Holland, 2014; Perry, Barron, & Cheng, 2013). Because of their secondary regression, tunicates are not included in this table.
See the appendix for table references, numbered 1–67 (p. 229).

	Chordates			
	Vertebrates			
		Tetrapods		
Fish	Amphibians	Reptiles	Birds	Mammals
carp[10] stingray[11] others[12,13] sea bream[14]	toad[15,16] others[17]	turtle[18] others[19]	pigeon[20]	rat[20] dog[20]
trout[24] goldfish[24,25]	toad[26] frog[27] salamander[28]	lizard[29]	sparrow[30]	rat[31] human[32]
trout[33]/—[34]	—[35]/—[34]	lizard?[36]/—[34]	pigeon[37]/ starling[38]	rat[39]/rat[39]
nine teleosts?[41,42]	—	crocodile?[42]	pigeon[43] chicken[44]	rat[45–49]
goldfish[55] zebrafish[56–61]	frog[62]	lizard?[63]	quail[64]	rat[65–66] monkey[67]

Table 8.4 Comparative neuroanatomy of positive and negative affects: Features for reward, nociception, and fearlike responses

		Chordates	
			Vertebrates
	Protostomes	Cephalochordates	Lampreys
1. Dopamine neurons for reward learning and seeking*	all Bilateria[1] C. elegans[2] Drosophila[3]	amphioxus[4-6]	river lamprey[7-9]
2. Nociceptors (needed for, but not alone sufficient for, pain)	C. elegans[12-14] Drosophila[12,13,15] leech, sea slug[12,13] squid[16]	—[17]	sea lamprey[18] silver lamprey[19]
2A. Nociceptive fiber types	(Not homologous to those of vertebrates)	—[17]	fibers, but no subtypes
3. Opioid receptors associated with relief of nociception	flatworm[35] scallop[36] octopus[37] (C. elegans: No[38] Drosophila: No[38] crayfish: uncertain[38])	amphioxus?[39]	sea lamprey[40](2)
4. Autonomic nervous system (visceral motor)	—	—	all[44-46] sea lamprey[45,46] (incomplete)
5. Spinoreticular, spinomesencephalic, and trigeminothalamic paths for nociception	—	amphioxus?[48]	all[49-52] Japanese lamprey[53]
6. Amygdala for fearlike avoidance, memory and learning, some generation of feelings?	—	—	all?[57-59]
7. Periaqueductal gray of midbrain, for panic, reproduction, aggression, calling, pain relief	—	—	sea lamprey[67]
8. Lateral septum, for social behavior, territoriality, and evaluating stimulus novelty, regulating affect ("mood," stress, anxiety)	—	—	sea lamprey[74-76] river lamprey[74,76]

*Dopamine signals the errors in predicted versus experienced rewards, which is the degree of "surprise" in appetitive outcomes: Rutledge et al. (2015); Schultz (2015).
Note: Features 7 and 8 also belong to the social behavior network of table 8.6.
See the appendix for table references, numbered 1–81 (p. 232).

| | | | Tetrapods | | |
| | | | | | Amniotes | |
Cartilage fish	Bony fish	Amphibians	Reptiles	Birds	Mammals
all[7]	all[7,10]	all[7,10]	all[7,10]	all[7,10]	all[7,10,11]
rays, shark[20-23]	carp[23,24] trout[23,25]	frog[26-28]	snake[29,30]	chick[31]	rat[32] human[33,34]
Aδ	Aδ/C C are sparse	Aδ/C	Aδ/C	Aδ/C	Aδ/C
all[40](4)	all[40-42](4)	all[40-43](4)	all[40](4)	all[40,42](4)	all[40-42](4)
all[44,47]	all[44]	all[44]	all[44]	all[44]	all[44]
shark[49-51,54]	all[49-51] goldfish[55]	all[49-51]	all[49-51]	all[49-51]	all[49-51] human[56]
shark?[60]	all[61-63]	all[61,62,64] newt[65]	all[61,62,64]	all[61,62,64]	all[61,62,64,66]
all[68] midshipman[69]	all[68] frog[70]	all[68] lizard[71]	all[68] finch[72]	all[68]	all[68,69] rat[73] human[73]
all[74,77]	all[74,77]	all[74,77-79]	all[74,77,79]	all[74,77,79]	all[74,77,79] rat[80,81]

Chordates — Vertebrates — Tetrapods — Amniotes

Table 8.5 Comparative neuroanatomy of positive and negative affects: Mesolimbic reward system (MRS), or mesolimbic dopamine (reward/aversion) system, with the functions listed for mammals

	Protosomes	Chordates	Chordates — Vertebrates
		Cephalochordates	Lampreys
1. Ventral tegmental area (VTA) for evaluating salience of stimuli: dopaminergic, role in reward, analgesia, aversion	—	—	—
1A. Diencephalic (or mesencephalic) structures that may correspond to VTA in anamniotes	—	amphioxus?[8]	all[9] sea lamprey[10] river lamprey[11]
2. Nucleus accumbens, where motivation translates to action, says approach or avoid stimulus; important in causing pleasure and displeasure (dread and fear)	—	—	—
3. Ventral pallidum, for reward processing in mediating motor output of behaviors	—	—?[22]	river lamprey?[23]
4. Laterodorsal tegmental nucleus of reticular formation, adjusts intensity of emotions, negative and positive, signals exploratory locomotion	—	amphioxus?[30] (has reticular formation)	all[31-33]
5. Striatum, for learning and reinforcing goal-directed behaviors	—	—?[38]	all[39,40] river lamprey[41]

	Chordates				
		Vertebrates			
			Tetrapods		
				Amniotes	
Cartilage fish	Bony fish	Amphibians	Reptiles	Birds	Mammals
shark[1]	—	—	all[2] gecko[3] turtle[3]	all[2] finch[4]	all[2] rat[5,6] mouse[7]
all[9] shark[12]	all[2,9,13] midshipman[14]	all[2,9] frog[15]	—	—	—
—	teleosts?[16]	all[17] frog[18]	all[17] lizard[19]	all[17] sparrow[20]	all[17,21] rat[21]
shark?[24]	zebrafish[25,26]	all[27] frog[28]	all[27]	all[27]	all[27] rat[29]
all[31]	all[31] zebrafish[34]	all[31]	all[31]	all[31]	all[31] rat, cat[35,36] mouse[37]
all[39,40] shark[42]	all[39,40]	all[39,40,43]	all[39,40,43]	all[39,40,43]	all[39,40,43]

Table 8.5 (continued)

| | | Chordates | |
| | | | Vertebrates |
	Protosomes	Cephalochordates	Lampreys
6. Pallial amygdala, for sensory-emotional association, emotional and fear learning, operant avoidance learning, negative and positive valence (called basolateral amygdala in amniotes)	—	—	lamprey?[44-46]
7. Hippocampus, for remembering experiences and spatial maps for operant learning	—	—	lamprey?[55,56]
8. Habenula, for coding negative reward, and suppressing movements (also during sleep)	—	—?[65]	all[66-68] river lamprey[69]
9. Neocortical parts: insula, ant. cingulate cortex, orbital and medial prefrontal cortex, for cognitive modulation and regulation of affects, for affect-influenced decisions	—	—	—[74]

Note: General references on the mesolimbic dopamine system O'Connell and Hofmann (2011); Goodson and Kingsbury (2013); Binder, Hirokawa, and Windhorst (2009); Butler and Hodos (2005); Navratilova et al. (2012).

See the appendix for table references, numbered 1–83 (p. 235).

| | | | | Tetrapods | | |
| | | | | | Amniotes | |
Cartilage fish	Bony fish	Amphibians	Reptiles	Birds	Mammals
shark?[47]	all[48,49]	all[48-50] newt[51]	all[48-50]	all[48-50]	all[48-50, 52-54]
sharks[57] stingray[58]	all[55,59-60] goldfish[61]	all[55,59] toad[62]	all[55,59] turtle[63]	all[55,59,64]	all[55,59,64]
all[66-68] shark[70]	all[66-68] zebrafish[71,72]	all[66-67]	all[66-67]	all[66-67]	all[66-67] rat[73]
—[74]	—[74]	—[74]	all?[74] (insula?)[75]	all[74,76-78] (insula?)[75]	all[74,79,80] rat[81] human[82,83]

Chordates — Vertebrates

Table 8.6 Adaptive behavior network ("social behavior network"), which signals behaviors necessary for survival; linked to the mesolimbic reward system, which motivates and rewards these adaptive behaviors

		Chordates	
			Vertebrates
	Protostomes	Cephalochordates	Lampreys
1. Pre-optic area of hypothalamus, plus paraventricular nucleus and vasotocin, for sexual behavior	—?[1]	amphioxus?[2–4]	sea lamprey[5–6]
2. Anterior hypothalamus, aggression and sex behavior	—	amphioxus?[13]	sea lamprey[14,15]
3. Ventromedial hypothalamus, sexual behavior, energy metabolism and food satiation, response to threats	—	amphioxus?[19]	sea lamprey[20]
4. Periaqueductal gray of midbrain, for panic, reproduction, aggression, calling, pain relief	—	—	sea lamprey[25] river lamprey[26]
5. Subpallial amygdala, for emotion-directed responses and actions	—	—	—?[33]
6. Lateral septum, for social behavior, territoriality, and evaluating stimulus novelty, regulating affect ("mood," stress, stress-coping anxiety)	—	—	sea lamprey[42–44] river lamprey[42,44]

Note: General references on adaptive behavior network (social behavior network): O'Connell and Hofmann (2011, 2012); Goodson and Kingsbury (2013). See Forlano and Bass (2011) on fish periaqueductal gray, hypothalamus, and social behavioral network in reproduction.

See the appendix for table references, numbered 1–49 (p. 239).

Finding Sentience

Chordates					
	Vertebrates				
			Tetrapods		
				Amniotes	
Cartilage fish	Bony fish	Amphibians	Reptiles	Birds	Mammals
shark[7-8] skate[7-8]	all[8-10] midshipman[11]	all[9]	all[9]	all[9]	all[9,10] mouse[12]
shark?[16] ray?[16]	all[17,18]	all[17,18]	all[17,18]	all[17,18]	all[17,18]
—	all[21,22]	all[21]	all[21]	all[21]	all[21] rat[23] mouse[24]
all[27]	all[27] midshipman[28]	all[27] frog[29]	all[27] lizard[30]	all[27] finch[31]	all[27,28] rat[32] human[32]
—?[34]	all[35-37]	all[35,37,38] salamander[39]	all[35,37,38]	all[35,37,38]	all[35,37,38,40] human[41]
all[42,45]	all[42,45]	all[42,45] frog[46]	all[42,45,47]	all[42,45,47]	all[42,45,47] rat[48,49]

mesolimbic reward system (table 8.5). More specifically, after the mesolimbic reward system evaluates the salience of sensed stimuli as positive or negative, it signals the ABN to carry out the appropriate adaptive response.[35] Furthermore, the reward system reinforces the adaptive behaviors performed by the ABN, so that the animal seeks more of the reinforcer. For these reasons, the existence of ABN structures in a vertebrate group can indicate that this group has affective conscious states. And ABN structures are simple for neuroscientists to find and identify because their neurons contain sex-hormone receptors, making them easy to label and locate in slices of brain tissue.

Table 8.6 shows the distribution of ABN structures in the brains of the various chordates, along with the affect-driven behaviors to which these structures contribute in mammals. All vertebrates have almost all parts of the ABN. For example, the table shows that lampreys have at least five of the six ABN traits that are present in mammals and in all the other bony vertebrates. This is further, though less direct, evidence that affective consciousness evolved in the first vertebrates.

This section of the chapter has answered the three central questions that we and others have asked about the origin of affective consciousness (end of chapter 7): Where are the core affects generated in the brain? And, when did they evolve in vertebrate history, both in absolute time and relative to exteroceptive consciousness? The core affects are subcortical, not cortical. They evolved in the first vertebrates between 560 and 520 mya,[36] in the same time interval as did exteroceptive consciousness. The *lamprey* brain and limbic system is the minimum level that provides affective consciousness in the living vertebrates.

When Did Interoceptive Consciousness Evolve in Vertebrates, and Do Fish Feel Pain?

General Interoception

Having dated the appearance of the affective aspect of consciousness at 560–520 mya, we will now consider the age of the related, interoceptive/pain aspect. As mentioned earlier and documented back in table 7.2, the interoceptive neural pathways seem to be similar in all the vertebrates and to project similarly to the limbic brain. This implies that interoception first appeared and influenced affective consciousness at the dawn of the vertebrates, 560–520 mya. That in turn fits the fact that interoceptive sensory neurons develop from the vertebrate-defining embryonic tissues, namely, the neural crest and ectodermal placodes (we know this because *all* the sensory neurons of vertebrates develop from crest and placodes: see chapter 5). Our conclusion that all vertebrates have the affective aspect of interoceptive consciousness is not airtight, however, because the interoceptive pathways are only incompletely known in the nonmammalian vertebrates.

By contrast, when did the *isomorphic* (somatotopic) aspect of interoceptive consciousness evolve? Did it evolve in the first vertebrates, with the appearance of the optic tectum, given the tectum's role in consciously perceiving so many other isomorphic senses (chapter 6)? Investigating this, we found fragmentary documentation that the tectum does receive interoceptive input with a somatotopic organization (but reported only in mammals),[37] and nociceptive input in sharks, teleost fish, and lampreys.[38] But this evidence is sparse and uncertain.

Pain

Better evidence is available, however, to date the origin of pain, and the result is surprising. Alone among the interoceptive affects, agonizing pain seems *not* to occur in all vertebrates. That is, fish seem to lack the pain associated with suffering. This theme has been repeatedly promoted by James D. Rose and his colleagues.[39] Yes, all of the jawed vertebrates could have fast, sharp pain, because all have Aδ nerve fibers. But the C fibers of slow and agonizing pain are rare or absent in fish (see table 8.4, structure 2A). More specifically, cartilaginous fish have no C fibers at all,[40] and in teleost fish only 4 or 5% of the axons in their nerves are C fibers,[41] which is "far below numbers seen in mammals (or even amphibians)."[42] It is also far below the 25% number in those rare people who are so insensitive to pain that they often hurt themselves without realizing it.[43]

Rose and coworkers studied the behavioral evidence that has been offered in favor of fish pain, especially of trouts rocking and rubbing their lips against the bottom of their aquarium,[44] after these lips were surgically injected with an irritating substance.[45] They concluded that such responses could have been artifacts of anesthetizing the fish before surgery or of other complicating factors. They presented counterevidence indicating no pain: fish often recover quickly from surgery or injury or from being hooked by fishermen. Rose and colleagues demonstrated that many of the studies claiming to show pain in fish showed only nociception instead.

To explain why bony fish have only the fast, sharp kind of pain, Rose and coworkers presented an interesting evolutionary hypothesis that says prolonged, suffering pain is situationally adaptive rather than universally adaptive.[46] That is, suffering pain warns a wounded animal to hide, rest, and take time to heal. For example, the first mammals and birds probably lived in forests or under cover where they could hide (chapter 6), and many other tetrapods hide out of sight (salamanders, frogs, small lizards). This could explain why tetrapods have more C fibers that signify pain. But some kinds of animals cannot afford to hide as they recover, because they must seek food continually in the open. Such animals would not benefit from prolonged pain, especially because pain-reflecting behaviors attract predators.[47]

Thus, a lack of agonizing pain need not preclude fish from experiencing the affective conscious states from the other interoceptive senses, and such pain may have been the final affect to evolve among vertebrate lineages. It would have appeared in the first tetrapods, 360 mya. We, like others,[48] originally assumed that agonized pain is the most basic type of negative affect, so it must have evolved earliest, but this is evidently not so. Suffering would have arrived last, not first. Therefore, studies that investigate pain to find the most primal affect or the primal kind of qualia could be misguided.

Genes and Affective Consciousness

In chapter 6, we named some genes that are associated with exteroceptive consciousness, thereby tracing consciousness all the way down to the level of molecules. Many genes also associate with affective consciousness. These genes code neuromodulator and neurotransmitter molecules that are released by neurons in the circuits of the limbic and autonomic nervous systems, and are involved with positive and negative affects, such as excitement, fear, and so on. As Damasio says, affective value "is expressed ... in the release of chemical molecules related to reward and punishment."[49] Such chemicals for which the genes are known include dopamine, serotonin, noradrenaline, acetylcholine, nitrous oxide, glutamate, GABA (gamma-aminobutyric acid), glycine, and other neurotransmitters.[50] Genes are also known for oxytocin and vasotocin, hormones that affect the brain and promote feelings of "liking."[51]

Conclusion

The behavioral and anatomic evidence both point to the same date of origin of sentience in vertebrates, and it was early (figure 8.1). At the time the earliest vertebrates were constructing their first mental maps of the external world from their exteroceptive sensors, they were also imbuing their interoceptive

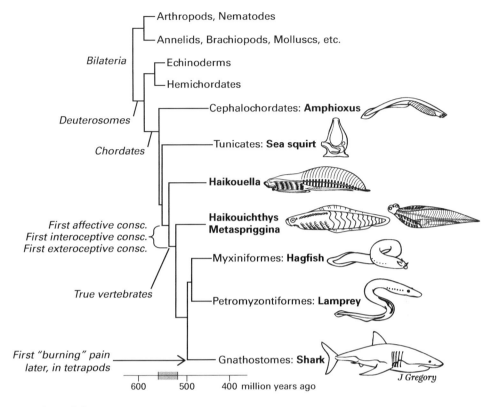

Figure 8.1
Phylogenetic tree showing that affective, interoceptive, and exteroceptive consciousness all existed in the first vertebrates of the Cambrian explosion. The suffering or "burning" kind of pain evolved later in the vertebrate line, with the tetrapods.

and exteroceptive sensations with affect, value, and therefore full sentience. The values were assigned in their limbic brain, but also at all the levels of the sensory neural hierarchies, according to which stimuli were beneficial and which were harmful. As Michel Cabanac and others have proposed, these affects efficiently directed the ensuing motor responses and decisions, by telling the animal which stimuli to approach and which to avoid.[52] The affects translated to the behaviors through the adaptive behavior network (table 8.6). Affect also marked those things that should be remembered and learned, to guide future behavior.[53]

The above line of reasoning implies that the first vertebrates performed conscious, purposive actions with goals. That is, because these animals experienced positive affects consciously, they felt pleasure and desire for rewards, so they had motivations and engaged in consciously directed reward-seeking.

This deduction comes from the work of neurophysiologist Wolfram Schultz,[54] and if true it goes a long way toward solving the difficult problem of mental causation (how the subjective can influence the objective world through purposive action: chapter 2).

Even though our figure 8.1 dates the first affects to the earliest vertebrates, we cannot rule out the possibility that affects evolved somewhat earlier than this, before the evolution of exteroceptive consciousness.[55] We say this because even early animals with crude senses could have felt the affective urge and motivation to forage in the environment to meet their homeostatic needs for food, safety, and mates. Only upon that "affective base" would improved senses and exteroceptive consciousness have evolved. This hints at the possibility of affective consciousness in some protostome invertebrates, which will be explored in the next chapter.

In summary, consciousness evolved earlier, and exists in more animals and in more diverse neural architectures, than neuroscientists previously suspected. The exteroceptive, interoceptive, and affective aspects of consciousness can be traced not only to the cerebral cortex of humans and other mammals, but also to the optic tectum of fish and amphibians (the isomorphic sensory consciousness) and to the subcortical limbic structures of all vertebrates (the affective consciousness).

9 Does Consciousness Need a Backbone?

Are invertebrates conscious?[1] We have concluded that amphioxus and tunicates are not, so now we are posing this question about the *protostome* invertebrates. These animals include the arthropods, molluscs, flatworms, nematode roundworms, and dozens of other phyla (figure 9.1). Many protostome animals have a central nervous system consisting of two longitudinal nerve cords with ganglia at intervals along the cords (a ganglion is an expanded mass of neurons). This is shown in figure 9.2. The cords have a ventral location that differs from the dorsal nerve cord and brain of the vertebrates. The tree in figure 9.1 was also presented in chapter 4, figure 4.2. There, we used it as evidence that the first bilaterian animal, the ultimate ancestor of invertebrates and vertebrates, had a simple, diffuse nervous system and no brain, and that most of the invertebrate clades never advanced far enough beyond this stage to obtain consciousness. But we did not test this deduction rigorously, and some scientists and animal-rights advocates claim that all animals with a central nervous system are conscious.[2] We should investigate this claim. In addition, several groups of protostome invertebrates have relatively complex brains and might have evolved consciousness independently of vertebrates and independently of one another. Here we will go back and test for invertebrate consciousness more carefully.

In this chapter we talk a lot about arthropods and molluscs, so we should familiarize the reader with the members of these two groups.[3] We will do so in simplified terms but in more depth than shown in figure 9.1. In the arthropods, the main subgroups are (1) insects, plus the related *crustaceans* such as lobsters, crabs, shrimps, barnacles, and krill; (2) myriapods (mostly centipedes and millipedes); and (3) chelicerates (spiders, horseshoe crabs, scorpions, and their relatives). For the molluscs, the main subgroups are (1) gastropods (for example, snails and slugs, including the often-studied sea hare, *Aplysia*); (2) cephalopods (squid, octopus, and cuttlefish, plus the nautilus of figure 9.1);

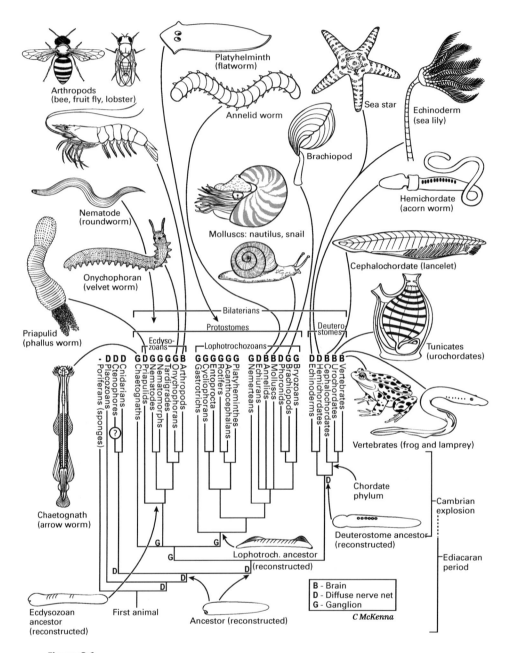

Figure 9.1
Relations of the living animal groups. Especially note the protostomes. As shown at the lower right, "B," "D," and "G" mark the complexity of their nervous systems.

Does Consciousness Need a Backbone?

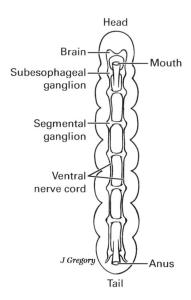

Figure 9.2
Plan of the central nervous system of most protostome invertebrates, as reconstructed in a presumed worm ancestor. Unlike in vertebrates, it is ganglion based and the nerve core is ventral.

and (3) bivalves (for example, clams, oysters, and scallops). Molluscs also include some smaller groups, the most familiar of which is the chitons, or coat-of-mail molluscs.

In the first eight chapters of this book, we worked out a set of criteria for recognizing phenomenal consciousness in animals; now we will use these criteria to judge if protostome invertebrates are conscious. This will be difficult because the neurobiology of invertebrates has not been studied as much as that of vertebrates, at least in those aspects that can reveal anything about consciousness. We will focus on the best-studied invertebrates or the ones with the best chance of having consciousness: octopuses, squids, and gastropods among the molluscs, and insects among the arthropods. But we will also consider flatworms (platyhelminths) and roundworms (nematodes), as protostome invertebrates with simpler nervous systems. Our set of criteria can test for both the affective consciousness of feelings and the isomorphic exteroceptive consciousness. Affective consciousness will be investigated first because the behavioral criteria are relatively easy to apply, and they will be familiar from the previous chapter.

Affective Consciousness in Invertebrates?

Table 9.1 lists five invertebrate clades that have been studied for affects, along with our criteria for judgment (from table 8.2). The clades are listed from left to right in order of increasing brain complexity, as judged by the total number of neurons in their brain, except that both the groups of molluscs were kept together. For these two kinds of molluscs, it must be emphasized that the cephalopods (squid, octopus, and cuttlefish) have much more complex nervous systems than the gastropods (snails, slugs, etc.), so they need not have similar degrees of consciousness or nonconsciousness.

Recall from the previous chapter that we chose these particular criteria for affective consciousness because each one suggests that animals can recognize, experience, and remember rewards versus punishments. Results show that the roundworm (with only about 140 neurons in its brain) and the flatworm (with about 1,000 brain neurons) fail to meet most of the criteria. The gastropod slugs and snails (with 5,000 to 10,000 brain neurons) meet two or three of the behavioral criteria, and the arthropods (with <100,000 to about 1 million brain neurons) meet almost all the criteria. The cephalopod octopuses and cuttlefish, the most neurologically complex of all invertebrates (50 million brain neurons), fit the criteria of operant learning and behavioral trade-off, but they apparently have not been tested for the other criteria. This unfortunate shortage of information hinders a decision on whether cephalopods have affects. However, cephalopods do engage in play, which is a potentially enjoyable experience. In summary, by our criteria and the evidence that is available, arthropods are the best candidates for having affective consciousness.

However, we emphasize that there is not enough direct evidence that *all* arthropods or all molluscs have affects, just that some do (for example, insects and some crustaceans: see table 9.1). Arthropods and molluscs are exceptionally large phyla with many species—especially the arthropods; and each contains diverse members with various body plans and behaviors—especially the molluscs.[4] Nor can we say anything about whether molluscs or arthropods experience the suffering type of *pain*, which is very difficult to recognize (see chapter 8).[5]

Do Invertebrates Have Exteroceptive Sensory Consciousness?

Table 9.2 applies seven criteria for exteroceptive sensory consciousness, criteria we developed in chapters 2, 5, and 6, to the best-studied of the protostome invertebrates. We put in the vertebrates for comparison (at far right). The

Table 9.1 Affective consciousness: Suggested behavioral evidence of positive and negative affect in protostome invertebrates

Animal Clade, with number of brain neurons	Nematode roundworms 140[1]	Platyhelminth flatworms 1,000[2,3]	Gastropod molluscs 5,000–10,000[4-6]	Cephalopod molluscs 50 million[7-8]	Arthropods <100,000–1 million[9-11]
Operantly learned responses to punishments or rewards[12]	C. elegans?[13-15]	—	sea hare[16-19] pond snail[19]	octopus[20-23] cuttlefish[24-25]	fruit fly[19] bee[26] crayfish[27] lobster and crab[28] jumping spider[29]
Behavioral trade-off	—	—	predatory sea slug[30-32]	octopus[33,34]	crayfish[32] cricket[35] crab[36] jumping spider[37]
Frustration/Contrast	—/—	—/—	—/—	—	fruit fly?[38]/—
Self-delivery of analgesics or rewards	—	—	land snail Helix?[39]	—	fruit fly[40]
Approaches reinforcing drugs/ conditioned place preference (CPP)*	C. elegans?[41]	flatworm[42-45]	—	—	fruit fly[45-47] bee[45] crayfish[45,48,49]

NOTE: The criteria for affective consciousness at left are the same as used in tables 8.2 and 8.3.

*Play may be an indicator of advanced affective behavior. It occurs in cephalopods (Mather and Kuba, 2013). Burghardt (2005) is good lead-in to that play literature. Also see Kuba et al. (2006), p. 189, for a comment on insect and crustacean play. See the appendix for table references, numbered 1–49 (p. 242).

Table 9.2 Sensory (exteroceptive) consciousness: Evidence for protostome invertebrates

Criteria	Arthropods (mostly insects)	Molluscs — Gastropods (snails, etc.)	Molluscs — Cephalopods (squids, etc.)	Vertebrates (for comparison)
1. Complexity: neuron number in brain (B) or in whole CNS (C) or entire nervous system (N)	*no?* 100,000 (C): crab[1] 100,000 (C): lobster[1] 100,000 (B): fruit fly[2] 250,000 (B): ant[3] 960,000 (B): bee[2] 1 million (B): roach[3]	*no* 20,000 (C): sea hare[3-4] 1000 (in cerebral ganglia of brain): sea hare[4] 5,000–10,000 (B): gastropods[5]	*yes* 300 million (N): octopus[6] 50 million (B): octopus[7,8]	*yes* 10 million (N): zebrafish[3] 16 million (N): frog[3] 10 million (B): frog[9] 200 million (N): rat[3] 85 billion (N): human[2,3]
2. Multiple sensory hierarchies, levels of neurons (n) before pre-motor center	*yes* visual: 5n[10,11] olfactory: 2 or 4n[12-17] taste: 2 or 4n[18] hearing: >2n[19,20] balance (gravity): >2n[21] touch-mechanosensory: >2n[22], 4n[23] lateral line-like (wind sense): ?n[22]	*few, short* chemosensory: 3n[24-27] mechanosensory: 1 or 2n[28-32] photoreception:?n[33-36]	*yes* visual: 3 or 4n+[37,38] olfactory: ?n[39] taste: ?n[40-41] hearing: ?n[42] balance: >1n[43] touch-mechanosensory: >3n[44-45] lateral line: ?n[44]	*yes* visual: >5n mammal, >4n fish* olfactory: >4n mammal and fish* taste: >4n mammal, >3n fish* hearing: >5n mammal, >4n fish* balance: >4n mammal, >4n fish* touch-mechanosensory: >4n mammal, >3n fish* lateral line: >4n fish*
3. Isomorphic organization	*yes* visual[46-49]: 5th level[46] olfactory: >2nd level[50,51] taste: >2nd level[52,53] hearing: >2nd level[54] touch-mechanosensory[55-58]: >2nd level[57,58]	*yes* mechanosensory: to >2nd or >3rd level[59,60]	*yes:* visual 3rd or 4th level[61] *no:* mechanosensory?[62-67] (other senses unknown)	*yes* all exteroceptive senses: see chapters 5 and 6

Table 9.2 (continued)

		Molluscs		
Criteria	Arthropods (mostly insects)	Gastropods (snails, etc.)	Cephalopods (squids, etc.)	Vertebrates (for comparison)
4. Reciprocal interactions	yes (few?)[68-70]	yes (in simple circuits)[71,72]	yes (but unstudied deep within brain)[73]	yes (extensive)
5. Multisensory convergence and possible sites of conscious unity	yes mushroom body: bee; other parts of protocerebrum: all insects[74-80]	yes? cerebral ganglia[81]	yes vertical and frontal lobes of brain: visual, tactile, more[82,83]; peduncle lobe: olfactory and visual input[84]	yes see chapters 5 and 6
6. Memory region	yes mushroom body[85-88]	yes? cerebral and buccal ganglia[89-93] (simple learning)	yes vertical (and frontal) lobes[94-96]	yes hippocampus and dorsal pallium: chapter 6
7. Selective-attention mechanisms	yes to a visual target[97-99]; to colors[100,101]; oscillatory binding in mushroom bodies; in medial protocerebrum and optic lobes[102,103]	yes to salient stimuli[104] arousal of feeding and swimming in cerebral and buccal ganglia[105,106]	yes limited to behavioral evidence[107-109]	yes see chapter 6

*For the vertebrates, these "n" values for number of levels of neurons before the premotor center were obtained by counting the neuronal levels to the cerebral cortex in figures 5.9 to 5.12 (mammals) and counting the neuronal levels to the optic tectum in figure 6.4 (fish). See the text for why we use the "greater than" symbol (>) here.
See the appendix for table references, numbered 1–109 (p. 244).

roundworm *C. elegans* is also well studied, but because its "brain" has only about 140 neurons, we judged it unlikely to be conscious and did not include it. For each invertebrate group in the table—arthropods, gastropod molluscs, and cephalopod molluscs—we will go through the seven criteria one by one.

But first we must comment on Criterion 2, which says consciousness requires multiple sensory hierarchies, each with a certain number of neuronal levels. Recall the last section of chapter 5 where, after much discussion, we concluded that the minimum number of levels allowing consciousness in the vertebrates is four for most of the senses, but could be three for the somatosensory pathway of fish and amphibians. In the vertebrates, however, we had the luxury of knowing the brain regions where consciousness may arise: the cerebral cortex in mammals or the optic tectum in the fish and amphibians. In invertebrate brains, by contrast, the regions where any sensory consciousness might emerge are unknown or highly uncertain; so for invertebrates we must find another way to frame the "number of levels to consciousness" criterion. We must frame it as the number of neuron levels before reaching the brain's *premotor* centers, because a premotor center defines where the sensory hierarchy ends and the motor system begins. But then, we have to check to be sure this "levels before premotor" parameter is comparable to the *vertebrate* criterion, which is actually "the number of levels to, and including, the conscious regions" (chapter 5).

In running this check, we see that the "conscious" cerebral cortex and tectum of vertebrates do project to premotor centers, namely to the basal ganglia and motor reticular formation, respectively;[6] so yes, the vertebrate and invertebrate numbers are indeed comparable. However, this ignores the fact that additional neuronal levels occur internally, within the vertebrate cortex and tectum. To compensate, we simply designate the numbers of levels in vertebrates with the "greater than" symbol (>) in table 9.2. The bottom line is that if an invertebrate has a sensory hierarchy with *three* or especially *four* levels, then it may signify consciousness, because that number approximates the vertebrate situation.

Arthropods

Of the arthropods, insects have been studied the most. In their ventral nerve cords, the head ganglia are enlarged into a three-part brain: the protocerebrum, deuterocerebrum, and tritocerebrum (figure 9.3A). Insects perform a wide range of behaviors, show behavioral flexibility, and they can learn.[7] But are they conscious of what they sense?

Does Consciousness Need a Backbone?   179

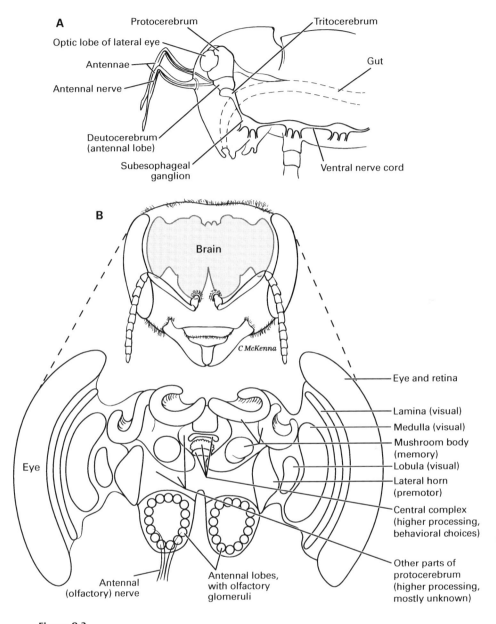

Figure 9.3
Insect and arthropod nervous system. A. Insect brain and nerve cord seen from the side. B. Bee brain in the head, in front view, with an enlarged brain below. C. Crab brain, labeled to show that it has the same parts as the brains of insects. D. Pathways for the various senses in insects: chains of neurons before the premotor centers, with the neurons numbered.

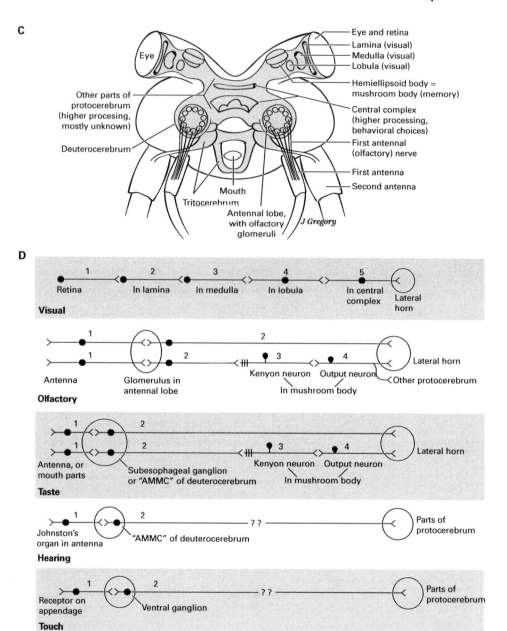

Figure 9.3 (continued)

Does Consciousness Need a Backbone?

Let us start with Criterion 1: complexity. Insects are small animals with tiny brains, and many investigators have wondered whether they have sufficiently many brain neurons to perform true cognitive functions or to be conscious. The <100,000 to 1,000,000 neurons in their brains are considerably fewer than the multimillions in the brains of vertebrates (see the "Vertebrates" column in table 9.2). Even after adjustment for insects' smaller body size, many insects have brains that are smaller than those of any vertebrate.

This latter point is shown in figure 9.4,[8] which graphs brain size against body size in various animals. Invertebrates are compared to the different groups of vertebrates, whose size ranges are shown as polygon boxes. The invertebrate points are circled, and so is the point for a lamprey, as a small-brained vertebrate. The dashed line at lower left is an extension of the average line in the "Bony fish and amphibians" box, extending the fish values as if fish were as small as insects. Notice that three of the five arthropod-brain values are below (smaller than) the vertebrate polygons and line.

Brain size has not been measured for many other arthropods, but spiders also have tiny brains (figure 9.4), and lobsters and crabs have even fewer neurons than insects do (fewer than 100,000: see table 9.2). Back in chapter 2, table 2.1, we adopted neural complexity as an important criterion for consciousness, and we already suspect that arthropods fail this criterion.

But let us see whether insects meet the other criteria, starting with Criterion 2: multisensory hierarchies with a certain number of neuronal levels. In the earlier chapters, we reasoned that sensory consciousness requires a wide variety of distance senses.[9] Insects certainly have such senses: high-resolution vision, olfaction, taste, hearing, balance or gravity sense, many kinds of touch mechanosenses, and even a sense of wind-detection that is comparable to the lateral-line sense of fish.[10] We emphasize the compound, image-forming eyes of insects and other arthropods,[11] having argued (in chapter 3) that such eyes are a key marker for exteroceptive consciousness in the vision-first theory of conscious evolution. Criterion 2 in table 9.2 lists the insect senses and gives references to the literature. Each kind of sense takes a hierarchical sensory pathway to the insect brain, fulfilling the hierarchy criterion for consciousness (see table 2.1). Some of the insect sensory hierarchies attain a fourth or fifth level before reaching the premotor forebrain (where they stop being *sensory* hierarchies). This is also shown in figure 9.3D, and it compares well with the vertebrate numbers in table 9.2. Note, however, that some of the insects' smell and taste pathways reach a premotor forebrain region in only *two* levels, perhaps reflecting the greater simplicity of the insect brain. Overall, insects fulfill this multisense/hierarchy Criterion 2 with a firm "yes."

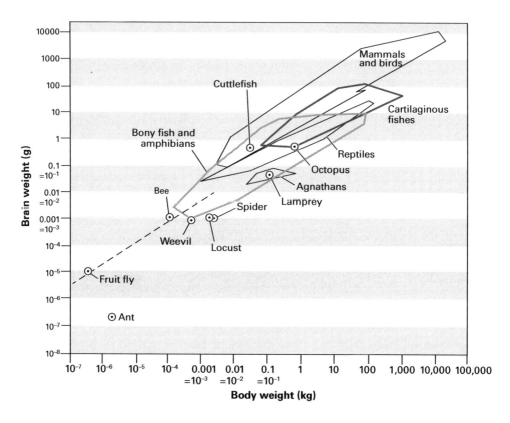

Figure 9.4
The relation between brain weight and body weight in various animals: vertebrates with some invertebrates.

Criterion 3 is isomorphic organization of the sensory pathways, and it is central to our own theory of exteroceptive sensory consciousness (chapters 2, 5, 6). Insects show such organization in all their sensory pathways: as visual retinotopy, hearing tonotopy, touch somatotopy, smell "odortypy," and so on.[12] In this way, insects match vertebrate isomorphism, point by point and sense by sense. Most importantly, Johannes Seelig and Vivek Jayaraman showed that retinotopy continues to at least the *fifth* level of the visual hierarchy in fruit flies,[13] into a brain region called the central complex[14] (figure 9.3B), which is involved in higher multisensory processing and behavioral decisions. This fifth-level retinotopy is rather astounding. In mammals, fifth-order processing is way up in the conscious cerebral cortex, well beyond the primary visual cortex. In fish and amphibians, by comparison, visual consciousness emerges at just the *fourth* retinotopic level, in the optic tectum (according to our hypothesis, discussed in chapter 5). Fifth-order retinotopy in the fruit fly is extremely impressive, *and it offers the strongest evidence for sensory consciousness in an arthropod.*

Also significant is that all groups of arthropods have not only the compound, image-forming eye, but also visual retinotopy in their brains. Such retinotopy has now been documented in insects, crustaceans, myriapods, and chelicerates.[15] Recall that we consider retinotopy to be the most important isomorphic marker for the existence of conscious mental images (chapter 5).

The fourth criterion for consciousness is the existence of reciprocal neural interactions, including interactions between the processing centers of the different levels of the sensory hierarchies (Criterion 4 in table 9.2; see also figure 6.5). This is a difficult criterion to apply, because all neuronal circuits in all animals show reciprocal interactions to some extent. Chittka and Niven worried that the small number of neurons in insect brains allows few reciprocal connections, thus limiting cognition in insects, which in turn would preclude their being conscious.[16] But these connections have not been counted, making this concern hard to validate. Insects do have the reciprocal connections so they must be marked as passing this criterion.

Our fifth criterion, well documented in the vertebrates, is that the pathways for different senses run separately through the nervous system and then come together somewhere in the brain, for a "merging of the senses" into an image (Criterion 5 in table 9.2). This merging is necessary for the *unity* of the experience of sensory consciousness. Chittka and Niven did not find such a multisensory "gathering place" in the insect brain, and they suggested the brain's small size did not allow this; but convergence sites are starting to be found in certain regions of the insect protocerebrum, whose connections and functions are largely a mystery.[17] Another multisensory convergence site has long been recognized as the especially large, memory-forming "mushroom bodies" of bees and some other insects (see figure 9.3B).

Our two final criteria for consciousness are Criterion 6, localized brain regions for memory, which is met by insects' memory-forming mushroom bodies (as just mentioned); and Criterion 7, mechanisms of selective attention to stimuli. Selective attention has been demonstrated in insects, especially in fruit flies trained to orient to a visual cue as they fly, and in bees that show they can tell between similar colors by "paying attention." The brain regions responsible for such selective attention have been identified in the protocerebrum,[18] and intriguingly, the attention process involves oscillatory signals in the gamma frequencies (see Criterion 7 in table 9.2). These are the same kind of waves that mark selective attention in vertebrates (see chapter 6).

Insects have met most of our criteria for sensory conscious. But nagging doubts remain, because their brains are so small with so few neurons. The literature contains phrases like "a fraction of the size of the head of a pin," and "the brains of some insects are smaller than a group of several vertebrate neurons."[19] Some investigators don't find this to be a problem. They have suggested that the miniaturized neurons of insects are each special supercomputers, because their complex axons and dendrites might be divided into functional segments. Each such segment would have its own inputs and outputs, making one insect neuron equal to dozens of typical vertebrate neurons.[20] Other investigators, by contrast, give evidence that insect neurons are nothing special. That is, insect neurons use the typical neurotransmitter chemicals and have the usual ways of making and carrying electrical signals, of connecting into networks, and so on.[21] The problem of potentially too few neurons remains, and it still brings insect sensory consciousness into question.

We seem to be at an impasse, but here is a possible way around it. In the end, the only thing that matters for sensory consciousness is forming mental images. If an insect can be shown to do that, then it has consciousness, period. Although many investigators speak of such images, their existence is very difficult to prove. But we are convinced by an elegant behavioral experiment on bumblebees by Karine Fauria and her associates.[22] The bees were trained to fly within a long box to reach a hole at one end that led to food. The bees had to learn which of several target or bull's-eye patterns surrounding the hole signified that the food was behind it. But they first had to pass through a gate on the way to the hole and to learn which of several striped patterns on a gate signified that the gate led to the food. That is, through visual inspection, the bees had to learn both the correct gate design and the correct target design. The key point was that the gate blocked the view of the target, so bees could not see the target until they had passed through the gate. Still, they successfully learned which combination of gate-stripe pattern and target-hole pattern led to the food. That could only mean that the bees successfully formed a mental *image* of the gate pattern, put that image into memory, and then related it to the correct target pattern. They had formed a visual image, implying consciousness—at least in bees.[23]

If insects have exteroceptive sensory consciousness, then members of the other groups of arthropods must also have it. Here is our reasoning. Recall that the noninsect arthropods are the crustaceans, which include crabs, lobsters, and crayfish; the centipedes and millipedes; and the spiders and their relatives. All these groups have the same distance senses and same brain regions as insects. This is exquisitely documented by Nicholas J. Strausfeld in his book *Arthropod Brains*,[24] and we illustrate the similarities by presenting a crustacean (crab) brain in figure 9.3C, for comparison with the insect brain in figure 9.3B. Furthermore, some fossil evidence is now available about the brains of basal arthropods and near-arthropods from the Cambrian.[25] The particular fossil animals are shown in figures 4.4T (an anomalocarid) and 4.6B (*Fuxianhuia*). As the authors who described these fossils emphasized, their brains had the same parts as in modern arthropod brains, especially the visual parts starting with compound, image-forming eyes. Importantly, these early arthropod brains were not simple or small, but were already full-blown. Thus, if living arthropods are conscious, the first arthropods probably were, too.

Our final decision has to be that, based on our criteria, all groups of arthropods probably have both exteroceptive and affective consciousness, but with the counterevidence of their small brains properly noted. If arthropods were conscious from the start—or at least always had their documented set of complex behaviors and cognitive abilities—then why did they not progress to large brains and higher consciousness like the vertebrates did?[26] The reason may be that all arthropods molt their outer body covering, or cuticle, as they grow.[27] This leaves them soft, physically unsupported, and vulnerable to predation for a time after molting. Giant, helpless molt stages would not be able to hide well and would be easily killed by predators. This limits the maximum size of an arthropod's body, and thus of its brain.

Molluscs
Gastropods

Most snails and slugs creep along slowly, though some water-dwelling species can swim. They have the paired ventral nerve cords so typical of invertebrates, without much enlargement of their head ganglia into a brain. This is illustrated in the sea hare, *Aplysia*, in figure 9.5. The behaviors of gastropods are comparatively simple, and scientists use them to study such simple activities as how they decide to approach, take in, or expel food; how a pond snail leaves oxygen-poor water to breathe air; and escape movements, including escape swimming.[28]

The snails and slugs that have been studied score poorly on several important measures of sensory consciousness (see "Gastropods" column of

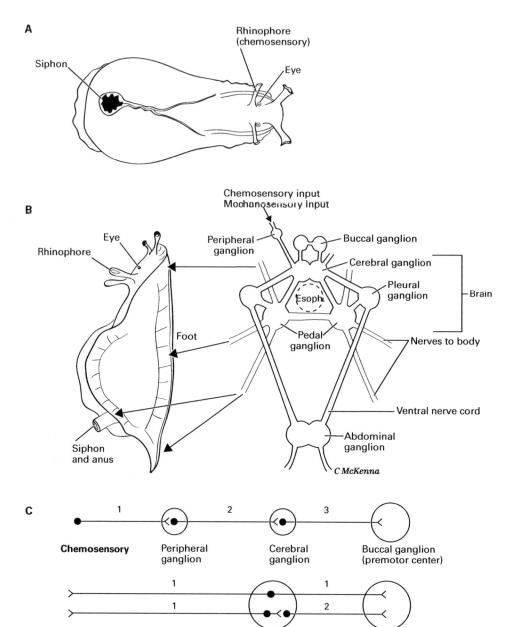

Figure 9.5

Gastropod nervous system. A. Sea hare, *Aplysia* viewed from above. B. Top view of the nervous system (right) of an *Aplysia*, with the whole animal shown in side view at left. C. The chemosensory and mechanosensory pathways (after figure 1 in Baxter and Byrne, 2006).

table 9.2). First is the complexity criterion. Their brain has fewer than 10,000 neurons (table 9.1), and its main ganglia (cerebral ganglia) can have fewer than 1,000.[29] Their senses are limited to mechanoreception, chemoreception, and photoreception.[30] The eyes are small, and in most gastropods, including those used in the studies of table 9.2, they are for light detection, not vision.[31] Overall, these are not the "distance senses" we have tied to consciousness, but the senses expected of the early, unconscious bilaterian animals (see chapter 4). The sensory pathways of gastropods are chains of just 1 to 3 neurons (figure 9.5C), forming shorter "hierarchies" than in vertebrates or arthropods.

But gastropods do have some features associated with consciousness, though always in basic form and within simple neural circuits. A clean, somatotopic arrangement has been found for touch neurons in the ganglia.[32] But these are only the first neurons in the pathway, with the higher levels unstudied for somatotopy. Chemosensory and mechanosensory inputs converge in the brain ganglia (figure 9.5B), and learning (memory) has been tied to two of these ganglia, the cerebral and buccal. Mechanisms of selective attention to salient stimuli, and arousal neurons, have been traced to the feeding and swimming circuits (table 9.2).

Our verdict is that gastropods probably do not have exteroceptive-sensory consciousness because they have few neurons, few senses, simple neural circuits, relatively small cerebral ganglia, and short sensory hierarchies. It is interesting that some of the conscious criteria they do meet (conditioned memory, attention to stimuli, and arousal) relate to the affective behaviors they show (table 9.1), namely to operant learning from rewarded or punished stimuli. This hints that affective consciousness can start to evolve without (or prior to) exteroceptive sensory consciousness.

Next we consider the cephalopod molluscs, who have expanded greatly on the basic mollusc nervous system seen in snails and slugs.

Cephalopods

Modern cephalopods include the coleoids, or octopuses, squids, and cuttlefish (figure 9.6), and the more distantly related *Nautilus* (figure 9.1). The behaviors of coleoids are complex, with much learning, behavioral flexibility, cognition, and foresight that leads to tool use and trickery. Details of their brain circuitry and brain function are poorly known, which is a big impediment to our analysis, but the features that are known suggest sensory consciousness (see the "Cephalopods" column in table 9.2). Their brains (see figure 9.6) are the largest among invertebrates. After being adjusted for body

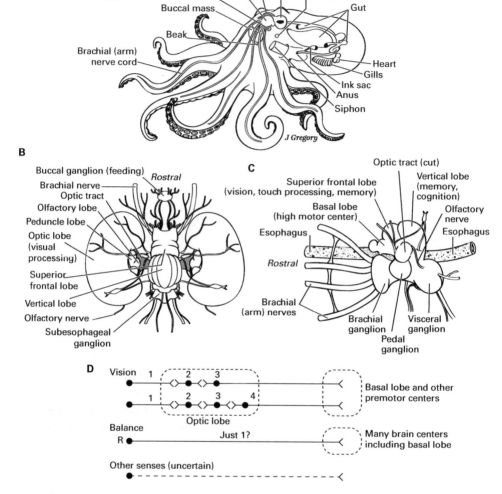

Figure 9.6

Cephalopod nervous system. A. An octopus, showing the position of its brain. B. Octopus brain in top view. C. Octopus brain in side view. D. Some sensory pathways. For more detail on the brain and especially on the pathways, see Shigeno and Ragsdale (2015; p. 1299, and fig. 10).

size, cuttlefish and octopus brains are in the size range of reptile, bird, and mammal brains (see figure 9.4). Octopuses, the best-studied cephalopods, have over 300 million neurons in their central nervous system. Two-thirds of these neurons are in their active and sensitive arms, but about 50 million—still a huge number—are in the brain (see Criterion 1 in table 9.2). This is more than in the entire nervous system of many fish and all amphibian vertebrates (table 9.2), so these cephalopods pass the "complexity" criterion.[33]

Cephalopods have the same variety of distance senses as vertebrates. The eyes of squids, octopuses, and cuttlefish are large, camera-like, and form sharp images, resembling the eyes of vertebrates despite having evolved independently.[34] They have an olfactory organ and a balance organ in the head, and taste receptors and many mechanoreceptors in the suckers on their arms. A lateral line along the side of their body not only detects vibrations in the water (as in fish), but may also "hear." All the senses project to the brain, so multineuron sensory pathways should exist, though these have not been traced systematically from start to end. The visual pathway, however, has been traced beyond third- or fourth-level neurons in the brain (figure 9.6D and Criterion 2 in table 9.2); and the *touch* pathway seems to have three or four.[35]

What of the isomorphism criterion? A real stumbling block for deciding whether cephalopods are conscious is a lack of evidence for topography in their sensory paths. The problem is an apparent absence of mechanosensory somatotopy in their brain, which has been sought but never found. This absence in turn has been related to the special arms of cephalopods in general and of octopuses in particular.[36] The eight flexible arms of an octopus can move in every possible direction, and such complex movements would seem too difficult to calculate, control, and coordinate from one center alone—from only the brain. So these arms are semi-independent of the brain, each arm having its own nerve cord, complex reflexes, neural-processing centers, encoded movement patterns, and even behavioral programs.[37] This has led to the claim that somatotopy does not exist in the octopus brain because the sensory information is processed out in the arms instead. This in turn would mean that the octopus does not unify the arms' sensory information in its brain, and without mental unity it cannot be a conscious animal. In short, the criterion of conscious "unity" would fail because the isomorphism criterion fails.

While accepting the semi-independence and remarkable behaviors of cephalopod arms, we do not believe this prohibits a unified, central consciousness in the cephalopod brain. We refute the claim in four ways. First, the studies that failed to find somatotopy recorded from the *motor* parts of the brain;[38] more recordings need to be done on the brain's *sensory*-processing regions. Second, the head and trunk senses of olfaction, balance, and lateral-line mechanosensation have never been examined for topographic organization in cephalopods. Third, the visual pathway from the cephalopod eye is retinotopic, all the way up to the third or fourth level in the brain, the highest level of the visual hierarchy that has been investigated.[39] As already emphasized, the existence of retinotopic visual maps supports the idea of sensory consciousness.

Fourth, the arms of cephalopods are not as independent as once thought. They send their sensory information on to the brain,[40] which allows the brain to direct the movements of all arms toward common goals. The best demonstration of this was by Tamar Gutnick and her colleagues, who showed that octopuses use vision to guide their arms through a transparent-walled maze to grab food that they could see (and vision is processed entirely in the octopus brain).[41] To do this successfully, the octopus brain must have had a unified "touch plus vision" picture of the situation. Thus, this experiment removes the objections to the possibility of mental unity in the octopus brain.

Nonetheless, the lack of positive evidence for isomorphic mapping of the nonvisual senses remains a difficulty for the idea of cephalopod consciousness. Further investigation is needed. And for now, this is as much a hurdle to cephalopod consciousness as small brains are to arthropod consciousness.

But cephalopods do fulfill the four final criteria for sensory consciousness (Criteria 4–7 in table 9.2). They have extensive reciprocal cross-talk among the (lower) levels of their visual pathway, and of their balance pathway.[42] Different sensory pathways converge in higher parts of the brain (Criterion 5), in regions called vertical, frontal, and peduncle lobes (figure 9.6B, C). The vertical lobe is the best candidate for the site of conscious unity in cephalopods, and brain-injury experiments suggest that it is involved in sensory memory (Criterion 6). Turning to the criterion of selective attention, the brain regions are unknown, but cephalopods are renowned for their ability to focus and concentrate on what they are learning.

Overall, our step-by-step analysis finds that cephalopods are good candidates for exteroceptive consciousness, despite the questions about isomorphic organization and the wide gaps in our knowledge of their brains. The coleoids have the key marker of retinotopy. Were cephalopods conscious from the start, from their first appearance in the Late Cambrian? We cannot tell, because today's nautilus cephalopods (see figure 9.1) live deep in the ocean where it is dim or dark, and they have bad vision from eyes with no lenses.[43] This implies that *Nautilus* lacks visual consciousness, but it is impossible to know whether this is the primitive state or a secondary loss that means the ancestral cephalopods once lived in the light and had good vision. On the other hand, the common ancestor of squids, cuttlefish, and octopus (first coleoid) did have good eyes, but it lived long after the days of the first, Cambrian cephalopods: 276 mya, versus about 490 mya.[44] Thus, the most we can say is that if any cephalopods have consciousness, they evolved it sometime between 490 and 276 mya.

How do cephalopods compare to the other animal groups that evolved consciousness—the vertebrates and probably the arthropods? All three groups have the full variety of well-developed distance senses, good cognitive ability, high mobility, and active lifestyles—traits that therefore could be consistently associated with consciousness in animals. Interestingly, cephalopods combine the key advantages of both vertebrates and arthropods, advantages that we presented at the end of chapter 4. Like the vertebrates (fish), they are fast swimmers with a streamlined body shape (best seen in squids, and powered by jet propulsion).[45] Like the arthropods, they have many limbs that can manipulate objects including food items. With this double advantage, along with their large and possibly conscious brains, should not cephalopods have outcompeted the fish and arthropods to dominate the world's oceans? Yet they did not. In fact, they have to hide and use camouflage, and very many of them are eaten by fish and whales.[46] Cephalopods' failure to outcompete may be because fish are still better swimmers (cephalopods jet *backwards*); because fish evolved good manipulative structures of their own (jaws); because fish have better armor in their scales and bone (early cephalopods had heavy, restrictive shells and modern coleoids are unarmored and vulnerable); because cephalopods' big arms must remain soft and unarmored for maneuverability, making them tempting morsels of meat to predators; or because cephalopods have short life spans so they cannot learn enough in a lifetime. But the real reasons can only be guessed at.

Conclusions

"Lower" invertebrates such as nematode roundworms and the flatworms show little evidence of either affective or exteroceptive consciousness (tables 9.1 and 9.2). The gastropod snails and slugs offer incomplete evidence for both types, perhaps more evidence for the affective, but they lack the brain complexity one would expect for consciousness. The vast majority of the other invertebrate clades have brains at or below this level of complexity and probably are not conscious either. These would be the many clades labeled "D" or "G" in figure 9.1.

That leaves arthropods and cephalopod molluscs as the main candidates for conscious invertebrates. Almost all arthropods, including the first, Cambrian ones, have brains of roughly the same complexity, they meet almost every criterion for affective consciousness (table 9.1) and also most criteria for sensory consciousness, including multiple levels of retinotopy (table 9.2). But we can only judge them as "probably conscious," the main problem being that their brains may be too small, with too few neurons for sensory consciousness. The large-brained cephalopods meet every criterion for exteroceptive consciousness except for nonvisual isomorphism (an important problem), and they meet the few criteria for affective consciousness for which they have been tested. But because of the relative lack of information about cephalopods, we have to call them "possibly" rather than "probably" conscious.

If arthropods and cephalopods should eventually prove to be conscious along with all the vertebrates, what would this say about the evolution of consciousness in the animal kingdom? Arthropods and vertebrates would have evolved consciousness in the Early Cambrian during their competitive "arms race" for improving distance senses, cognition, and locomotion (chapter 4). Cephalopod molluscs did so some time later. In all three cases, consciousness offered the adaptive advantages of (1) forming mental reconstructions of the world, images to which a conscious animal refers to guide its behaviors (in exteroceptive sensory consciousness); and (2) the advantage of revealing which behavioral actions to take or to avoid, based on learned feelings toward cues (in affective consciousness).

But in the long run, consciousness seems to have served the vertebrates better than it could have served the invertebrate groups, at least in the sense that vertebrates include more of Earth's largest animals and largest brains, and have many complex behaviors. It is difficult to argue that any invertebrate made the leap to the heavily memory-enhanced consciousness that we attribute to the large-brained birds and mammals (chapter 6) or achieved self-consciousness or a "theory of the mind" of others. Contributing to these later advances in vertebrate consciousness were some early advantages that were not mental at all. From the start, vertebrates had an excellent locomotory and swimming mechanism; they soon evolved the strongest jaws; and their internal skeleton meant that they did not have to molt (unlike arthropods), so they could evolve large body sizes. Future studies may show that octopuses and squids do have mammal-like levels consciousness and refute some of these ideas. But for now, we have taken a conservative approach: We've presented the evidence that arthropods and cephalopods have only the most basic, phenomenal consciousness, while admitting to reservations even to this much.

This chapter used objective criteria, laid out in tables 9.1 and 9.2, to raise the possibility of invertebrate consciousness, in arthropods and cephalopods. If one or both of these invertebrates should prove to be conscious, then they would reinforce the theme that consciousness is more widespread on Earth than is generally believed, is generated by more diverse neural substrates, and evolved independently several times.

10   Neurobiological Naturalism: A Consilience

In this book, we have analyzed the "mystery" of consciousness from multiple viewpoints. We can now update and summarize our theory of neurobiological naturalism as the most scientifically plausible account of the relationship between brain and mind. Although the theory could be expanded and divided into innumerable elements and synthetic generalizations, here we reduce it to three major postulates that encapsulate this point of view (table 10.1).

Postulate 1: Sensory Consciousness Can Be Explained by Known Neurobiological Principles

Neurobiological naturalism says that consciousness is consistent with generally accepted and known scientific laws, so no new "fundamental," or quantum-level, or "mysterious" properties are required to explain it (chapter 2). Sensory consciousness across all animal species that possess it and in all its varieties results from several essential features that in combination distinguish it from anything else in nature. Drawing upon the general and special features presented in table 2.1, we can summarize them as follows: sensory consciousness is an *emergent* characteristic of a *living, neural,* and *complexly hierarchical* brain.[1]

Living

The seeds of ontological subjectivity go back to the origins of life itself. We have already made the point that in nature, both life and consciousness are *embodied* features of individual animals and thus their existence is ontologically specific to a particular life (chapter 2). Because sensory consciousness is an emergent and embodied feature of a particular life, it follows that consciousness must also be an ontologically subjective feature of that specific life.

Table 10.1 The three postulates of neurobiological naturalism

Postulate 1: Sensory consciousness can be explained by known neurobiological principles.
Postulate 2: Sensory consciousness is ancient and widespread in the animal kingdom, and diverse neural architectures can create it.
Postulate 3: The philosophical issues of ontological subjectivity, neuroontological irreducibility, and the "hard problem" can be explained by the nondissociable confluence of neurobiological and adaptive neuroevolutionary events.

Evan Thompson expresses the continuity between the emergence of life and the emergence of consciousness:

> I have argued that the standard formulation of the hard problem is embedded in the Cartesian framework of the "mental" versus the "physical," and that this framework should be given up in favor of an approach centered on the notion of life or living being. Although the explanatory gap does not go away when we adopt this approach, it does take on a different character. The guiding issue is no longer the contrived one of whether a subjectivist concept of consciousness can be derived from an objectivist concept of the body. Rather, the guiding issue is to understand the emergence of living subjectivity from living being, where living being is understood as already possessed of an interiority that escapes the objectivist picture of nature. It is this issue of emergence that we need to address, not the Cartesian version of the hard problem.[2]

According to this view, the roots of ontological subjectivity go far back in evolution, to the point when living things first emerged from nonliving matter. And therefore, even a living thing with no nervous system at all, like a bacterium, possesses emergent features that are ontologically specific to that individual life.

However, life and all the general biological features of living things we enumerated in chapter 2 cannot alone explain consciousness. A one-cell bacterium is embodied and alive, but it is not conscious. Therefore, additional features must be present.

Emergent

Just as consciousness is an emergent property of life and living systems, it is also an emergent feature of neural interactions. A single retinal cell of the eye does not "see" anything; nor does one neuron in the auditory cortex "hear." Having a conscious sensory experience—any conscious experience—is an emergent process requiring many interacting neural cells arranged in particular patterns with input from other, nonsensory sources such as the memory and reticular-activating systems.

But emergence is still not enough to explain consciousness. All living systems, with or without consciousness, are replete with emergent features. Digestion emerges from the digestive system, circulation from the circulatory system, but these non-nervous systems are not conscious. A colony of ants building an anthill is a system or "society" in which individual ants do not intend to build the hill, yet their collective behavior emergently creates it.[3] But this is emergence without any system consciousness. Ironically, even if individual ants may be conscious (see chapter 9), their societies—the collective groups of individuals—are not. The question is, why not?

The reason why an ant colony can create "intelligent" emergent features the way a conscious brain does, yet not be conscious, is obvious. It is because the individual nervous systems within a colony of ants are not collectively woven together into a single functional and structural unity. An ant colony cannot be in a "state of red" or a state of anything because the ant colony cannot create a system-wide embodied neural state that is required for consciousness. This is why complex emergence is an insufficient (though necessary) condition for consciousness.

Neural

In presenting this neural feature of consciousness, we mean communicating chains and networks of neurons, which evolved long ago as short reflex arcs. Neural reflexes are not in and of themselves conscious, but they are an essential ingredient of the neural substrate that makes consciousness possible (table 2.1). The reasons for this are many. Unlike the cells of any other organ system, neurons are capable of tight and rapid sensory processing of numerous stimulus types, and more neurons can be added to form complex chains and circuits, and hence do more data processing. Thus, neurons uniquely enable the emergence of sensory consciousness. But as we have seen, while neurons and simple reflexes are the "royal road to consciousness," they do not in and of themselves create conscious. Therefore, something else—something relating to increasing complexity—must be the "missing ingredient" that creates consciousness.

Neural Hierarchies and the Special Neurobiological Features

The special neurobiological features that we enumerated in chapter 2 (table 2.1) and throughout this book must also be present. Only a unified brain that has evolved elaborate neural hierarchies creates the unique neural–neural interactions that mark the transition from nonconscious reflexes to subjective phenomenal consciousness. Multiple hierarchies, based on a wide variety of sensory modalities and receptors, may also be essential (vision, hearing, smell, touch: table 9.2, Criterion 2). The hierarchies require reciprocal interconnections and oscillatory binding for mental unity, and, for the distance senses, isomorphic representations that allow mental images. All these requirements are nicely illustrated by the huge architectural differences we found between the small, unconscious brains of the invertebrate chordates with short, simple sensory hierarchies (chapter 3) and the conscious brains of every vertebrate, with their long, complex neural hierarchies (figures 5.9–5.12, 6.4).

Based on the evidence we have uncovered, we can assemble the neurohierarchical features that allow sensory consciousness into one statement: Consciousness stems from unique neural interactions within and between discrete chains of neurons, for many specialized sensory modalities, each chain usually requiring four or more levels, and with the chains arranged as hierarchically organized, modality-specific pathways that merge to form a "neural map" that simulates the real world, or else feed into affective circuits, and ultimately serve sustained processing and behavior. Such hierarchies are documented and illustrated, in a nutshell, in figures 6.4, 6.5, 7.1, and 9.3D. Attention-directing and arousal mechanisms must also emerge, and memory centers must contribute (chapter 6).

To restate Postulate 1: the factors of life, emergence, and unique, complex neural hierarchies in combination can explain ontological subjectivity without invoking any unknown or mysterious principles.[4]

Postulate 2: Sensory Consciousness Is Ancient and Widespread in the Animal Kingdom, and Diverse Neural Architectures Can Create It

Sensory Consciousness Is Ancient

Figure 10.1 shows and dates the key events in our reconstructed history of consciousness. Its timeline emphasizes that consciousness must be an evolutionary continuum, from ancient nonconsciousness organisms to animals with primary consciousness to those with higher kinds of consciousness. A theme of this book is that exteroceptive-sensory consciousness began over 520 mya, when the incipient vertebrates and arthropods each evolved better distance senses in response to the onset of animal predation during the Cambrian explosion, and when their brains processed the new sensory stimuli that that event produced (chapter 4). Rapidly improving vision was most

Neurobiological Naturalism

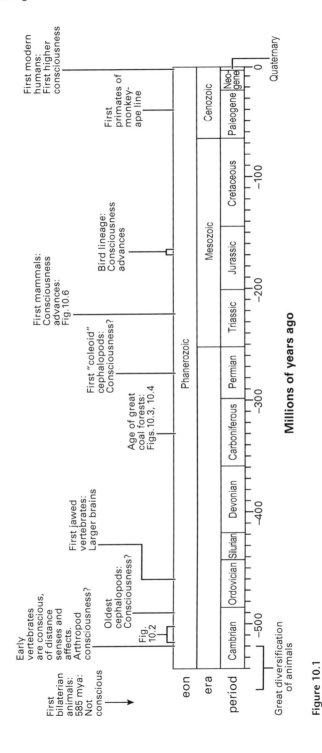

Figure 10.1
Timeline showing the major events in the history of consciousness.

important (chapter 5). We also found good evidence that the affective type of consciousness evolved long ago, in or before the first vertebrates (tables 8.3–8.6). The fact that roundworms and flatworms show little evidence for affective behaviors (table 9.1) indicates that the ancestral bilaterian worms from about 585 mya had no affective consciousness, nor did they have exteroceptive consciousness (chapter 9).[5] To emphasize the importance of the Cambrian events in the timeline, figure 10.2 re-presents our illustration of the Cambrian animals from chapter 4, this time marking all the animals that we interpret to have been conscious. That is, figure 10.2 uses shading to show the conscious vertebrate and the "probably conscious" arthropods.

Next, we follow the timeline in figure 10.1 to just after the Cambrian explosion. At 490 mya, the oldest cephalopod molluscs appear in the fossil record. As we concluded in chapter 9, if cephalopod consciousness exists, it would have appeared sometime between the dates of these first cephalopods and the last common ancestor of octopuses and squid (the first coleoids at 276 mya).[6]

Vertebrate consciousness probably did not undergo any qualitative advances for the first half of its long existence, from 520 to 220 mya (see chapter 6). However, brain size and probably cognition increased when the first jawed vertebrates evolved about 460 mya. Two panoramas from the age of the coal forests show the conscious vertebrates that lived in the middle of this time interval, at 330 mya (figures 10.3 and 10.4). Again, the pictures mark these vertebrates with shading, along with the probably conscious arthropods and the possibly conscious cephalopods.

Following the timeline ahead to 220 mya, vertebrate sensory consciousness advanced in the first mammals to become more informed by learned and remembered imagery so that it better interpreted its mental images (Step 2 in chapter 6). Figures 10.5 and 10.6 show the animals of this time period, both the conscious and nonconscious ones, with the tiny mammals looking rather insignificant but possessing this more advanced consciousness (figure 10.6). The same advance occurred somewhat later in the dinosaur lineage leading to birds (170–165 mya: figure 6.8G). By comparison, few or no arthropods ever made this second advance, though note that the memory regions of bee, wasp, and termite brains are enlarged (these are the mushroom bodies, covered in chapter 9). Perhaps the affective aspect of consciousness also became more complex in the first mammals and birds.[7] For example, these are the only animals that show behavioral negative contrast (table 8.3), which is degraded behavior when a learned reward unexpectedly stops and could therefore indicate "disappointment."

Figure 10.2
Animals in the Cambrian ocean (520–505 mya). Based on figure 4.3B. Shading marks the animals we found to be conscious (the vertebrate, Q) or probably conscious (the arthropods). Animals that are not conscious (the other invertebrates) are unshaded. Because the consciousness of arthropods is less certain and only probable, we marked each of their identifying letters with an asterisk. A. sponge, *Vauxia*. B*. Anomalocarid, *Amplectobelua*. C. Jellyfish of family narcomedusidae. D. Sea gooseberry, ctenophore *Maotianoascus*. E*. Near-vertebrate *Haikouella*. E. Sponge, archaeocyathid. G. Brachiopod, *Lingulella*. H. Sponge, *Chancelloria*. I*. Arthropod trilobite, *Ogygopsis*. J. Priapulid phallus worm, *Ottoia*. K. mollusc relative, hyolithid. L*. Arthropod, *Habelia*. M*. Arthropod, *Branchiocaris*. N. Brachiopod, *Diraphora*. O. Hemichordate worm, *Spartabranchus*. P. Annelid worm, polychaete *Maotianchaeta*. Q. Vertebrate, *Haikouichthys*. R*. Arthropod, *Sidneyia*. S. Varied worm trails within the sediment. T*. Arthropod trilobite, *Naraoia*. U. Sea anemone, *Archisaccophyllia*. V. Lobopodian worm, *Microdictyon*.

Figure 10.3
Animals in the seas of the Carboniferous Period (330 mya), the age of the great coal forests. Shading indicates the animals we found to be conscious (vertebrates), probably conscious (arthropods with asterisks *), and possibly conscious (cephalopod with dagger †). A. Ray-finned fish, *Kalops*. B. Shark, *Ctenacanthus*. C*. Mantis shrimp, *Tyrannophontes*. D. Shark, *Cobelodus*. E. Lamprey, *Mayomyzon*. F*. Proetid trilobite. G. Hagfish, *Myxinikela*. H. Coral, *Acrocyathus*. I. Echinoderm sea lily, *Agaricocrinus*. J. Echinoderm sea lily, *Actinocrinus*. K. Bryozoan colony. L. Mollusc clam, *Aviculopecten*. M. Brachiopods. N. Mollusc snail, *Pleurotomaria*. O†. Mollusc cephalopod ammonite *Anthracoceras*. P. Lobe-finned fish, coelacanth, *Hadronector*.

Figure 10.4
Animals on land in the Carboniferous Period (330 mya), the age of the great coal forests. Shading and asterisks indicate animals with consciousness and probable consciousness, as in the previous two figures. A. *Sigillaria* tree. B. Lycopsid tree, *Bothrodendron*. C. Lycopsid scale-tree, *Lepidodendron*. D. Tree fern, *Psaronius*. E. Horsetail-plant relative, *Sphenophyllum*. F. Horsetail-plant relative, *Calamites*. G. Amphibian, temnospondyl. H. Fern, *Mariopterus*. I*. Spider relative, *Eophrynus*. J. Lobe-finned fish *Strepsodus*, washed up on the shore of a stream. K*. Giant millipede, *Arthropleura*. L*. Eurypterid ("sea scorpion"), *Hibbertopterus*. M. Early amniote, *Paleothyris*. N. Early synapsid (mammal-like reptile), *Eothyris*. O. Freshwater shark, *Xenacanthus*.

Figure 10.5
Animals in the seas in the Triassic Period (220 mya), the age of reptiles. Shading, asterisks, and daggers indicate animals with consciousness, probable consciousness, and possible consciousness, as in the previous three figures. A*. Shrimp, *Schimperella*; B. Ichthyosaur reptile, *Mixosaurus*. C. Shark, *Hybodus*. D†. Mollusc cephalopod, straight-shelled ammonite. E†. Mollusc cephalopod, curve-shelled ammonoid, *Ticinites*. F. Sea weed, *Codium*. G. Mollusc gastropod, snail *Wannerospira*. H. Various corals. I. Various gastropod snails. J. Mollusc gastropod snail, *Palaeonarica*. K. Echinoderm sea star. L. Mollusc clam, *Enteropleura*. M. Ray-finned fish, *Kyphosichthys*, which is related to today's gars. N. Ray-finned teleost fish, *Pholidophorus*.

Figure 10.6
Animals on land in the Triassic Period (220 mya), the age of reptiles. Shading and asterisks are used to indicate animals with consciousness and probable consciousness, as in previous four figures. A. Ginkgo tree. B. Gymnosperm plant, *Glossopteris*. C. Pterosaur (winged reptile), *Caviramus*. D. Early dinosaur, *Plateosaurus*. E. Cycad tree, *Leptocyas*. F. Conifer (pine) tree, *Voltzia*. G. Late mammal-like reptile, *Placerias*. H*. Bee relative, of Xylidae family. I. Seed fern, *Dicrodium*. J. Early dinosaur, *Coelophysis*. K*. Beetle, *Carabus*. L. Early mammals, *Hadrocodium*, with their nest and eggs. M. Thecodont reptile, *Ornithosuchus*. N. Early turtle, *Odontochelis*.

A final advance, to self-consciousness, language, and the realization that other individuals have minds, occurred in the primate line of mammals, perhaps when the first modern humans evolved about 200,000 years ago (figure 10.1).[8]

Sensory Consciousness Is Widespread

Much of this book involved determining which animals have sensory consciousness by identifying the criteria and then applying them systematically to the clades of vertebrates and invertebrates (tables 8.3–8.6, 9.1, 9.2, and 2.1). Notice that tables 9.1 and 9.2 use our own new findings to flesh out these criteria and to present them in more depth than in chapter 2. More specifically, our original criteria (table 2.1) mostly involved complex neural hierarchies, and our new findings added such criteria as multisensory convergence in the brain, a role for memory, and the behavioral criteria for affective consciousness.

The findings indicate that consciousness is much more widespread than previously suspected. Few investigators have thought that all vertebrates are conscious and in all three basic ways (figure 8.1). That is, our criteria say that the first vertebrates had (1) exteroceptive, (2) affective, and probably (3) interoceptive consciousness (though they did not suffer pain), but the invertebrate chordates like amphioxus are not conscious, and neither are most other invertebrates. Arthropods and cephalopods meet many of the criteria for consciousness, although each raises uncertainties (chapter 9). If these two clades of protostome invertebrates should prove to be conscious, as are vertebrates, then consciousness originated independently in three distantly related clades of animals (figure 10.7): in the first vertebrates, the first arthropods, and in the early or later cephalopod molluscs. By convergent evolution, consciousness could have evolved three different times, once in each of three great groups of bilaterian animals: in deuterostomes, Lophotrochozoa, and Ecdysozoa.

Figures 10.2 to 10.6 show the types of conscious animals in various periods in Earth's history. Although these figures are by no means random or unbiased samples of the kinds of animals existing at the times, they do reveal that consciousness became more widespread through the ages. That is, they show how the conscious vertebrates became increasingly abundant and varied after the Cambrian, and how conscious animals gradually radiated and invaded more kinds of environments in the sea and on land. It is indeed surprising how many species were conscious (the vertebrates), or potentially conscious

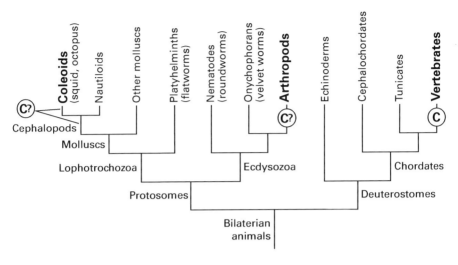

Figure 10.7
Phylogenetic relations of the animals that show evidence of consciousness. "C" indicates a conscious clade, and "C?" indicates the possibly or probably conscious clades. At left, the "C?" and the leaders extending from it mean that consciousness could have arisen either in the first cephalopods or in the first coleoids, but not in any other molluscs such as gastropods or clams (chapter 9). Condensed from the tree in figure 4.2.

(the arthropods and cephalopods), throughout the past half-billion years. The vertebrates seem to have had special advantages, perhaps stemming from better locomotion (chapters 4, 9), and they seem to have derived more of their success from their conscious state. Additionally, the figures illustrate the many kinds of invertebrates that existed and were common in Earth's ecosystems and yet *never* evolved consciousness. These were mostly immobile and armored or shelly types such as clams, corals, sea lilies, and bryozoans. Finally, the figures show the higher proportion of land animals than sea animals that have always been conscious.

The illustrations end with the Age of Reptiles, 220 mya, when the broad groups of conscious animals already were established and in their ecological roles. To be complete, however, we will verbally describe similar scenes from the present day. Modern animals are familiar so no illustrations are needed, and readers can use their imaginations to make the modern-versus-ancient comparisons. Starting with today's equivalent of the Triassic land scene of figure 10.6, imagine the grasslands of present Southern Africa, teeming with conscious animals. Mammals and birds now fill most of the former reptile niches. The flying pterosaur reptile of figure 10.6 (animal C) is replaced by any of a number of birds, the long-necked dinosaurs (D) have been replaced by giraffes, the hairy reptile (G) by a hippopotamus, the ostrich-like dinosaur (J) by a real ostrich, the crocodile-like thecodont reptile (M) by a true crocodile, the earliest mammals (L) by any of a number of small rodents or shrews of today. Turtles (N) still persist, and the insect groups (H, K) are much the same as today. A difference between the Triassic and today would be the presence of humans, so imagine today's grassland scene including a band of Kung! or San people ("bushmen"), who show some genetic and cultural traits of the earliest modern humans.[9]

The animals in today's shallow ocean are also comparable to those in the Triassic sea, shown in figure 10.5. The nonconscious kinds of molluscs (G, I, J, L) still are very abundant (clams, oysters, many kinds of gastropods), and nonconscious corals, echinoderms (sea stars, sea urchins), and worms still inhabit the modern ocean floor. The potentially conscious invertebrates in today's seas include the marine arthropods such as lobsters and the cephalopod groups of octopuses, squids, and cuttlefish. Conscious vertebrates thrive in the seas. Though sharks (as in C) are still important carnivores and predators, advanced bony fish (teleosts) are more abundant and varied now than in the Triassic, and they are the dominant vertebrates in the ocean (think of cod, anchovies, tuna, eels, reef fish, barracudas, and the like). Sea turtles and sea snakes survive today, but the fishlike reptiles in the Triassic scene (B) have been replaced by the cetacean mammals—dolphins and whales. Dolphins are thought to have levels of consciousness and self-awareness akin to those of apes.[10] Seals and their relatives are other marine mammals, and penguins are marine birds. With the many vertebrates, arthropods, and cephalopods in today's seas, consciousness in the ocean has held its own since the Triassic, or has even expanded.

Sensory Consciousness Is Diverse
In addition to consciousness being widespread among animals, we found that it is *diverse* regarding the different neural architectures that can create it. One example is that the advanced consciousness of mammals depends on a rather different architecture of the cerebral pallium than that of birds (figure 6.9). Another, more striking example of such diversity is that the exteroceptive-sensory consciousness of fish and amphibians is based on a different neural substrate (the optic tectum) than in mammals and birds (the cerebral cortex and equivalent dorsal pallium), for all the nonolfactory senses (chapter 6).

Tectal versus corticothalamic consciousness

Many experts would disagree with our claim for such diversity because they assign consciousness to just one, corticothalamic, system. That is, most investigators claim that consciousness requires the cerebral cortex of mammals (or the homologous dorsal pallium of birds), and they would not agree that the optic tectum of fish and amphibians generates it. Their reasoning is based in part on the observation that humans and other higher primates lose critical aspects of sensory consciousness after suffering cortical lesions. For instance, humans lose visual awareness (become "cortically blind") after brain lesions that destroy the cortical visual areas. Therefore, we know that humans do not have conscious tectal vision. From this, the "corticocentric theorists" of consciousness conclude that an animal without a cortex or elaborated dorsal pallium—an animal like a lamprey—cannot have conscious vision based on the tectal visual system.[11]

But we have built the case that a basally arising vertebrate like a lamprey has image-forming eyes, excellent vision, a topographic mapping of vision, and all the other distance senses in its tectum, and thus that lampreys have mapped, mental images without having a substantial pallium (chapter 6). We looked to the fish themselves, rather than indirectly deducing their state from humans and other mammals. In our view, the elaborate senses of fish—along with their intimate interactions with their stimulus-rich environments[12]—indicate that the pallium is *not necessary* for sensory consciousness rather than proving that the pallium *is necessary* for all sensory consciousness.

Our view faces the challenge of *nonconscious perception* of visual and other sensory stimuli. In presenting this challenge, advocates of the rival corticothalamic hypothesis point out that human sensory hierarchies, which meet all our criteria for consciousness in tables 2.1 and 9.2, can perceive stimuli without conscious knowledge, largely subliminally, when the very highest levels of the cerebral cortex are not engaged. Experimentally demonstrated examples of such nonconscious perception in primates include blindsight, subliminal perception during masking, the attentional blink, binocular rivalry, implicit cognition, inattentional preconsciousness, mismatch negativity, response inhibition, nonconscious error detection, and conflict resolution.[13] By using some version of such nonconscious perception, the corticothalamic advocates reason, the sensory systems of fish and amphibians could be functional but without any consciousness.

The problem with this challenge is that the types of nonconscious perception that have been detected in humans and monkeys are so weak and incomplete that they effectively amount to inattention, and any fish relying on them could not sense dangers well enough to survive in nature. These nonconscious perceptions are always studied along with mammalian consciousness and cannot be dissociated from the latter, nor can they be distinguished from a degraded type of consciousness. By contrast, the exteroceptive systems of lampreys and other anamniotes are not like that. Fish process stimuli to a high level that can make the fine-grained sensory distinctions that are necessary for survival. Any human who was dependent on the nonconscious, low-level sensory discriminations that are demonstrated in the laboratory would be almost "senseless" and could not survive if left to his or her own devices.[14]

Next, our view faces the critical challenge of *unconscious hierarchies*. Regarding our claim that having "complex neural hierarchies of isomorphic representations" is a fundamental feature of sensory consciousness (table 2.1), critics could argue that this is wrong if such a hierarchy exists for any *un*conscious sense. But we do not claim that *all* the isomorphic sensory hierarchies are conscious. In humans, the hierarchies for balance and for some interoceptive and proprioceptive senses have unconscious aspects.[15]

The proprioceptive senses illustrate this best, proprioception being the monitoring of one's own body movements by sensing the stretch of muscles, tendons, and joint capsules during such movements. Proprioception has both an unconscious system with isomorphic projections to the cerebellum, and a conscious system that has isomorphic projections to the cerebral cortex. The conscious, cortical system is unproblematic and has the same neural architectonics as all conscious sensory systems (table 5.3). But why is the (possibly older) cerebellar proprioceptive system isomorphic yet unconscious? We have an answer. Theoretical considerations suggest that it is far more important to devote consciousness circuits to exteroceptive senses that monitor the outside world, which is much more dangerous, unpredictable, and varied than proprioceptively sensing the changes resulting from one's own movements.[16] The same reasoning holds for other inner-body changes, which are already being kept within safe and narrow limits by hormones and other automatic homeostatic mechanisms.

Tononi and Koch present another reason why not all of the complex brain regions give rise to consciousness, specifically why the proprioceptive cerebellum does not.[17] The cerebellum's many neuronal modules act more independently of one another than do those of the (conscious) cerebral cortex, with fewer interactions, so they have less integration and no consciousness. The number of organized, integrated interactions is the key.

Finally, rival investigators would challenge us by claiming that the cerebral cortex is necessary and responsible for *affective* consciousness. We refuted this in chapters 7 and 8 with evidence that the basic affects are not cortical but subcortical, stemming from brain regions that exist not only in mammals but in all vertebrates (tables 8.4 and 8.5). Also, we were able to assemble and apply a justifiable set of *behavioral* criteria, well beyond the level of reflexes, for judging which vertebrates have affective consciousness (table 8.2); and all vertebrates pass these criteria whether they have a cerebral cortex or not (table 8.3).

It is most important to recognize and emphasize that subcortical affects and selective attention are indicators of primary consciousness. It is not easy to tell whether an animal is experiencing a "private" mental image of the world in its brain,[18] but it is simple to see that even a fish pays attention to stimuli and shows valenced preferences.

Within-brain diversity

Continuing our theme of the great diversity of conscious systems, neuroarchitectonic diversity also exists within each single vertebrate brain. A good example is how, in every vertebrate, the conscious pathway for every sense relays through the thalamus—*except* for olfaction (chapters 5 and 6; figures 5.9 through 5.12, 6.4). Here, the diversity is indicated by the two different kinds of paths, for olfaction without the thalamus and for the other senses with direct thalamic projections. Actually, the olfactory path does have a transthalamic branch, but it is of secondary importance and olfactory consciousness does not depend on it.[19] In the main olfactory path, information is transferred by gamma oscillations that are much like—but generated independently from—the thalamocortical oscillations that bind the other senses of vision, hearing, balance, and touch in mammals into conscious percepts for those senses.[20] In mammals, the olfactory percepts would next be bound with the percepts for the other senses over large areas of the prefrontal cortex, but the key point is that until then they are two distinct systems of gamma-bound projections, as if they'd evolved separately for (1) olfactory consciousness versus (2) consciousness of the other senses. We repeat: the two different systems indicate diversity.

Additionally, diversity exists in the form of three types of consciousness: exteroceptive, interoceptive, and affective. Here are the most important comparisons among these types, summarized from chapters 7 and 8. Let us start by comparing the two extremes, the exteroceptive and the affective. One difference is that the exteroceptive comes from the brain's dorsal pallium or optic tectum whereas the affective is from the subcortical, limbic core. Exteroceptive and affective consciousness also differ in their amount of isomorphic mapping: much isomorphism for the exteroceptive and little for the affective. Interoceptive consciousness is the third type, adding to the diversity even though it shares features with both exteroceptive consciousness (isomorphism) and affective consciousness (more input to the limbic system, more valence in its qualia).

Figure 10.8 provides more in-depth comparisons of the three types of consciousness, and of the brain regions responsible for each type. The figure shows how sensory inputs from exteroceptors and interoceptors (the two thick arrows in the lower left quadrant) go to the brain (the two big boxes in the middle), leading to exteroceptive, interoceptive, and affective consciousness (the three boxes at the top). Despite the brain's schematic look in this figure, it is physically set in space. Its *dorsal* part (the pallium and tectum at left) does the most processing of sensory inputs, especially of isomorphically mapped inputs leading to conscious mental images of the outside world and inner body ("Exteroceptive images" and "Interoceptive images, feelings," in the boxes at the top). The brain's *ventral* part at right, especially the deep limbic core of the brainstem, uses sensory input to generate the basic affective feelings of good and bad. The brain's *caudal-rostral* axis is important because sensory processing becomes more complex rostrally, so that consciousness emerges at the top (far rostrally). As shown by arrows, the "Distance senses, from exteroceptors" project heavily to the thalamus and cerebral cortex, or to the tectum, yielding exteroceptive images. The distance senses also project to the limbic core, inducing affects. The "Body senses, from interoceptors" project even more to the limbic core, to produce basic affects. In mammals, they also reach the cerebrum's insula and anterior cingulate cortex (ACC), both to produce images and to adjust affects. In the core, the reticular activating system (RAS) receives both interoceptive and exteroceptive inputs, activating (arousing) both affects and exteroceptive awareness. The sensory input to the core can also activate the mesolimbic reward system, whose parts are marked by dashed ovals. For more about the limbic structures, see tables 8.5 and 8.6. At the bottom of the figure is a key to the abbreviations used for some limbic structures. At top, the two darker arrows signify the isomorphically organized images, and the three light arrows signify affects.

To summarize the diversity theme so far: rather than there being a single emergent process that explains all of sensory consciousness—as most authors

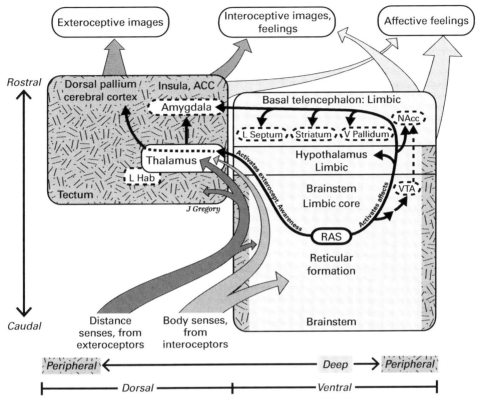

Figure 10.8
Relationships among the three types of primary consciousness in vertebrates: exteroceptive, interoceptive, and affective. See p. 212 for a full explanation.

insist—we see that many "emergences" contribute to subjectivity, each of which can be explained when considered in the appropriate evolutionary context.

Commonality within diversity

Given that so much diversity exists, the question arises: What common factors are shared by all types of sensory consciousness, both within and across species? We find that except in a very broad sense, it is almost impossible to discover any *single* neuronal network that could account for all three types of sensory consciousness (exteroceptive, affective, interoceptive) in all vertebrates. In the mammals, a well-known commonality does exist in the form of extensive interactions and connections between (1) the exteroceptive and interoceptive sensory areas of the cerebral cortex and (2) the subcortical and cortical limbic structures for affects.[21] But this applies only to mammals, and these particular connections are unlikely to exist in the brains of fish and amphibians, which lack a cerebral cortex.

On the other hand, olfactory-to-limbic connections are well documented in all vertebrates, and memory-related hippocampus-to-limbic connections are known in fish and tetrapods[22] These are genuine exteroceptive-affective commonalities. Furthermore, we recognize another structure that exists throughout the vertebrates and is tied to both the exteroceptive and affective types of consciousness: the reticular activating system (RAS in figure 10.8), specifically its laterodorsal tegmental nucleus (LDN), which is located near the junction of the hindbrain with the midbrain (table 8.5). This LDN adjusts the intensity of the emotions, both positive and negative.[23] The LDN exists in every vertebrate group that has been studied including lampreys.[24] Given that the reticular activating system is also involved with exteroceptive arousal (chapter 6), this makes its LDN a common link between exteroceptive and affective consciousness throughout vertebrates.[25]

A further commonality, which might link the different consciousnesses of *invertebrates* and vertebrates, could exist in the genes. Most promising are the genes that universally code for neuromodulator chemicals of affective consciousness (chapter 8). For example, genes for dopamine, the neurochemical that relates to reward, appetitive surprise, and motivation, exist in many animal phyla (molluscs, arthropods, vertebrates, and so on: table 8.4).[26]

Finally, there is the commonality of hierarchy. Our analysis of consciousness has shown how incredibly versatile the widespread feature of the neurohierarchy is, with its reciprocal "top down–bottom up" interactions (figure

6.5). From this one, basic, neural bioplan that evolved over 520 million years ago, the vertebrate brain can create an amazing diversity of qualia, ranging from the image of a pink sunset to the smell of a rose; from the pleasant taste of sugar to the pain of a pinprick. The analogous hierarchies of arthropods and cephalopods might do the same. All the diverse qualia are built on the same principles of complex neurohierarchy, with the diversity of the subjective experiences resulting from differences in the structures and neurotransmitters of the various neurons involved and in their particular connections. Surely this is one of the most astounding features of consciousness.

A continuum of consciousness, and the higher types

We found a diversity of primary consciousness within individual brains, across different species and brain organizations, and across animals with different amounts of brain complexity. However, we wish to reemphasize what was said in chapter 1: we make this claim only for primary sensory consciousness, the most basic type, for "something (anything) it is like to be," and for consciousness at its early stages in evolution. We are not claiming that all aspects of consciousness, and especially of "higher consciousness," exist in fish or are of equal complexity in fish, amphibians, reptiles, mammals, birds, and humans. Rather, our position is that while the basal form of sensory consciousness exists in all vertebrates, there also was an evolutionary progression of consciousness, with the more complex brains having attained additional, higher kinds (e.g., the memory-enhanced consciousness called Step 2 in chapter 6). That said, we argue that nowhere in this continuum has any "mysterious" ingredient been added, but rather it all was built on naturalistic changes in brain complexity that resulted in increasing intelligence and especially increasing self-awareness.

In contrast to the primary consciousness that we have examined in this book, full-blown *self-awareness* implies many additional features, such as meta-awareness (awareness of being aware), thoughts about the self, thoughts about thoughts, the ability to identify oneself in a mirror, theory of mind, verbal self-report, or other related concepts.[27] Researchers have proposed several hierarchical models of the self that apply here. Damasio, for instance, suggests that there are three levels of the self that advanced through evolution via increasing brain complexity, the most basic of these being a "proto-self."[28] According to Damasio, this proto-self registers the current state of the organism, including its moment-to-moment perceptions, but is said to be unconscious (in contrast to our view, which would assign such perceptions to primary consciousness). Next, at a higher level, he posits a "core-self" that is conscious and represents a second-order awareness of the proto-self. Finally, the highest level is an "autobiographical self" that also is conscious, but additionally uses autobiographical memory and the anticipation of the future. It exists in humans, great apes, and dolphins. None of Damasio's levels exactly corresponds to primary consciousness and "something it is like to be," but the point is that *above and beyond* the evolutionary stages of primary consciousness that we have discussed, ample opportunity arose over millions of years of evolution for an increase in the complexity of the self and self-consciousness until the highest stage of an "autobiographical self" was realized.

Memory researcher Endel Tulving has proposed a different model of self and consciousness that assigns a key role to memory. Like Damasio, Tulving claims that there are three hierarchically arranged levels of self and consciousness, these being distinguished by the type of memory function each uses. The lowest level is *anoetic* (nonknowing) consciousness that occurs in animals at the very least "capable of perceptually registering, internally representing, and behaviorally responding to aspects of their present environment, both external and internal."[29] It seems this level would correspond to primary consciousness. Anoetic consciousness is supported by *procedural memory*, which is the simplest form of memory function and requires only the ability to acquire, retain, and retrieve cognitive, perceptual, and motor skills. The next hierarchical level of consciousness to evolve, according to Tulving, was *noetic* (knowing) consciousness. Animals with noetic consciousness possess *semantic memory* that includes knowledge of the world and allows the animal to be aware of and act upon objects and events that were previously experienced but are not immediately present.[30] Tulving believes that many nonhuman species, particularly mammals and birds, developed semantic memory systems and that these animals may therefore possess noetic consciousness. The noetic stage roughly matches our memory-enhanced Step 2 attained by the first mammals and birds (discussed in chapter 6).

The highest level within Tulving's hierarchy is *autonoetic* (self-knowing) consciousness, which he believes is achieved only by humans. This level most closely corresponds to Damasio's autobiographical self. The key to the evolution of autonoetic consciousness is the development of *episodic memory*, the most recently evolved, latest-developing, and advanced form of memory function. Note that this idea differs from our views in that Tulving assigns the evolution of episodic memory to a later stage than we do, because newer evidence shows that many amniotes and some fish remember specific experiences.[31] However, there is little doubt that episodic memory is most highly developed in humans. Autonoetic consciousness makes possible "mental time travel," allowing one to reexperience, through autonoetic awareness, one's own previous experiences. Tulving claims that this ability to envisage one's past and mentally reexperience specific life episodes allows the emergence of identity, self, and self-awareness. We accept that humans have these things, but so may great apes, dolphins, gray parrots, and elephants.[32]

In conclusion, although we do not agree with Damasio and Tulving in every detail, we do agree that there were gradual advances in self-awareness that went beyond basic sensory and bodily awareness, beyond the primary consciousness that is the theme of our book. Our model of the evolutionary origins of primary consciousness is compatible with the later evolution of higher consciousness, self-awareness, and intelligence as built upon primary consciousness.

Postulate 3: The Philosophical Issues of Ontological Subjectivity, Neuroontological Irreducibility, and the "Hard Problem" Can Be Explained by the Nondissociable Confluence of Neurobiological and Adaptive Neuroevolutionary Events

Our study found that the irreducible, ontologically subjective aspects of sensory consciousness in vertebrates—and the origins of the "hard problem"— came into being at a critical evolutionary branch point over 520 million years ago, when reflexes evolved into subjective experiences that are unified, referred, and qualitative, and that exert mental causation (table 1.1). Postulate 3 holds that to understand ontological subjectivity, one must integrate the neurobiological, neuroevolutionary, and philosophical perspectives, and that in fact these perspectives *cannot be dissociated.* As we proposed in chapter 1, the integration of these perspectives—the points of view of the "four blind men pondering the elephant"—is the key to answering the hard problem of consciousness.

But if we plan to use evolution to tackle the hard problem, we first must establish as rigorously as possible that consciousness is a real, adaptive phenomenon that is of evolutionary survival value to the conscious organism. This is the topic of the next section.

Consciousness and Its Neuroontologically Subjective Features Are Adaptive
The four mysteriously subjective features of consciousness—unity, referral, mental causation, and qualia (table 1.1)—are readily understood as adaptations that resulted from natural selection. Mental *unity* must be adaptive,

because choosing one's behavior from a *disjointed* mental-map of reality would be inefficient and a waste of time. Failing to synthesize and unify all the sensory information about a stalking predator would be fatal. Likewise, a disunity of affective feelings would lead to hesitation and death in times of danger. Conscious *referral* is adaptive because to survive one must deal with the external world or with one's body outside the brain, not with the irrelevant inner workings of one's own neurons. Mental *causation* is adaptive because it allows the conscious animal to interact with the outside world proactively rather than just reactively. *Qualia* are adaptive because they discriminate different stimuli or qualities from one another, the difference between experiencing one electromagnetic wavelength as blue and another as red. That discrimination would not have evolved were it not vital for the animal—say, a vulnerable river guppy—to distinguish a predatory blue heron from a red herring. Discriminations of qualia are valuable not only to vision, but to all the other senses as well: distinguishing a threat call from a mating call, or fresh food from rotten food by taste. Here is another reason why sensory qualia are adaptive: being simulations, they allow one to predict how the real objects they represent will move in the immediate future, thus helping to avoid attackers and catch prey (chapter 6). Finally, *affective* qualia are adaptive: positive and negative affects tell the organism which stimuli to approach if helpful and which to avoid if harmful.

But even more fundamentally, consciousness itself must be shown to be adaptive at a deeper level, if our entire thesis is to have any validity. After William James, Gerald Edelman, and others,[33] we have argued throughout this book that the conscious features of pleasure, displeasure, representational mental images, and selective attention are helpful and aid survival. But that is not enough. We also must answer the counterargument that consciousness is just an inert or useless by-product (epiphenomenon) of some other brain function that itself is adaptive, such as increased cognition, more learning capacity, more intelligence, or more information processing by larger brains.[34] In chapter 2, we presented Nichols and Grantham's rebuttal to the idea of consciousness as by-product: they say that consciousness relies on too many complex neural and expensive structures to be useless.[35] To this, we add the argument of Mark Bradley.[36] He pointed out that specific experiences (qualia) are *too tightly correlated* to specific stimuli (such as sensing a predator), to specific brain structures and neural activities (such as activating the amygdala),[37] and to specific responses (escape) to be coincidental by-products with no causal role.

At first glance, Bradley seems to be claiming that "correlation implies causation," which is a notorious logical fallacy,[38] but strong or perfect correlation does make causality more likely. Look at it this way. If the other view—that consciousness is a useless by-product—were true, then natural selection for consciousness would be relaxed and the correlation would continually be loosening. *This* particular quale would not need to be associated with *this* particular stimulus for the animal to survive. It would be fine for someone often to react to danger with pleasure and attraction, or to run away in terror from helpful, needed food. This does not happen.

If consciousness were not adaptive, then some members of a population of humans or other vertebrates—healthy members with normal lifespans in nature—should be missing at least one of these three features of consciousness: (1) attention (the animal pays no particular attention to salient stimuli); (2) affect (the animal shows no preference for helpful over harmful stimuli); or (3) sensory isomorphism (the sensory neural pathways are not mapped isomorphically in this member of the species). We know of no such individuals, at least none that could survive on their own.

Further attesting to the adaptive importance of sensory consciousness is that natural selection has sustained it over the past half-billion years by making the vertebrates' subjective images and affects match reality closely, or at least as closely as needed for survival. As we have pointed out elsewhere, "the mad, the persistently hallucinating, or any other animals with distorted sensory perceptions and inappropriate emotional responses quickly die in the 'tooth and claw' state of nature."[39] Subjectivity is thus kept "accurate" and grounded in reality.

That said, consciousness is not a universally beneficial adaptation, because many animals did not evolve it. The great complexity of consciousness makes it energetically expensive to maintain, and it never evolved in the majority of invertebrates that have cheaper modes of survival (see chapter 4). Examples are clams and the other immobile invertebrates that are protected by hard shells, as shown in figures 10.2, 10.3, and 10.5.

Has consciousness ever lost its adaptive value and disappeared in a clade of animals that once had it? To the best of our knowledge, no adult vertebrate clade has lost sensory consciousness. That is, no known clade of fish, amphibians, or amniotes is without multiple distance senses or is immobile like a clam or a hookworm. When a vertebrate clade has lost a distance sense, it has compensated with more acute versions of the other senses: blind cave fish and aquatic cave salamanders, for example, each have a highly sensitive lateral line for detecting objects they encounter.[40] But who can say whether young larval vertebrates with very simple brains, such as the youngest feeding stages of zebrafish and frog tadpoles, are conscious? Perhaps they use reflexes and their consciousness develops later in life.

And, if we can consider arthropods to have consciousness, then interesting examples of its loss come to mind. Many myriapods (the centipede and millipede group) have secondarily reduced eyes and heads and relatively small brains,[41] so some of them might have lost visual images and consciousness. But the best example is the arthropod group of rhizocephalans,[42] which are related to barnacles. They parasitize the organs inside the abdomens of crabs. They have a multistage life cycle in which most stages resemble a worm or a set of branching roots and have no nervous system at all. Their final life-stage is just a giant egg-sac, with no possibility of consciousness. This suggests that consciousness can be lost in parasites, with their protected habitat, easy access to food, low mobility, and their simple, stable, "unstimulating" sensory environment.[43]

Subjectivity Evolved Such That It Cannot Be Objectively "Experienced": Auto- and Allo-ontological Irreducibility

Having established that consciousness is adaptive, we can now get down to "solving" the problem of subjectivity, by approaching it from a particular philosophical perspective. So far, we've considered the general and special features of consciousness (tables 2.1, 9.1, 9.2), which are *positive* features. To understand ontological subjectivity more fully, however, we must bring in two *negative* features that prevent the scientific reduction of consciousness. Gordon G. Globus sets the stage for these features in the following quote:

The ontological claim that mental events are strictly identical with neural events unfortunately coalesces the perspectives of both subjective (S) and objective (O) observers. The term "mental events" implies the perspective of S who has the mental events immediately given by direct acquaintance (without inference), whereas the term "neural events" implies the perspective of O who is presumably the brain of S. Thus O cannot have S's mental events by direct acquaintance because they are private to S; for example, O cannot experience S's pain. For S, there are no neural events by direct acquaintance—that is, that his own mental events are physically embodied is not directly known by S, unless he observes his own brain in the manner of O using Feigl's imaginary "autocerebroscope." Mental events contain no information about any neural embodiments, for example, S's pain does not have the typical characteristics of physical objects in that S cannot see his pain or touch it. Nor is there anything about pain which seems at all like neurons. ... It does not appear that the brain in any way codes or represents in any way its own structure. (The nervous system has no sensory apparatus directed to its own structure.)[44]

This subjective-objective divide is expressed in the idea of *auto-ontological* and *allo-ontological irreducibilities* (figure 10.9).[45] Auto-ontological irreducibility means that the subject cannot experience the workings of his or her own

Neurobiological Naturalism

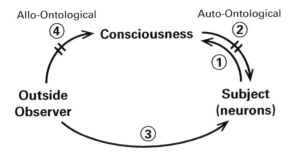

Figure 10.9
The problem of auto- and allo-ontological irreducibilities of consciousness. (1) Subject has access to his or her conscious experiences. (2) Auto-ontological irreducibility: subject lacks access to his or her own objective neurons. (3) Observer has access to subject's material neurons. (4) Allo-ontological irreducibility: observer lacks access to subject's experience.

neurons, and allo-ontological irreducibility means that an outsider cannot access the subject's experiences.

Let's consider auto-ontological irreducibility first. This conundrum stems from the fact that the neural processes that create sensory consciousness refer all feeling states away from the brain itself to something else. This forms a gap between *being* a brain that it is in a feeling state and *observing* or examining a brain in that feeling state.[46]

Neuroscientists and medical doctors use several kinds of scanning machines to crudely "observe" the living brain. But such neuroimaging can never close the auto-ontological gap, as demonstrated by an imaginary machine called the *autocerebroscope*, a device that would insert a probe through your skull and let you see and measure the firing of neurons in your own brain:

> This, for the time being, of course, must remain a piece of science fiction (conceived in analogy to the doctor's fluoroscope) with the help of which I would be able to ascertain the detailed configurations of my cortical nerve currents while introspectively noting other direct experiences, such as the auditory experiences of music; or my thoughts, emotions, or desires.[47]

But even when you observe your neurons as they are in the act of creating the experience, you cannot "feel" neurons in action, so the observation does not explain your experience. In other words, the autocerebroscope merely makes one into a third-person observer of one's self, with all the limitations that vantage point entails.

Actually, this fact that the brain and its consciousness cannot objectively experience the neurons that create it is not mysterious when considered from the perspective of how it evolved: survival depends on neuronal networks referencing the outer world and the body, where the dangers arise against which consciousness helps to protect. These networks should not waste effort in consciously perceiving the detailed firings of their own neurons. That would be wasteful because these neurons are already protected by other homeostatic mechanisms (for example, by physiological mechanisms that maintain the constancy of the body fluids that bathe and sustain the neurons). The manager of a sports team leaves the playing to the players, instead duplicating their effort.[48]

Allo-ontological irreducibility is the opposite of the auto-ontological. Just as the subject lacks objective access to his or her brain that generates experiences, the outside observer lacks access to these experiences as they are felt by the subject (see "4" in figure 10.9). Only the subject has that access. In summary, from the outside viewpoint the brain is observable but not the experience, whereas from the inside the experience is observable but not how the brain constructs that experience.

Here is how an understanding of evolutionary history and neurobiological principles can help solve the dilemmas of auto- and allo-ontological irreducibility. Basal forms of sensory responsiveness were reflexive and innate, and they characterized the early, nonconscious bilaterian animals. Reflexes can be fully described in an objective way. But when the increasing brain complexity of the evolving vertebrates first turned reflexive responses into mental images and mental states before 520 mya, *both kinds of irreducibility were abruptly created because subjectivity appeared.* That is, the specific manner in which consciousness evolved created the brain–mind distinction, and this occurred within an entirely natural biological framework as a historical event. For us to solve the philosophical difficulties of these irreducibilities it was necessary to bring in the evolutionary and neurobiological perspectives, with each perspective informing the others.

The same basic idea and same solution to the hard problem would also apply if primary consciousness actually originated in the first mammals and first birds, as held by those who adhere to the corticothalamic hypothesis.

The Transition from the General to the Special Features: Bridging the Gaps

Although adding the special features to the general features and reflexes (tables 2.1 and 9.2) created the transition to consciousness, the basic general features of life were preserved. These general features form the foundation of,

Neurobiological Naturalism

and set the direction for, the special features, and the special features are integrated with them. Using the general and special features, we can now explain the four perplexing neuroontologically subjective features of consciousness (table 1.1). We next consider each one of these NSFCs and how the general, reflexive, and special (or gen-refl-spec) features join to create its subjective aspect.

The Transition to Referral

How does the progression through the gen-refl-spec features explain referral—the referral of brain activity onto the body (in itch, skin pain), into the body (hunger pangs) or out to the world (in vision, hearing)? In a 2012 paper, coauthor Feinberg[49] saw referral as the most representative and informative NSFC in this context, because so many of the gen-refl-spec features obviously contribute to it. Starting with its general features, he saw that referral is a *living process* and a *system feature* of an *embodied* organism. Its *embodiment* feature is seen in the fact that referral distinguishes qualia in the body from those outside the body. Importantly, referral also requires the general feature of *hierarchical organization*, as elaborated into the special feature of *complex neurohierarchies* with their related features of extensive *neural interactions*, *attention mechanisms*, and so on (table 2.1). Among the many gen-refl-spec features that contribute to referral, a few others are noteworthy: *reflexes*, because their inputs and responses automatically and non-consciously relate to/refer to the outside world (see pp. 25–26); *isomorphic mapping*, which simulates the objects to which consciousness is referred; and the *adaptation* aspect of *auto-ontological irreducibility* that avoids wasting energy by uselessly referring to the firing of one's own neurons (see p. 222).

Thus, referral is a multidetermined process based on both the general and special features. All neural levels—even the most reflexive—are critical to the origins of referral, but the neural architecture required for sensory consciousness does not appear until the special neurobiological features are added to the general features. The transition was smooth and uninterrupted, owing to the continuous increase in neurohierarchical complexity.

We conclude that referral is an ontologically subjective and irreducible process, yet at the same time it has a scientific explanation that does not require any "mysterious" or "fundamental" physical properties that would create a "gap" between the brain and the experience. The referral of a stimulus comes from a unique *constellation* and *interaction* of natural features.

The Transition to Mental Unity

The neurobiological basis of mental unity, like that of referral, is multi-level and multi-determined by the gen-refl-spec features (table 2.1). As an example of a general biological feature on which it relies, mental unity is a *process*, as is life, and not a material thing that can be located in space, because there is no single place in the brain where consciousness is *physically* unified (chapter 2, p. 30). Unity is also, as is referral, a feature of an *embodied hierarchical system*, whose smooth evolutionary transition to the special features of *complex neurohierarchies* was critical. We explained this in chapter 2, by showing that the conscious brain acts as a *nested hierarchy* that progressively unifies sensory precepts (assembles the parts of a falling apple: p. 30) and also that higher brain-levels impose the unifying control of downward *constraint* on lower levels of the hierarchy (of which cyclopean vision is an example: figure 1.2).

Of the many gen-refl-spec features contributing to mental unity, two contribute especially closely. First, as discussed on pages 217–218, unity is highly *adaptive* because fragmented sensory maps or feelings would be inefficient or fatal. Second, unity has strong ties to the special feature of *attention* because most things to which we attend are within our unified experience, and most unattended things are not (see the Attention section in chapter 2).

Finally, mental unity is auto-and allo-ontologically irreducible. The conscious subject (the "auto-") experiences the embodied, systemic, neurohierarchical processes as nested and unified, although the objective observer (the "allo-") knows that they are not unified but distributed and "gappy." Recall our conclusion from the previous section that these auto- and allo-irreducibilities are real but biologically explainable.

The Transition to Mental Causation

Full-blown mental causation, in which a conscious animal changes the world through intricate goal-directed behaviors, involves all the gen-refl-spec features of table 2.1. These features, again culminating in *complex neurohierarchies*, process many kinds of stimuli to set up varied and context-specific behaviors. But causation has quite primitive roots. Causative actions can be *reflexive* or even "prereflexive" because any organism that moves (in) water, air, or soil affects its surroundings. An example of this is water-pumping by sponges, which have no nervous system at all. Therefore, causation may have been the first NSFC to start evolving.

Among the many gen-refl-spec features that contribute to real, mental causation, *embodiment* or embodied system is especially important: As Feinberg put it in his 2012 paper, causal efficacy exists in my *own* brain *system*, from which the action commands are issued.[50] The general feature of *self-organization* is also important for causation, in the following way: Mayr convincingly showed that all living things have programmed, goal-directed processes within this self-organization feature[51]—and mental causation is highly goal-directed. Finally, the *memory* feature is important because aspects of action and motor programs must be learned.

The Transition to Qualia

All four NSFCs interact with each other, but qualia is the most difficult to untangle from the rest and to analyze independently. This is because qualia relate so closely to unity (which is all the qualia united into a whole) and to referral (qualia projected outside of the brain). Even so, this book has already characterized qualia so fully that we can analyze it as one phenomenon. Chapters 2 and 5 through 8 covered how qualia arose from the general features of *systems hierarchy* and *adaptive* evolution. In brief, qualia are the result of a unique, multifactorial neurobiological substrate and recursive interactions within and between higher and lower neurohierarchical levels. As with all the other NSFCs, the key is the unbroken transition from simple to complex neurohierarchies.

Some gen-refl-spec features that seem especially important for qualia are the multisensory *isomorphic maps*, which construct exteroceptive qualia, plus all the other aspects of *complex sensory hierarchies* from their receptors on up. Evolutionary *adaptation* is a vital feature because qualia serve to discriminate among a wide range of sensory stimuli and valences, as is needed for survival (see p. 218). And qualia also relate closely to the features of auto- and allo-ontological irreducibility: all qualia are experienced by the brain, and none are reachable objectively from outside that embodied brain. If our scientific explanation of the auto- and allo- irreducibilities is correct (see p. 222), then these two features could bring qualia into the realm of science.

To summarize this section, by smoothly linking the general-life features and reflexes to the special features of consciousness for each of the four NSFCs, we have philosophically bridged many of the explanatory gaps between natural neural processes and ontological subjectivity, while rooting consciousness firmly in the scientific field of biology. This could be the holy grail of consciousness studies, an unbroken continuity between subjectivity and the explainable, tractable life processes.

The "Character of Experience": Do You Believe in Magic?

Let us consider our naturalistic answer to the question of consciousness from one more angle. Our integrated philosophical, neurobiological, and evolutionary approach can solve a very difficult aspect of the hard problem: the *character of experience.* Someone who is skeptical about neurobiological naturalism might argue that, even if we have correctly identified when, how, and why consciousness evolved, does our theory really explain why "red" feels exactly like red does? Why does pain "hurt"? Isn't that magic? Isn't that beyond scientific explanation? Surely that's a "fundamental" or "radically emergent" phenomenon? Chalmers states the problem of the character of experience as follows:

> Why do individual experiences have their particular nature? When I open my eyes and look around my office, why do I have *this* sort of experience? At a more basic level, why is seeing red like *this*, rather than like *that*? It seems conceivable that when looking at red things, one might have had the sort of color experiences that one in fact has when looking at blue things. Why is the experience one way rather than the other? Why, for that matter, do we experience the reddish sensation that we do, rather than some entirely different sensation, like the sound of a trumpet?[52]

We reply that our *integrated* approach that combines the neurobiological, neuroevolutionary, and neurophilosophical domains is necessary to answer this seemingly impossible question. If we ask you: "Why does red subjectively feel 'red,' and not 'hurt'?" what would you say? First, you could argue that we know that the neurobiology of color processing and of pain processing are quite different, so they shouldn't feel the same (a neurobiological answer). Second, you could say that the distinction between the feeling of red and pain evolved because there is strong adaptive value in a response to harm that differs from a response to color (a neuroevolutionary answer). Third, to answer the "why subjectivity?" part of the question, you could say that these qualia and feelings are ontologically subjective because the subject has a different point of view than the observer, so no objective feature will ever totally replace, be reduced to, or fully explain the subjective qualities (the neurophilosophical answer). Note that all three explanations are correct, but that *no single* explanation is sufficient, just as the blind men cannot individually understand the elephant. The solution lies in the confluence of the neurobiological, neuroevolutionary, and neurophilosophical domains. And all three domains of consciousness came into being in unison at approximately the same time in the Early Cambrian.

Conclusion

Francis Crick once famously pronounced:

> The Astonishing Hypothesis is that "You," your joys and your sorrows, your memories and your ambitions, your sense of personal identity and free will, are in fact no more

than the behavior of a vast assembly of nerve cells and their associated molecules. As Lewis Carroll's Alice might have phrased it: You're nothing but a pack of neurons.[53]

Was Crick correct? We would answer "yes and no." The spirit of Crick's proposal is that we find nothing more "fundamental" in the conscious brain than we find in biology in general, and with this we agree. But what about your being "nothing but a pack of neurons"? Well, in our view, this sounds too strictly reductionist, oversimplifying the many emergent features of the brain and subjectivity that make consciousness the most complex and unique of all natural phenomena. And it is these unique features that create the difficult gaps between experience and the brain.

But can we completely close all the "explanatory gaps" between the physicochemical properties of an organism and consciousness? On the one hand, we propose that our list of general and special neurobiological features, along with the principles of auto- and allo-ontological irreducibilities, do indeed close the explanatory gaps between the brain and subjective experience. These ideas provide a seamless reconstruction of the nature and origins of consciousness. This "case" can be considered, in all its essential aspects, "closed."

On the other hand, for many of the same reasons, the "ontological gap" between first-person subjective experience and third-person objective knowledge cannot be closed in the same way. The objective–subjective divide is a real ontological barrier, for the numerous reasons we have described. Because the brain creates experiences that are uniquely embodied, personal, and auto- and allo-ontologically irreducible, no manner of explanation can eliminate its ontological subjectivity. However, because we have explained how the brain creates experience, we therefore conclude that our theory of neurobiological naturalism bridges and naturalizes the ontological gap in a manner that is consistent with normal biological science. Thus, we argue that this gap can now be "bridged," but not closed.

Finally, we also conclude that, in contrast to theories that focus on a single approach to the "mystery of consciousness," a satisfying and complete explanation of primary consciousness requires a *confluence* of points of view, necessarily including neurobiological, evolutionary, and philosophical arguments, each contributing important answers to the "hard problem." Perhaps one reason no one has solved it before is that it requires all three perspectives, including what happened over half a billion years ago.

Appendix: Table References

Table 8.3
Distribution of the behavioral evidence for positive and negative affect across animals

A. Operant learning

See Perry, Barron, and Cheng (2013) for extensive documentation of the invertebrates.

1. Qin and Wheeler (2007) suggest yes for *C. elegans*.
2. Zhang et al. (2005) suggest yes for *C. elegans*, although poisonous food is an extremely powerful conditioner. That is, aversion to poison may be so important that it is hard-wired as an innate reflex.
3. Perry, Barron, and Cheng (2013) say no for *C. elegans*.
4. Brembs (2003b), for sea hare *Aplysia*.
5. Nargeot and Simmers (2011), for sea hare *Aplysia*.
6. Nargeot and Simmers (2012), for sea hare *Aplysia*.
7. Kandel (2009), for sea hare *Aplysia*.
8. Brembs (2003a), on pond snail *Lymnaea* and *Aplysia*.
9. Casimir (2009) for no known learning behavior in cephalochordates (amphioxus).
10. Mesquita (2011), for carp.
11. Kuba et al. (2010), for stingray.
12. Flood et al. (1976), for various teleosts.
13. Chandroo et al. (2004), for older literature on teleosts.
14. Millot et al. (2014), for sea bream.
15. Papini et al. (1995), for toad *Bufo*.
16. Schmajuk et al. (1980), for toad.
17. Burghardt (2013), for amphibians.

18. Weiss and Wilson (2003), for tortoises.
19. Burghardt (2013), for reptiles.
20. http://en.wikipedia.org/wiki/Operant_conditioning for bird, rat, dog.

B. Behavioral trade-off

21. Gillette et al. (2000), for predatory sea slug, *Pleurobranchaea*.
22. Hirayama et al. (2012), for *Pleurobranchaea*.
23. Herberholz and Marquart (2012), on trade-offs in predatory sea slug, crayfish, and primates.
24. Dunlop et al. (2006), for trout and goldfish.
25. Millsopp and Laming (2008), for goldfish.
26. Daneri et al. (2011), for toad.
27. Bennett et al. (2013), for leopard frog tadpoles *Lithobates*.
28. Gabor and Jaeger (1995), for salamander *Plethodon*.
29. Balasko and Cabanac (1998), on iguana and all amniotes.
30. Pilastro et al. (2002), for sparrow *Petronia*.
31. Cabanac and Johnson (1983), on rats.
32. Cabanac (1995), on human.

C. Frustration/Contrast

33. Vindas et al. (2014), for trout.
34. Papini (2002), says contrast supression in mammals only.
35. Shibasaki and Ishida (2012), for no frustration in newt.
36. Forster et al. (2005): possible frustration in lizard because aggression elevates in a second fight?
37. Azrin et al. (1966), for frustration in pigeon.
38. There has been difficulty showing 'contrast' in birds, but Freidin et al. (2009) seem to have shown it in starlings.
39. Papini and Dudley (1997), for frustration and contrast in rat and other mammals.

D. Self-delivery

40. Balaban and Maksimova (1993), for electrodes in brain of *Helix* snail, including brain areas for sex; it is unclear whether these areas can really be called pleasure centers.

Appendix 231

41. Campbell (1968), for teleost fish, reward of mild electric shock.
42. Campbell (1972), for teleost fish and crocodile, reward of mild shock.
43. Delius and Pellander (1982), rewarding brain stimulation in pigeon.
44. Danbury et al. (2000), for chicken, analgesic.
45. Weissman (1976), for rat, analgesic.
46. Martin et al. (2006), for rat, analgesic.
47. Colpaert et al. (1980), for rat, analgesic.
48. Solinas et al. (2006), for rat, cocaine reward.
49. Martin, Kim, and Eisenach (2006), for rat, opioid delivery.

E. Approach drugs, CPP

50. Mori (1999) found that nematode *C. elegans* is attracted to alcohol, but it is attracted to a wide variety of volatile chemicals, so uncertain if the alcohol attraction is specific or if it reflects reward.
51. Kusayama and Watanabe (2000), for planarian.
52. Pagán et al. (2013), for planarian.
53. Raffa et al. (2013), for planarian.
54. Søvik and Barron (2013), their table 2, for planarian (and arthropods).
55. Lett and Grant (1989), for goldfish.
56. Ninkovic and Bally-Cuif (2006), for zebrafish.
57. Webb et al. (2009), for zebrafish.
58. Mathur et al. (2011), for zebrafish.
59. Klee et al. (2011), for zebrafish.
60. Kily et al. (2008), for zebrafish.
61. Kalueff et al. (2014), for zebrafish.
62. Presley et al. (2010), for frog.
63. Farrell and Wilczynski (2006), on anolis lizard, conditioned place preference to a location of the reward of an aggressive encounter, not to a drug reward.
64. Levens and Akins (2001), for quail *Coturnix*.
65. Rutten et al. (2011), for rat.
66. Huston et al. (2013), for rats, other vertebrates, and flies.
67. Wang et al. (2012), for monkey.

Table 8.4

Comparative neuroanatomy of positive and negative affects: Features for reward, nociception, and fearlike responses

1. Dopamine neurons

1. Barron et al. (2010), for all bilaterians such as nematodes, flatworms, gastropod *Aplysia*, and mammals.
2. Mersha et al. (2013), for nematode worm, *C. elegans*.
3. Waddell (2013), for *Drosophila*.
4. Moret et al. (2004), for amphioxus.
5. Candiani et al. (2005), for amphioxus.
6. Candiani et al. (2012), for amphioxus.
7. Wullimann (2014), for every group of vertebrates.
8. Pierre et al. (1997), for lamprey.
9. Robertson et al. (2012), for lamprey.
10. O'Connell and Hofmann (2011), for all bony fish and tetrapods.
11. Berridge and Kringelbach (2013), for rat.

2. Nociceptive fibers

12. Smith and Lewin (2009), for the distribution of nociceptors in bilaterians: in some ecdysozoans (nematode worm *C. elegans* and fruitfly *Drosophila*), and in some lophotrochozoans (leech *Hirudo* and sea slug *Aplysia*).
13. Elwood (2011), for nociceptors across bilaterians.
14. Carr and Zachariou (2014), on nematode.
15. Im and Galko (2012), for nociceptors in *Drosophila*.
16. Crook et al. (2013), for squid *Loligo*.
17. Wicht and Lacalli (2005), for amphioxus. Note amphioxus has many types of receptors, some of which may be nociceptive (unproven), but none are from neural crest so cannot be homologous with those of vertebrates.
18. Matthews and Wickelgren (1978), on sea lamprey *Petromyzon*.
19. Rovainen and Yan (1985), on silver lamprey *Ichthyomyzon*.
20. Snow et al. (1993), for stingray, shovelnose ray, and black-tip shark.
21. Sneddon (2004), for elasmobranchs.
22. Braithwaite (2014), see p. 329 on chondricthyes.
23. Rose et al. (2014), on all kinds of fish.

Appendix

24. Roques et al. (2010), for carp.
25. Sneddon (2002), for trout.
26. Hamamoto and Simone (2003), for frog *Rana*.
27. Stevens (2011), on amphibians.
28. Machin (1999), on amphibians.
29. Hisajima et al. (2002), on snake *Agkistrodon*.
30. Liang and Terashima (1993), on crotaline snakes.
31. Gentle et al. (2001), on chick.
32. Schmalbruch (1986), on rat.
33. Guo et al. (2003), on human.
34. Hall (2011), on human.

3. Opioid receptors

35. Raffa et al. (2013), for flatworm *Planaria*.
36. Guo et al. (2013), for scallop.
37. Sha et al. (2013), for octopus.
38. See Søvik and Barron (2013) for absence of opioid receptors in *C. elegans* and *Drosophila* and uncertainty about their presence in crayfish and in most invertebrates.
39. Nordström et al. (2008), Fig. 1, for cephalochordates. But McClendon et al. (2010) could not find any in cephalochordates.
40. McClendon et al. (2010), for the two receptors (2) in lamprey and the four receptors (4) in all classes of jawed vertebrates.
41. Dreborg et al. (2008), for teleosts, frog, newt, and mammals.
42. Sundström et al. (2010), for teleost osteichthyans, frog, birds and mammals.
43. Machin (1999), on amphibians.

4. Autonomic nervous system with parasympathetic and sympathetic divisions

44. Kardong (2012), pp. 644–645, for all vertebrates.
45. It is incomplete in lampreys: Häming et al. (2011).
46. It is incomplete in lampreys: Glover and Fritzsch (2009).
47. Nilsson (2011), for sharks.

5. Spinoreticular, spinomesencephalic, and trigeminothalamic pathways (including nociception)

48. Holland and Yu (2002) show a somatosensory path from body surface up to the brain's reticular-formation region in amphioxus. Not known if this is fully the same as the vertebrate spinoreticular pathway. The latter depends on neurons from neural crest, and amphioxus lacks neural crest.
49. Butler and Hodos (2005), p. 553: all vertebrates have a spinoreticular tract and spinal trigeminal nucleus.
50. Butler (2009b), p. 1419, for these sensory paths existing across vertebrates.
51. Nieuwenhuys et al. (1998), for the trigeminothalamic pathway of lamprey and other vertebrates.
52. Nieuwenhuys and Nicholson (1998), pp. 425–426, for the spinoreticular tract (includes spinomesencephalic) in lamprey.
53. Koyama et al. (1987), for spinal trigeminal nucleus in lamprey.
54. Anadón et al. (2000), for spinal trigeminal nucleus in shark.
55. Puzdrowski (1988), for spinal trigeminal nucleus in goldfish.
56. DaSilva et al. (2002), for spinal trigeminal nucleus in human.

6. Amygdala

57. Watanabe et al. (2008), for possible amygdala in lamprey (but unproven).
58. Martínez-de-la-Torre et al. (2011), for possible amygdala in lamprey (but unproven).
59. Maximino et al. (2013), for possible amygdala in lamprey (but unproven).
60. None is certain in cartilaginous fish, but a possible amygdala is called Nucleus A: see Vargas et al. (2012); Northcutt (1995).
61. Butler and Hodos (2005), for amygdala in bony fish and tetrapods.
62. Wulliman and Vernier (2009b), see p. 1424 for bony fish and tetrapods.
63. Hurtado-Parrado (2010), for teleosts.
64. González and Moreno (2009), for tetrapods.
65. Moreno and González (2007), for salamanders.
66. LeDoux (2007), for mammals.

7. Periaqueductal gray

67. Freamat and Sower (2013), for lamprey.
68. Forlano and Bass (2011), for teleosts and all jawed vertebrates.

Appendix

69. Kittelberger et al. (2006), for midshipman fish and mammals.
70. Brahic and Kelley (2003), for frog *Xenopus*.
71. Ten Donkelaar and de Boer-van Huizen (1987), for lizard *Gekko*.
72. Kingsbury et al. (2011), for finch.
73. Quintino-dos-Santos et al. (2014), for rat and human.

8. Lateral septum, septal nuclei

74. Butler and Hodos (2005), for all vertebrates.
75. Freamat and Sower (2013), on lateral septum in sea lamprey.
76. Robertson et al. (2007), for river and sea lampreys.
77. Northcutt (1995), for all jawed vertebrates.
78. Endepols et al. (2005), for frog.
79. Lanuza and Martinez-Garcia (2009), for all tetrapods.
80. Trent and Menard (2013), for rats.
81. Singewald et al. (2010), for rats.

Table 8.5
Comparative neuroanatomy of positive and negative affects: Mesolimbic reward system (MRS), or mesolimbic dopamine (reward/aversion) system, with the functions listed for mammals

1. Ventral tegmental area, or equivalent

1. Carrera et al. (2012), for VTA in elasmobranch shark *Scyliorhinus*.
2. O'Connell and Hofmann (2011), for VTA in all tetrapods, and an equivalent called posterior tuberculum is in amphibians and bony fish.
3. Domínguez et al. (2010), for VTA in gecko and turtle.
4. Hara et al. (2007), for VTA in zebra finch.
5. Kender et al. (2008), for rat.
6. Schifirnet et al. (2014), for rat.
7. Lammel et al. (2012), for mouse.

1A

8. For diencephalic structures in amphioxus that may include a distant precursor, see Wicht and Lacalli (2005).
9. Rink and Wullimann (2001): posterior tuberculum is in all fish.

10. Ryczko et al. (2013), for posterior tuberculum in sea lamprey.
11. Pombal et al. (1997), for posterior tuberculum in river lamprey.
12. Ferreiro-Galve et al. (2008), for VTA and posterior tuberculum in elasmobranch sharks.
13. Goodson and Kingsbury (2013), for VTA equivalent of teleost fish.
14. Petersen et al. (2013), for teleost fish, midshipman.
15. Chakraborty and Burmeister (2010), for tungara frog.

2. Nucleus accumbens

16. O'Connell and Hofmann (2011), say teleost fish may have an equivalent structure called dorsal ventral telencephalon, or Vd.
17. O'Connell and Hofmann (2011), for all the tetrapods.
18. Hoke et al. (2007), for tungara frog.
19. Guirado et al. (1999), for lizard.
20. Earp and Maney (2012), for sparrow.
21. Berridge and Kringelbach (2013), on pleasure- and dread-causing functions, rat and other mammals.

3. Ventral pallidum

22. For forebrain structures in amphioxus that may include a distant precursor of the pallidum structures, see Wicht and Lacalli (2005), esp. pp. 145–147.
23. Lampreys have a pallidum part of their striatum, so a ventral pallidum may exist: Ericsson et al. (2013).
24. Chondrichthyes have a pallidum part of their striatum, so a ventral pallidum may exist: Quintana-Urzainqui et al. (2012).
25. O'Connell and Hofmann (2011) discuss the uncertain situation in teleosts.
26. Teleosts have a pallidum part of their striatum, so a ventral pallidum may exist: Ganz et al. (2012), on zebrafish.
27. O'Connell and Hofmann (2011), for amphibians, reptiles, birds and mammals.
28. Sánchez-Camacho et al. (2006), for frog.
29. Berridge and Kringelbach (2013), for rat.

Appendix

4. Laterodorsal tegmental nucleus

30. Wicht and Lacalli (2005), for reticular formation in amphioxus.
31. Rodríguez-Moldes et al. (2002), for laterodorsal nucleus in all vertebrate groups.
32. Pombal et al. (2001), for lampreys.
33. Ryczko et al. (2013), for lamprey mesencephalic locomotory region including this nucleus.
34. Mueller et al. (2004), for zebrafish.
35. Brudzynski (2014), for mammals, negative and positive emotions.
36. Brudzynski (2007), for mammals, reward.
37. Lammel et al. (2012), for mouse, reward.

5. Striatum

38. For forebrain structures in amphioxus that may include a distant precursor of the striatum structures, see Wicht and Lacalli (2005), esp. pp. 145–147.
39. Grillner et al. (2013), for a striatum in all vertebrates.
40. Ericsson et al. (2013), for a striatum in all vertebrates.
41. Robertson et al. (2012), for river lamprey.
42. Quintana-Urzainqui (2012), for shark.
43. O'Connell and Hofmann (2011), for amphibians, reptiles, birds and mammals.

6. Pallial amygdala

44. Watanabe et al. (2008), for possible amygdala in lamprey (but unproven).
45. Martínez-de-la-Torre et al. (2011), for possible amygdala in lamprey (but unproven).
46. Maximino et al. (2013), for possible amygdala in lamprey (but unproven).
47. None is certain in cartilaginous fish, but a possible amygdala is called Nucleus A: see Vargas et al. (2012); Northcutt (1995).
48. Butler and Hodos (2005), for amygdala in bony fishes and tetrapods.
49. Wulliman and Vernier (2009b), see p. 1424 for bony fish and tetrapods.
50. González and Moreno (2009), for tetrapods.
51. Moreno and González (2007), for salamander (newt).
52. LeDoux (2007), for mammals.

53. Roozendaal (2009), for mammals; also see Janek and Tye (2015).
54. Namburi et al. (2015), for reward and avoidance learning in mouse.

7. Hippocampus

55. Bingman et al. (2009), in all bony vertebrates and including lamprey possibility.
56. Watanabe et al. (2008), for lampreys' hippocampus-identifying developmental genes, Lhx1/5 and 2/9.
57. Fuss et al. (2014), for shark hippocampus (functional evidence from brain ablation).
58. Schluessel and Bleckmann (2005), for stingray (functional evidence from brain ablation).
59. O'Connell and Hofmann (2011), for bony fish, amphibians, reptiles, birds and mammals.
60. Demski (2013), for teleost hippocampus (and amygdala).
61. Portavella et al. (2002), for goldfish.
62. Muzio et al. (1994), for toad.
63. López et al. (2003), for turtle.
64. Allen and Fortin (2013), for birds and mammals.

8. Habenula

65. For diencephalon structures in amphioxus that may include a precursor of habenula and pineal structures, see Wicht and Lacalli (2005), esp. pp. 145–147.
66. Butler (2009a), for all vertebrates.
67. Hikosaka (2010), all vertebrates and its function.
68. Rodníguez-Moldes et al. (2002), for lampreys, elasmobranchs and all fish.
69. Stephenson-Jones et al. (2012), for river lamprey.
70. Giuliani et al. (2002), for elasmobranch shark.
71. Okamoto et al. (2012), for zebrafish.
72. Kalueff et al. (2014), for zebrafish.
73. Yang et al. (2014), for rat, and habenula's function.

9. Neocortical parts

74. All vertebrates have the dorsal pallium, which forms the neocortex in mammals: Medina (2009). See Murakami and Kuratani (2008) for lampreys. In the fish and amphibians, however, the dorsal pallium is often small and has nothing obviously like these mammalian subdivisions.
75. See Butler et al. (2011), for possible insula in reptiles and birds.
76. Butler and Cotterill (2006), on bird cingulate and prefrontal cortex as dorsolateral corticoid area and nidopallium caudolaterale.
77. Veit et al. (2014), on bird cingulate and prefrontal cortex as dorsolateral corticoid area and nidopallium caudolaterale.
78. Kirsch et al. (2008), on bird cingulate and prefrontal cortex as dorsolateral corticoid area and nidopallium caudolaterale.
79. Berridge and Kringelbach (2013), for mammals (gives the functions too).
80. Craig (2010), on insula in mammals.
81. Uylings et al. (2003): rats have a prefrontal cortex.
82. Heimer and Van Hoesen (2006), on limbic lobe of humans.
83. Barrett et al. (2007), on humans.

Table 8.6
Adaptive behavior network ("social behavior network"), which signals behaviors necessary for survival; linked to mesolimbic system, which motivates and rewards these adaptive behaviors

1. Pre-optic area with vasotocin neurons

1. Minakata (2010), for vasotocin/oxytocin and gonadotrophin-releasing hormone (GnRH) neurons in many bilaterians.
2. Wicht and Lacalli (2005), esp. pp. 145–147, for the possible location of a pre-optic area in amphioxus brain.
3. Roch et al. (2014), on GnRH neurons in amphioxus cerebral vesicle. See also Abitua et al. (2015), for these cells in tunicates.
4. Kubokawa et al. (2010), on amphioxus paraventricular-vasotocin neurons in infundibular organ.
5. Tobet et al. (1996), for GnRH neurons in preoptic area of sea lamprey.
6. Kavanaugh et al. (2008), on GnRH cells in anterior and preoptic hypothalamus of sea lamprey.

7. Sherwood and Lovejoy (1993), for GnRH cells in preoptic area of chondrichthyes, shark and skate.
8. Forlano and Bass (2011), for teleosts and all jawed vertebrates, see their table 1 on preoptic area and paraventricular nucleus homologue of teleosts.
9. O'Connell and Hofmann (2011), for preoptic area in teleosts, amphibians, reptiles, birds, and mammals.
10. Balment et al. (2006), for vasotocin action in fish and mammals.
11. Goodson and Bass (2000), for reproductive function in midshipman fish.
12. Lee and Brown (2007), for function of preoptic nucleus in mouse.

2. Anterior hypothalamus

13. For the simple hypothalamus in amphioxus see Wicht and Lacalli (2005), esp. pp. 145–147.
14. Freamat and Sower (2013), on sea lamprey, anterior hypothalamus.
15. Kavanaugh et al. (2008), on GnRH cells in anterior and preoptic hypothalamus of sea lamprey.
16. Forlano et al. (2000) report an anterior hypothalamus in elasmobranchs, although this just may be part of the preoptic area.
17. O'Connell and Hofmann (2011), for anterior hypothalamus in teleosts, amphibians, reptiles, birds, and mammals.
18. Forlano and Bass (2011), for teleosts and all jawed vertebrates, see their table 1 on anterior hypothalamus homologue in teleosts.

3. Ventromedial hypothalamus

19. For the simple hypothalamus in amphioxus see Wicht and Lacalli (2005), esp. pp. 145–147.
20. Freamat and Sower (2013), on possible ventromedial hypothalamus in sea lamprey.
21. O'Connell and Hofmann (2011), for ventromedial hypothalamus in teleosts, amphibians, reptiles, birds, and mammals.
22. Forlano and Bass (2011), for teleosts, see table 1 on ventromedial hypothalamus homolog in teleosts.
23. Borszcz (2006), for role in response to threat and pain, rat.
24. Lee, H., et al. (2014), for role in sexual and aggressive behavior in mouse.

Appendix

4. Periaqueductal gray

25. Freamat and Sower (2013), for sea lamprey.
26. Sten Grillner, July 30, 2014, personal communication, for river lamprey.
27. Forlano and Bass (2011), for teleosts and all jawed vertebrates.
28. Kittelberger et al. (2006), for midshipman fish and mammals.
29. Brahic and Kelley (2003), for frog *Xenopus*.
30. Ten Donkelaar and de Boer-van Huizen (1987), for lizard *Gekko*.
31. Kingsbury et al. (2011), for finch.
32. Quintino-dos-Santos et al. (2014), for rat and human.

5. Subpallial (extended) amygdala

33. Amygdala may exist in lamprey (Watanabe et al., 2008), but whether it has a subpallial subpart is uncertain.
34. Amygdala may exist in cartilaginous fish (Vargas et al. 2012; Northcutt, 1995), but whether it has a subpallial subpart is uncertain.
35. O'Connell and Hofmann (2011), for bony fish and tetrapods.
36. See Maximino et al. (2013), for bony fish.
37. Wulliman and Vernier (2009b), p. 1424, for bony fish and tetrapods.
38. González and Moreno (2009), for tetrapods
39. Moreno and González (2007), for salamanders.
40. LeDoux (2007), for mammals.
41. Heimer and Van Hoesen (2006), for human.

6. Lateral septum, septal nuclei

42. Butler and Hodos (2005), for all vertebrates.
43. Freamat and Sower (2013), on lateral septum in sea lamprey.
44. Robertson et al. (2007), for river and sea lampreys.
45. Northcutt (1995), for all jawed vertebrates.
46. Endepols et al. (2005), for frog.
47. Lanuza and Martinez-Garcia (2009), for all tetrapods.
48. Trent and Menard (2013), for rats.
49. Singewald et al. (2010), for rats.

Table 9.1

Affective consciousness: Suggested behavioral evidence of positive and negative affect in protostome invertebrates

Numbers of brain neurons

1. Schrödel et al. (2013), for *C. elegans*, calculated from the numbers on p. 1016.
2. Agata et al. (1998), for planarian flatworms.
3. Bailly et al. (2013), for flatworms.
4. Lillvis et al. (2012), for gastropods.
5. See http://neuronbank.org/wiki/index.php/Aplysia_cerebral_ganglia, for sea hare, *Aplysia*.
6. See http://en.wikipedia.org/wiki/List_of_animals_by_number_of_neurons for gastropods.
7. Zullo and Hochner (2011), for octopus.
8. Hochner (2012), for cephalopods.
9. Chittka and Niven (2009), for insects.
10. Giurfa (2013), for insects.
11. See http://www.theguardian.com/world/2005/feb/08/research.higher education, for crabs and lobsters.

1. Operant conditioning

12. Perry, Barron, and Cheng (2013), for all the invertebrate documentation.
13. Qin and Wheeler (2007) suggest yes for *C. elegans*.
14. Zhang et al. (2005) suggest yes for *C. elegans*, although poisonous food is an extremely powerful conditioner. That is, averting poison may be so important that it is hard-wired as an innate reflex.
15. Perry, Barron, and Cheng (2013) say no for *C. elegans*.
16. Brembs (2003b), for sea hare *Aplysia*.
17. Nargeot and Simmers (2011), for sea hare *Aplysia*.
18. Nargeot and Simmers (2012), for sea hare *Aplysia*.
19. Brembs (2003a), on pond snail *Lymnaea*, *Aplysia*, and *Drosophila*.
20. Papini and Bitterman (1991), for octopus.
21. Packard and Delafield-Butt (2014), for octopus.
22. Crancher et al. (1972), for octopus.
23. Gutnick et al. (2011), for octopus.

Appendix

24. Andrews et al. (2013), for cuttlefish.
25. Cartron et al. (2013), for cuttlefish.
26. Kisch and Erber (1999), for honey bee.
27. Kawai et al. (2004), for crayfish.
28. Tomina and Takahata (2010), for lobster and Abramson and Feinman (1990), for crab.
29. Jackson and Wilcox (1998) for jumping spider, learning predatory strategy by trial and error.

2. Behavioral trade-off

30. Gillette et al. (2000), for predatory sea slug, *Pleurobranchaea*.
31. Hirayama et al. (2012), for *Pleurobranchaea*.
32. Herberholz and Marquart (2012), on trade-offs in predatory sea slug, crayfish, and primates.
33. Anderson and Mather (2007), for octopus.
34. Mather and Kuba (2013), for octopus.
35. Stevenson and Schildberger (2013), for cricket, *Gryllus bimaculatus*.
36. Elwood and Appel (2009) and Appel and Elwood (2009) for crab.
37. Jackson and Wilcox (1993) for jumping spider, detouring to reach prey and thus trading short-term chances for better long-term chances.

3. Frustration/Contrast

38. Pain (2009), for *Drosophila*.

4. Self-delivery of analgesics

39. Balaban and Maksimova (1993), for electrodes in brain of *Helix* snail, including brain areas for sex; but unclear whether these areas can really be called pleasure centers.
40. Søvik and Barron (2013), for *Drosophila*.

5. Approach drugs, CPP

41. Mori (1999) found that nematode *C. elegans* is attracted to alcohol, but it is attracted to a wide variety of volatile chemicals, so uncertain if the alcohol attraction is specific or if it reflects reward.

42. Kusayama and Watanabe (2000), for planarian flatworm.
43. Pagán et al. (2013), for planarian.
44. Raffa et al. (2013), for planarian.
45. Søvik and Barron (2013), for planarian and the arthropods.
46. Shohat-Ophir et al. (2012), for *Drosophila*.
47. Huston et al. (2013), for *Drosophila*.
48. Huber et al. (2011), for crayfish.
49. Huber (2005), for crayfish.

Table 9.2
Sensory (exteroceptive) consciousness: Evidence for protostome invertebrates

1. Complexity

1. Peter Fraser, University of Aberdeen, quoted in Adan (2005), for crabs and lobsters.
2. Chittka and Niven (2009), for fruit fly, bee, and human.
3. See http://en.wikipedia.org/wiki/List_of_animals_by_number_of_neurons, for ant, cockroach, sea hare *Aplysia*, octopus nervous system, zebrafish, frog, rat, human.
4. See http://neuronbank.org/wiki/index.php/Aplysia_cerebral_ganglia, for sea hare *Aplysia*.
5. Lillvis et al. (2012), for gastropods.
6. Hochner et al. (2006), for octopus nervous system.
7. Zullo and Hochner (2011), for octopus.
8. Hochner (2012), for octopus brain.
9. See http://www.neurocomputing.org/Amphibian.aspx for brain of frog, *Rana esculenta*.

2. Hierarchy, levels

Arthropods

10. Visual: Seelig and Jayaraman (2013), for insect, fruit fly.
11. Visual: Strausfeld (2013), for insects.
12. Olfactory (2 for direct path and 4 for indirect path): Farris (2005), for insects.
13. Olfactory (2 for direct path and 4 for indirect path): Giurfa (2013), for insects.

Appendix

14. Olfactory (2 for direct path and 4 for indirect path): Jacobson and Friedrich (2013), for fruit fly.
15. Olfactory (2 for direct path and 4 for indirect path): Liu et al. (1999), for fruit fly.
16. Olfactory (2 for direct path and 4 for indirect path): Perry, Barron, and Cheng (2013), for bee.
17. Olfactory (2 for direct path and 4 for indirect path): Strausfeld (2013), for insects.
18. Taste (2 for direct path and 4 for indirect path): Giurfa (2013), for bee.
19. Hearing (number of levels is uncertain): Kamikouchi (2013), for fruit fly.
20. Hearing (number of levels is uncertain): Strausfeld (2013), for insects.
21. Balance: Kamikouchi (2013), for fruit fly.
22. Touch-mechanosensory, and wind sense: Strausfeld (2013), for insects.
23. Touch-mechanosensory (and nociceptive): Ohyama et al. (2015), for larval *Drosophila*.

Gastropods
24. Chemosensory: figure 1 in Baxter and Byrne (2006), for *Aplysia*.
25. Chemosensory: Bicker et al. (1982), for *Pleurobranchaea*.
26. Chemosensory: Wertz et al. (2006), for *Aplysia*.
27. Chemosensory: Yafremava and Gillette (2011), for *Pleurobranchaea*.
28. Mechanosensory: figure 1 in Baxter and Byrne (2006), for *Aplysia*.
29. Mechanosensory: Bicker et al. (1982), for *Pleurobranchaea*.
30. Mechanosensory: Hirayama et al. (2012), for *Pleurobranchaea*.
31. Mechanosensory: Yafremava and Gillette (2011), for *Pleurobranchaea*.
32. Mechanosensory: Walters et al. (2004), for *Aplysia*.
33. Photoreception but no vision: Noboa and Gillette (2013), for *Pleurobranchaea*.
34. Photoreception but no vision: Zhukov and Tuchina (2008), for *Lymnaea*.
35. Photoreception but no vision. But for a few gastropods with lensed eyes, see Nilsson (2013).
36. Photoreception but no vision. But for a few gastropods with lensed eyes, see Ruppert et al. (2004).

Cephalopods
37. Visual: Young (1974) documents three or four levels; Shigeno and Ragsdale (2015, fig. 10) document more.

38. Visual: Williamson and Chrachri (2004), for cephalopods.
39. Olfactory: Mobley et al. (2008), for squid. (See also Messenger, 1979.)
40. Taste: Grasso (2014), for octopus.
41. Taste: Mather (2012), p. 120, for cephalopods.
42. Hearing: Mather (2012), p. 120, for cephalopods.
43. Balance: Williamson and Chrachri (2004), for cephalopods. (See also Messenger, 1979.)
44. Mechanosensory: Grasso (2014) and Mather (2012), for octopus.
45. Mechanosensory: Shigeno and Ragsdale (2015; fig. 10).

3. Isomorphism

Arthropods

46. Visual retinotopy: Seelig and Jayaraman (2013), for insect, fruit fly.
47. Visual retinotopy: Harzsch (2002) for crustaceans such as lobster and crab, and for chelicerate horseshoe crabs (p. 14).
48. Visual retinotopy: Strausfeld (2009) for malacostracan crustaceans (lobster, crab, etc.).
49. Visual retinotopy: Sombke and Harzsch (2015) for a centipede.
50. Olfactory odortypy: Jacobson and Friedrich (2013), for fruit fly.
51. Olfactory odortypy: Strausfeld (2013), for insects.
52. Taste: Newland et al. (2000), for locust.
53. Taste: Wolf (2008), for spider.
54. Hearing tonotopy: Strausfeld (2013), for insects.
55. Mechanosensory somatotopy: Schmidt, Van Ekeris, and Ache (1992), for lobster.
56. Mechanosensory somatotopy: Schmidt and Ache (1996), for lobster.
57. Mechanosensory somatotopy: Strausfeld (2013), for insects.
58. Mechanosensory somatotopy: Newland et al. (2000), for locust.

Gastropods

59. Mechanosensory somatotopy: Xin et al. (1995), for *Aplysia*.
60. Mechanosensory somatotopy: Walters et al. (2004), for *Aplysia*.

Cephalopods

61. Visual: Young (1974). (See also Messenger, 1979.)
62. No mechanosensory somatotopy: Godfrey-Smith (2013), for octopus.

Appendix 247

63. No mechanosensory somatotopy: Grasso (2014), for octopus.
64. No mechanosensory somatotopy: Hochner (2012), for octopus.
65. No mechanosensory somatotopy: Hochner (2013), for octopus.
66. Zullo et al. (2009), for octopus.
67. No mechanosensory somatotopy: Zullo and Hochner (2011), for octopus.

4. Reciprocal interactions

Arthropods

68. Few: Chittka and Niven (2009), for insects.
69. Connections documented: Giurfa (2013), for fruitfly and bee.
70. Connections documented: Strausfeld (2013), for insects.

Gastropods

71. Baxter and Byrne (2006), see Figure 1, in *Aplysia*.
72. Hirayama et al. (2012), in *Pleurobranchaea*.

Cephalopods

73. In balance and vision circuits: Williamson and Chrachri (2004), for cephalopods.

5. Multisensory maps, possible unity

Arthropods

74. In mushroom body and other parts of protocerebrum: Farris (2005), for insects.
75. In mushroom body and other parts of protocerebrum: Giurfa (2013), for insects.
76. In mushroom body and other parts of protocerebrum: Strausfeld (2013), chapters 6 and 7, for various arthropods.
77. In ventolateral lobes of protocerebrum: Anton et al. (2011), for moth.
78. In ventolateral lobes of protocerebrum: Nebeling (2000), for cricket.
79. In ventolateral lobes of protocerebrum: Ostrowski and Stumpner (2010), for cricket.
80. For a counterview, see Chittka and Niven (2009), for insects.

Gastropods

81. Baxter and Byrne (2006), figure 1, in *Aplysia*.

Cephalopods

82. Graindorge et al. (2006), for cuttlefish.
83. Hochner (2013), for octopus.
84. Williamson and Chrachri (2004), for cephalopods.

6. Memory site

Arthropods

85. Farris (2005), for insects.
86. Giurfa (2013), for insects.
87. Perry, Barron, and Cheng (2013), for bee.
88. Strausfeld (2013), for all arthropods.

Gastropods

89. Brembs (2003b), for *Aplysia*.
90. Brembs (2003a), for *Lymnaea* and *Aplysia*.
91. Hirayama et al. (2012), for *Pleurobranchaea*.
92. Nargeot and Simmers (2011), for sea hare *Aplysia*.
93. Nargeot and Simmers (2012), for sea hare *Aplysia*.

Cephalopods

94. Graindorge et al. (2006), for cuttlefish.
95. Hochner (2013), for octopus.
96. Shomrat et al. (2011, 2015), for octopus and cuttlefish.

7. Selective-attention mechanisms

Arthropods

97. Visual target: Giurfa (2013), for fruit fly.
98. Visual target: van Swinderen and Andretic (2011), for fruit fly.
99. Visual target: van Swinderen and Greenspan (2003), for fruit fly.
100. Color differences: Giurfa (2013), for bee.

Appendix

101. Color differences: Spaethe et al. (2006), for bee.
102. Oscillatory binding: van Swinderen and Greenspan (2003), for fruit fly.
103. Oscillatory binding: Tang and Juusola (2010), for fruit fly.

Gastropods

104. Attention to salient stimuli: Hirayama et al. (2012), for *Pleurobranchaea*.
105. Arousal: Hirayama et al. (2012), for *Pleurobranchaea*.
106. Arousal: Jing and Gillette (2000), for *Pleurobranchaea*.

Cephalopods

107. Fiorito and Scotto (1992), for octopus.
108. Gutnick et al. (2011), for octopus.
109. Mather and Kuba (2013), for cephalopods.

Notes

1 The Mystery of Subjectivity

1. Nagel (1974), p. 436. For a broad overview of the topic of consciousness in animals, and the many different ideas that have been proposed on this subject, see the entry in the *Stanford Encyclopedia of Philosophy* at http://plato.stanford.edu/entries/consciousness-animal/.

2. Nagel (1974).

3. Revonsuo (2006), p. 37. For further descriptions of primary consciousness, see Allen and Bekoff (2010); Edelman (1989); Revonsuo (2010).

4. Access consciousness is consciousness that is accessible for verbal report, reasoning, and the like. Block (1995, 2009) has emphasized the importance of distinguishing access consciousness from basic, phenomenal consciousness. See also Boly and Seth (2012).

5. See Griffin (2001), chap. 1.

6. Chalmers (1995), p. 201. Note that Chalmers includes within the "hard problem" of consciousness the puzzle of "something it is like to be," and the subjective "feeling" of perception, mental images, and emotions. See also Chalmers (1996, 2010). The hard problem, as it addresses how the brain produces consciousness, is called the "world knot" by Globus (1973) after A. Schopenhauer (1813).

7. Levine (1983); see also Block (2009). In essence, the "hard problem" of consciousness is the problem of explanatory gaps.

8. "Jainism and Buddhism," Udana 68–69: Parable of the Blind Men and the Elephant. http://www.cs.princeton.edu/~rywang/berkeley/258/parable.html.

9. In his most recent book, Antonio Damasio (2010, p. 16) proposed his own set of approaches. His first avenue is "the direct-witness perspective on the individual mind, which is personal, private, and unique to each of us" or "introspective, first-person inspection." His second is the behavioral perspective, the third is the brain perspective, and the fourth is the evolutionary perspective. These approaches differ somewhat from

ours. Damasio focuses on the self much more than we do, as we are unsure that primary consciousness has to include any real sense of self (Edelman, 1992, 1998; Seth, Baars, & Edelman, 2005, pp. 131–132). Damasio's neuroevolution is centered more on the *human* brain. Our view is that while Damasio's first perspective is related to the philosophical issues that we focus on in this book, it is not the same as the thorny philosophical problems posed by Searle's "first-person ontology" or "ontological subjectivity," Chalmers's "hard problem," or Nagel's "something it is like to be," which do not address self-related introspection or first-person inspection. Our goal is to solve primary consciousness, not to explain the self. In chapter 10, we propose our theory of neurobiological naturalism to explain how our various approaches can be integrated into a solution to the hard problem.

10. For a primer on ontology, see http://en.wikipedia.org/wiki/Ontology.

11. Searle, (1997), p. 212. Nagel (1974) makes the same argument that since subjectivity is inherently connected to a "single point of view," and any objective theory of consciousness must necessarily abandon that point of view, therefore subjectivity is irreducible to any objective account of its nature. See also Nagel (1989).

12. Feinberg (2012); Feinberg and Mallatt (2016a). In these earlier works, we called the NSFCs the "neuroontologically irreducible features of consciousness" (NOIFs).

13. For the definition of projicience, see Sherrington (1947).

14. For additional discussions of referral of conscious sensations, see Feinberg (1997, 2009, 2012); Velmans (2000, 2009).

15. Sellars (1963, 1965); see also Meehl (1966). For other discussions of mental unity, see Baars (1988, 2002); Bayne (2010); Bayne and Chalmers (2003); Cleeremans (2003); Feinberg (2009, 2011, 2012); Metzinger (2003); Teller (1992).

16. Feinberg (2001). Descartes (1649) reasoned that because the mind and the soul were unified entities derived from a singular "inside" point of view, and all the sense organs are "double," that the single pineal gland was the likely candidate to pull together the two hemispheres of the brain and thereby unify consciousness. We now know that the pineal gland has the more modest function of releasing a hormone that helps our body prepare for its nighttime activities (viz., melatonin: Mescher, 2013).

17. Crick and Koch (2003), p. 119. See also Crick (1995). For more on the philosophy of qualia, see Churchland (1985); Churchland and Churchland (1981); Dennett (1988, 1991); Edelman (1989); Flanagan (1992); Jackson (1982); Kirk (1994); Metzinger (2003); Revonsuo (2006, 2010); Tye (2000).

18. The first quote is from Kim (1998), p. 29, and the second from Revonsuo (2010), p. 298.

19. Further discussions of mental causation can be found in Dardis (2008); Davidson (1980); Heil and Mele (1993); Kim (1995); Jackson (1982); Popper and Eccles (1977); Walter and Heckmann (2003).

Notes 253

20. The various types of reduction are explained in Ruse (2005).

21. Searle (1992), p. 113.

22. Searle (2002).

23. Actually, epilepsy may not be so well understood after all, as it may relate more to how the brain processes sugar than was previously recognized (Sada et al., 2015). Still, scientists remain confident that epilepsy is solvable by the reductionist approach.

24. For differences between types of emergence theories, see Bedau (1997); Chalmers (2006); Searle (1992). For more discussion of the role of emergence in biology, science, and philosophy, see Beckermann, Flohr, and Kim (1990); Clayton and Davies (2006); Mayr (2004).

25. Sperry (1977), p. 119.

26. Schrödinger (1967), p. 154. While Sperry, Schrödinger, and others say that consciousness is a radically emergent process that is absolutely irreducible, still others claim that it is irreducible only to conventional scientific explanations. That is, they propose that consciousness is explained by fully natural but as yet incompletely understood effects of quantum mechanics, magnetic fields, microtubules in cells, or single neurons. See Hameroff and Penrose (2014); Penrose (1994); Sevush (2006). For critiques, see Smith (2006, 2009) and Llinás (2001), p. 209.

27. This quote is from Chalmers (2010), p. 17.

28. Here are the basics of evolutionary theory. Charles Darwin (1859) revolutionized the science of biology in the nineteenth century by explaining evolution through *natural selection*, based on this line of logic: (1) All organisms produce more offspring than can survive; (2) the offspring vary in their traits; (3) offspring with the most favorable traits are more likely to survive and pass these traits on to their own offspring. In this way, organisms become better adapted to their environments and species can survive when the world changes. The nervous system has evolved in this way ever since it first appeared in animals.

Although natural selection is the dominant mechanism, we do not mean to imply that it is the only thing that influences evolution. Other factors include mutation, genetic drift, and the newly recognized processes of niche construction and developmental bias. For a lead-in to the literature on modern views of evolution, see Laland et al. (2014).

29. For example, see Butler (2000, 2008a); Cabanac, Cabanac, and Parent (2009); Casimir (2009); Damasio (2010); Denton (2006); Fabbro et al. (2015); Griffin (2001); Jonkisz (2015); Lindahl (1997); Llinás (2002); MacLean (1975); Merker (2005); Nichols and Grantham (2000); Packard and Delafield-Butt (2014); Panksepp (1998); Tsou (2013).

30. We began this in an earlier paper (Feinberg & Mallatt, 2013), and greatly expand on it here.

31. These works, which will be treated in this book, include Allen and Fortin (2013); Binder, Hirokawa, and Windhorst (2009); Erwin and Valentine (2013); Holland (2014); Lacalli (2015); Lamb (2013); Mather and Kuba (2013); Murakami and Kuratani (2008); Nilsson (2013); O'Connell and Hofmann (2011); Parker (2009); Rowe, Macrini, and Luo (2011); Šestak et al. (2013); Šestak and Domazet-Lošo (2015); Strausfeld (2013); Trestman (2013); Vopalensky et al. (2012).

32. Boly et al. (2013); Butler (2008a). Higher-order thought theories are discussed by Block (2009).

33. The proposal that cognition and consciousness are roughly correlated is from Block (2009), as is the consideration of the human baby. For more evidence of consciousness in human infants, see Kouider et al. (2013).

34. For explorations of which nonhuman animals are conscious, see Butler (2008a); Edelman and Seth (2009); Griffin (2001). Diverse criteria for judging which animals have consciousness can found in Seth, Baars, and Edelman (2005); see also Edelman, Baars, and Seth (2005). For the conclusion that all birds and mammals are conscious, see Århem et al. (2008); Boly et al. (2013); Butler and Cotterill (2006); Butler et al. (2005); Edelman et al. (2011). For investigators who still say only humans are conscious, see Carruthers (2003); Dennett (1995); LeDoux (2012).

35. Searle (1992), p. 1.

36. Feinberg (2012); Feinberg and Mallatt (2016a). Many other scientists also say consciousness emerges only from natural processes; see, e.g., Damasio (2010), p. 15; Edelman (1992); Griffin (2001), chap. 1; Llinás (2001).

2 The General Biological and Special Neurobiological Features of Conscious Animals

1. This idea is derived from systems theory, in which higher and later developed levels are added to lower and earlier levels in evolutionary hierarchies. For further discussion, see Buss (1987); Salthe (1985); von Bertalanffy (1974). The idea of the stages being represented by general and special features of consciousness was previously presented in Feinberg and Mallatt (2016a), and it is updated here.

2. Mayr (1982), p. 53. This is not to imply that Mayr was puzzled or at a standstill. In a later book, he provided his own list of the characteristics of life: "Owing to their complexity, biological systems are richly endowed with capacities such as reproduction, metabolism, replication, adaptedness, growth, and hierarchical organization. Nothing of the sort exists in the inanimate world" (Mayr, 2004, p. 29).

3. Luisi (1998).

4. Schopf and Kudryavtsev (2012).

5. Knoll et al. (2006) give the date of the oldest fossil eukaryote cells as 1.8 billion years. Spang et al. (2015) may have identified the closest living sister group of eukaryotes as a clade of complex archaeans called Lokiarchaeota.

Notes

6. Lane and Martin (2010) discuss the evolution of mitochondria as symbiotic energy providers.

7. Thompson (2007), p. 225. This states his main theme that the border between the inside and the outside, between the subjective and the objective, is not really at the brain surface but at the body surface.

8. Mayr (1982), p. 74. "Reification" means treating an abstraction as if it were a concrete, material thing.

9. James (1904), p. 478.

10. Camazine et al. (2003), p. 8.

11. Mitchell (2009).

12. Salthe (1985); Simon (1962, 1973).

13. Feinberg and Mallatt (2016a). Buss (1987), pp. 183–188, also notes this tendency. Ellis (2006, p. 82) gives a similar reason: "At the topmost level, each type of emergence is characterized by adaptive selection in interaction with the physical and social environment, which are the boundary conditions for the system." It is akin to Ernst Mayr's argument that selection does not operate at the lower level of genes but at the higher level of the phenotype, which is expressed by the genes (Mayr, 2004, chap. 8).

14. Kim (1992); Salthe (1985); Simon (1962, 1973).

15. Kim (1992), pp. 121–122.

16. Knoll et al. (2006).

17. Butler and Hodos (2005).

18. Nomura et al. (2014).

19. Ahl and Allen (1996); Allen and Starr (1982); Campbell (1974); Salthe (1985); Simon (1962, 1973).

20. Pattee (1970), p. 119.

21. On teleonomy, see Monod (1971); see also Feinberg (2012) and Mayr (1982, 2004). See Deacon (2011) for a discussion of teleonomy, emergence, and constraint from the standpoint of thermodynamics. Ernst Mayr argues that teleonomy describes programmed, goal-directed processes Mayr (2004, pp. 53–59). Unlike us, he resists equating teleonomy and adaptation, saying that teleonomic processes are programmed and involve movements, whereas adaptations are stationary and do not involve movements. But Mayr admits that teleonomy and adaptation are both goal-directed and related to one another.

22. Nichols and Grantham (2000), p. 648.

23. Baxter and Byrne (2006); Dickenson (2006); Grillner et al. (2005); https://en.wikipedia.org/wiki/Central_pattern_generator.

24. Simon (1962), p. 468.

25. Mitchell (2009).

26. Tononi (2004, 2008); Tononi and Koch (2008, 2015).

27. Jabr (2012); Kandel et al. (2012); Strausfeld (2013); Underwood (2015); Zeisel et al. (2015).

28. Invertebrate neurons are a bit different from those of vertebrates, with their dendrites sometimes sending signals as well as receiving them, and their axons sometimes receiving signals as well as sending them (Matheson, 2002).

29. Sommer (2013).

30. Brodal (2010); Mescher (2013); Zeisel et al. (2015).

31. Ahl and Allen (1996); Allen and Starr (1982); Feinberg (2000, 2001, 2011, 2012); Feinberg and Mallatt (2016a); Pattee (1970); Salthe (1985).

32. Barlow (1995); Fried, MacDonald, and Wilson (1997); Gross (2002, 2008); Gross, Bender, and Rocha-Miranda (1969); Gross, Rocha-Miranda, and Bender (1972); Kreiman, Koch, and Fried (2000); Quiroga et al. (2005, 2008).

33. Quiroga et al. (2005, 2008).

34. Beckers and Zeki (1995); Engel et al. (1999); Engel, Fries, and Singer (2001); Engel and Singer (2001); Roskies (1999); Singer (2001); Uhlhaas et al. (2009); von der Malsburg (1995); Zeki and Marini (1998).

35. Dennett (1991); Feinberg (2012).

36. Kaas (1997); Kandel et al. (2012).

37. Gottfried (2010); Kandel et al. (2012); Shepherd (2012).

38. For example, Damasio (2000, 2010); Edelman (1989).

39. Sherrington (1947), p. 325.

40. Edelman (1992), p. 112.

41. Damasio (2000, 2010). For further discussion of mental images, see Feinberg (2009, 2011); Feinberg and Mallatt (2013). Llinás (2001, chap. 5) also talks of sensory images and their importance, in his otherwise movement-based theory of the evolution of primary consciousness.

42. Feinberg (2009, 2011).

43. Crick and Koch (2003).

44. James (1890), pp. 403–404.

45. For an introduction to the current literature on attention, see Tsuchiya and van Boxtel (2013).

46. Block (2012); Tononi and Koch (2008), pp. 241–242.

47. For more on this debate, see the overview by Tsuchiya and van Boxtel (2013). Studies that argue for a very close association between consciousness and attention, or even that they are identical, include Baars (1988); Baars, Franklin, and Ramsoy (2013); Chica et al. (2010); Cohen et al. (2012); De Brigard and Prinz (2010); Mole (2008). By contrast, studies arguing that consciousness and attention are dissociated include Boly et al. (2013); Dehaene and Naccache (2001); Dehaene and Changeux (2005, 2011); Kouider and DeHaene (2007); Tallon-Baudry (2011); Tononi and Koch (2008); and van Boxtel, Tsuchiya, and Koch (2010). The current debate is so centered on humans, the cerebral cortex, and cognitive top-down attention that it is difficult to know how much of it applies to distantly related animals such as fish and insects, which certainly do attend to stimuli but have no cerebral cortex. The debate does not affect our ideas, as long as one can assume a correlation between attention and consciousness.

48. Van Boxtel, Tsuchiya, and Koch (2010); Chica et al. (2010); Talsma et al. (2010).

49. Chica et al. (2010), p. 1205.

50. Denton (2006), p. 105.

3 The Birth of Brains

1. For more information on chordates and their basic neural anatomy, many general anatomy and neurobiology texts are helpful. For example, see Brodal (2010); Butler and Hodos (2005); Kandel et al. (2012); Kardong (2012); Marieb, Wilhelm, and Mallatt (2014).

2. Ruppert, Fox, and Barnes (2004).

3. Holland (2015), p. 2.

4. General neurobiology texts with more information include Brodal (2010), Hall, Kamilar, and Kirk (2012), and Kandel et al. (2012).

5. For the best overall treatments of the anatomy and biology of tunicates, see Burighel and Cloney (1997), Kardong (2012), and Ruppert, Fox, and Barnes (2004).

6. Berrill (1955); Garstang (1928); Romer (1970). See also Cameron, Garey, and Swalla (2000); Lowe et al. (2015); Stach et al. (2008); Wada (1998). *Neoteny* is defined by Gould (1977).

7. For a review of the evidence that larvaceans and the other tunicates are highly specialized, see Holland (2014). See also Niimura (2009).

8. Delsuc et al. (2006, 2008).

9. The main studies on tunicate brains are by Burighel and Cloney (1997), Glover and Fritzsch (2009), Lacalli and Holland (1998), and Mackie and Burighel (2005).

10. Lacalli and Holland (1998).

11. Here is a possible reason for the differential brain reductions in the different groups of tunicates: adult ascidians feed but do not locomote, so they only retain the part of the brain that controls feeding and inner-body functions. This is the midbrain and anterior part of the hindbrain (the shaded part in figure 3.4A). Larval ascidians swim but do not feed, so they have the locomotory and sensory parts of the brain, including most of the hindbrain and a simple, non-image-forming eye and a balance organ ("ocellus" and "otolith" in figure 3.4B). But their brain regions for feeding (the shaded "Ganglion") are rudimentary and undeveloped. More examples of the inconsistent presence of a forebrain, midbrain, or hindbrain are shown in figure 3.4C–E.

12. For the best overall treatments of the anatomy and biology of cephalochordates, see Kardong (2012), Ruppert (1997), and Ruppert, Fox, and Barnes (2004).

13. Delsuc et al. (2008); Holland (2014); Holland et al. (2013); Lowe et al. (2015); Mallatt (2009); Mallatt and Holland (2013); Northcutt (2005); Putnam et al. (2008).

14. For the size of the brain of adult amphioxus, see Butler and Hodos (2005) and Ruppert (1997).

15. The many papers by Lacalli and his colleagues on the anatomy and evolutionary interpretation of the brain of larval amphioxus include Lacalli (2008, 2013, 2015), Lacalli and Kelly (2003), Lacalli and Stach (2015), Vopalensky et al. (2012), Wicht and Lacalli (2005), and Wicht et al. (2013).

16. Vopalensky et al. (2012); Lacalli (2013).

17. For statements that amphioxus has no telencephalon, see Medina (2009) and Wicht et al. (2013). But Thurston Lacalli writes (personal communication, May 6, 2015): "the figures I drew mapping amphioxus CNS onto vertebrate brain are all based on the young larval stages, which have almost no differentiated nerve cells in regions corresponding to where the dorsal telencephalon ought to be (i.e., forward of the pineal). However, by the late larval stages this region has expanded and differentiated quite a lot. Linda [L. Z. Holland] ... concludes that amphioxus lacks a telencephalon based on the gene expression and structure, but all of that data is based on embryos and early larvae. The truth is that we really don't know what might be present later on." In other words, older amphioxus may have a telencephalon, but this has not been examined carefully.

18. Wicht et al. (2013).

19. Castro et al. (2006); Mazet and Shimeld (2002).

20. The developmental biologists Ariel Pani, Christopher Lowe, and their colleagues would take issue with some of these identifications, arguing that the amphioxus brain is not so easily compared to that of vertebrates (Pani et al., 2012). They have emphasized that the embryonic brain of amphioxus lacks some of the key genes that pattern the formation of the vertebrate brain into its three divisions of forebrain, midbrain, and hindbrain. From this shortcoming, they conclude that the amphioxus brain is even more reduced and secondarily simplified than that of tunicates. However, closer

Notes 259

analysis by Linda Holland and her coworkers (Holland et al., 2013) has shown that amphioxus expresses many other of the brain-patterning genes, enough to reveal three landmarks that subdivide its brain in a fully vertebrate manner. These landmarks are the midbrain-hindbrain boundary (MHB), a mid-diencephalon marker called the zona limitans intrathalamica (ZLI), and the anterior neural ridge (ANR), which is the rostral border of the forebrain (see figure 3.5). The ZLI of vertebrates divides the forebrain's hypothalamus from another major region called the thalamus. This raises the possibility that the amphioxus brain also has a thalamus, at least a rudimentary one.

The above relates to a disagreement about whether amphioxus has a midbrain: Pani et al. (2012) and Suzuki et al. (2015a, p. 259) say no, while Lacalli and Holland say yes (Holland, 2014, p. 344; Holland et al. 2013). More specifically, the debate is over whether the gene markers show amphioxus has a distinct midbrain, or a combined "hypothalamus plus midbrain," or if it has none and the midbrain evolved anew in prevertebrates from the posterior forebrain. We favor Lacalli's and Holland's interpretation that a real midbrain is present in amphioxus, because the anatomic landmarks say it is (figure 3.5B). That is, as in vertebrates, the midbrain must be the region of the amphioxus brain between the pineal (lamellar body at the back of the forebrain) and the hindbrain.

21. Lacalli (2008, 2015); Wicht and Lacalli (2005).

22. Nieuwenhuys, Veening, and Van Domburg (1987); see also Nieuwenhuys (1996).

23. Lacalli (2008, 2015); Lacalli and Stach (2015).

24. For a paper on how larval amphioxus feeds by sinking through the plankton, see Webb (1969).

25. For more on vertebrates' diencephalic locomotor region for behavioral control and its similarities to the postinfundibular neuropile of larval amphioxus, see Candiani et al. (2002); Canteras et al. (2011); Ménard and Grillner (2008); Vopalensky et al. (2012); Yáñez et al. (2009).

26. Although Lacalli's ideas were carefully constructed and creatively applied, two objections could be raised against his considering the brain of larval amphioxus as the platform from which the vertebrate brain evolved. First, just because this larval brain is simple does not automatically mean that it is primitive or reflects the prevertebrate brain, which could have been simple but different. Between the years 1860 and 1950, many evolutionary biologists believed that "ontogeny recapitulates phylogeny," meaning the order in which structures arise in a developing embryo and larva is the same as the order in which they were added during evolutionary history (the "biogenetic law"). Although this is true in some cases, these scientists were too often proven wrong by other cases where embryos did not resemble ancestors. Embryonic traits themselves evolve and change, and they can diverge away from the primitive condition. Lacalli did answer this larval-is-not-primitive objection, however, by carefully arguing that the amphioxus structures are homologous to those in vertebrate brains based on similar locations and connections, and by making a case that amphioxus has

only those brain structures that are essential for survival. Essential structures, by definition, must have been present from the very start of chordate and vertebrate history and therefore must be ancestral. A good book on the history and invalidity of the idea that ontogeny recapitulates phylogeny is Gould (1977).

The second objection claims that the brain of amphioxus became secondarily simplified, possibly when their free-swimming ancestors started burrowing and living in the sand, a dark, protected, and cognitively unchallenging habitat that led to brain degeneration (see, e.g., Conway Morris & Caron, 2012). If secondarily simplified like this, the amphioxus brain would be uninformative about original chordate brains. This objection can be answered. As discussed above, the evidence of Pani et al. (2012) that was said to support the objection—the evidence that amphioxus has lost key genes for brain development—was countered by showing that amphioxus expresses enough genes to pattern the full, tripartite brain (Holland et al., 2013). In addition, Martin Sestak and Tomislav Domazet-Loso (Sestak & Domazet-Loso, 2015) recently searched the entire genome of amphioxus, and although their findings are complicated by some puzzling aspects, they found that amphioxus has many of the genes that are expressed in the vertebrate forebrain, midbrain, and hindbrain. Therefore, extensive gene loss and secondary simplification of the amphioxus brain seem unlikely.

For Lacalli's responses to the claims that the brain of larval amphioxus is not primitive, see Lacalli (2008, 2015) and Wicht and Lacalli (2005).

27. Lacalli (2015).

4 The Cambrian Explosion

1. Much of the information about the Cambrian Period, the Cambrian explosion, and the time leading up to it can be found in Erwin and Valentine (2013), which gives many of the important papers and references.

2. For a timeline of the important events in the history of life, see the History of Earth page in Wikipedia: http://en.wikipedia.org/wiki/History_of_the_Earth.

3. For accounts of Darwin's dilemma, see Conway Morris (2006) and Gould (1989).

4. For descriptions of the modern phyla of animals, see Erwin and Valentine (2013) and Ruppert, Fox, and Barnes (2004). For interrelationships between the phyla, see Philippe et al. (2011) and Northcutt (2012a).

5. For the ways in which animal relations are reconstructed—by comparing their anatomic structures, functional characteristics, and genes—see Wägele and Bartolomaeus (2014).

6. These Cambrian and Ediacaran fossil localities are described in Erwin and Valentine (2013), chaps. 1 and 5.

7. For more on Wiffle-ball sponges from around 635 mya, see Maloof et al. (2011); on a more convincing Ediacaran sponge from around 600 mya, see Yin et al. (2015); on a

cone-shaped Ediacaran sponge candidate from before 580 mya, see Sperling, Peterson, and Laflamme (2011). For a more complete description of sponge biology, and of the biology of all the other invertebrate animals, see Ruppert, Fox, and Barnes (2004).

8. For more on the importance of microbial mats in Ediacaran times, see Gingras et al. (2011).

9. Pecoits et al. (2012) report on these earliest worm trackways that reveal simple behavior.

10. For more on deuterostomes, protostomes, lophotrochozoans, ecdysozoans, and their subdivisions and phyla, see Erwin and Valentine (2013), Philippe et al. (2011), and Ruppert, Fox, and Barnes (2004).

11. For more about the Cambrian fossil animals, see Conway Morris (1998), Conway Morris and Caron (2012), Chen (2012), Erwin and Valentine (2013), and Gould (1989).

12. For anomalocarids, see Paterson et al. (2011) and Van Roy, Daley, and Briggs (2015). For vetulicolians, see Garcia-Bellido et al. (2014).

13. Erwin and Valentine (2013) cover the complex-simple worm debate, but see also Holland et al. (2013).

14. Northcutt (2012a).

15. As a lead-in to the large literature on how Earth's atmosphere gained oxygen, see Planavsky et al. (2014). See also Erwin and Valentine (2013); Knoll and Sperling (2014); Sperling et al. (2013, 2015).

16. For more about the importance of sediment bioturbation and predation in structuring the Cambrian explosion and in maintaining ecosystems, see Meysman, Middelburg, and Heip (2006). For the increasing complexity of the trails and burrows in the Cambrian, see Erwin and Valentine (2013), chap. 5.

Although predation is considered central, still other factors could have contributed to the Cambrian explosion. These include the evolution of developmental and genetic pathways capable of building the complicated body of bilaterians, changes in ocean chemistry, and widespread glaciation before and during the Ediacaran ("snowball Earth"). See Erwin and Valentine (2013).

17. Plotnick, Dornbos, and Chen (2010).

18. For descriptions of the senses of vertebrates and invertebrates, see Kardong (2012); Ruppert, Fox, and Barnes (2004); and Shubin (2008), chaps. 8–10.

19. For more on the theory that complex nervous systems and brains are "expensive tissues" and will not evolve or will regress unless selected for, see Aiello and Wheeler (1995) and Kotrschal et al. (2013). For more about the many invertebrate phyla that did not become more active or improve their sensory systems, and about their basic biology and survival strategies, see Ruppert, Fox, and Barnes (2004).

20. Ruppert, Fox, and Barnes (2004) describe the features of arthropods; Kardong (2012) describes those of vertebrates. Trestman (2013) discusses the different advantages of arthropods versus vertebrates in the Cambrian Period.

21. Bambach, Bush, and Erwin (2007).

22. For the idea that arthropods evolved from lobopodians, and for more information on *Fuxianhuia*, see Erwin and Valentine (2013). See also Cong et al. (2014); Vannier et al. (2014); Van Roy, Daley, and Briggs (2015).

23. Shu et al. (2010).

24. Mallatt (1982).

25. For more information about the locomotion and senses of fish and amphioxus, see Kardong (2012).

26. Genome duplications occurred in the first vertebrates, but not in any invertebrate clades: Dehal and Boore (2005); Holland (2013); Kuraku, Meyer, and Kuratani (2009); Putnam et al. (2008). Here is more explanation for the evolutionary importance of these duplications. Many of the new gene copies could mutate and take on new adaptive functions while the original genes kept up the original functions. The new functions meant more body complexity in vertebrates. This is covered at https://en.wikipedia.org/wiki/Gene_duplication. Underscoring the uniqueness of this event in vertebrates, no such duplication occurred in cephalopod molluscs, the invertebrates with the largest bodies and brains (Albertin et al., 2015).

27. Arthropods are more abundant and have more species than vertebrates, but vertebrates' success is also shown by the fact that they attained the largest bodies and largest absolute brain sizes. The average body size of marine vertebrates has increased markedly since the Cambrian, whereas that of marine arthropods has not and may even have declined (see Heim et al., 2015, esp. their fig. 1).

5 Consciousness Gets a Head Start: Vertebrate Brains, Vision, and the Cambrian Birth of the Mental Image

1. Most of the information in this Vertebrate Brains section and in table 5.1 can be found in the textbooks by Brodal (2010), Butler and Hodos (2005), Kardong (2012), and Marieb, Wilhelm, and Mallatt (2014).

2. For more information on lateral-line canals and electroreceptive senses by which fish can detect objects in the water around them, such as approaching predators, see Baker, Modrell, and Gillis (2013); http://en.wikipedia.org/wiki/Lateral_line; http://en.wikipedia.org/wiki/Electroreception; and Shubin (2008), chap. 10.

3. Butler and Hodos (2005), pp. 192, 244.

4. Nieuwenhuys (1996).

Notes

5. Lacalli (2013); Vopalensky et al. (2012).

6. Wicht and Lacalli (2005).

7. Wicht and Lacalli (2005); see also Niimura (2009). For possible chemoreceptor or olfactory cells in tunicates and amphioxus, see Abitua et al. (2015) and Satoh (2005).

8. Šestak et al. (2013); Niimura (2009); Satoh (2005).

9. Parker (2003); Trestman (2013).

10. For how eyes evolved in animals, see Nilsson (2009, 2013), Lamb (2013), chap. 5 in Llinás (2001), and http://en.wikipedia.org/wiki/Evolution_of_the_eye.

11. Nilsson (2013).

12. Plotnick, Dornbos, and Chen (2010). Others who advocate smell-first are Lucia Jacobs (Jacobs, 2012), who says the building of smell maps of environmental space came first, and James Kohl (Kohl, 2013), whose model says chemical ecology is the main driver of adaptive evolution.

13. Erwin and Valentine (2013).

14. Jacobs (2012).

15. The following references document the three lines of physical evidence that vision evolved before smell. For homology of the amphioxus and vertebrates eyes, see Vopalensky et al. (2012). For the early evolution of vision genes, see Šestak et al. (2013). For information on eyed, fossil prevertebrates, see Mallatt and Chen (2003). Amphioxus has neurons that express olfactory receptor-like genes, but these neurons are rather broadly scattered in the skin of the snout, not concentrated in any olfactory organ (Satoh, 2005).

16. For good descriptions of the neural crest and placodes, see Green, Simoes-Costa, and Bronnwer (2015), Hall (2008), Patthey, Schlosser, and Shimeld (2014), Schlosser (2014), and Sommer (2013). Gans and Northcutt (1983) is the classic paper on the importance of crest and placodes for the evolution of the vertebrate head.

17. It is becoming more evident that the neural crest and placodes originated at or very near the time of the first vertebrates, in a real revolution. For decades, scientists have looked for homologous cells in amphioxus and tunicates with experimental techniques of ever increasing sophistication, yet with limited success. Researchers have proposed many candidates, especially among ectodermal and pigment cells, but these are all debated (Abitua et al., 2012; Green, Simoes-Costa, and Bronner, 2015; Ivashkin & Adameyko, 2013). Little is clear, suggesting the protochordates have only the simplest rudiments of the cell sources, and only parts of the gene pathways, for crest and placodes. Šestak et al. (2013) suggest that placodes evolved earlier than the neural crest, because tunicates express more of the placode-associated genes, but such a claim is hindered by protochordates' lack of crest-derived or placode-derived body structures (e.g., no lens equivalent, no dorsal root ganglia).

Finally, Abitua et al. (2015) have used genetic identification to find a credible homologue of the olfactory placode in tunicates. But it is rudimentary and its neurons more demonstrably secrete reproductive hormone than participate in olfaction, a sense that still has not been demonstrated in tunicates. The neurons might be olfactory but they differ in certain ways from vertebrate olfactory neurons, and no conventional odorant receptors have been found in the tunicate genome. Finding this elementary placode homologue was a master achievement, but it emphasizes the vast gulf between vertebrates and tunicates.

18. http://en.wikipedia.org/wiki/Lateral_line.

19. Abitua et al. (2015).

20. Lacalli (2008).

21. Butler (2000); see also Butler (2006).

22. Chen et al. (1995); Chen, Huang, and Li (1999); Hou, Ramsköld, and Bergström (1991); Mallatt and Chen (2003).

23. Chen, Huang, and Li (1999); Mallatt and Chen (2003).

24. The *Haikouella* eyes are nearer in size to those of larval lampreys (figure 4.6D), whose reduced eyes measure a third of a millimeter and are said not to form images but to be "nondirectional or broadly directional photoreceptor organs" (Suzuki et al., 2015b). This is presumably because these worm-shaped larvae burrow and live in the turbid waters of stream bottoms. *Haikouella* was not a wormy burrower, but the beds of shale (fossilized mud) in which it is found suggest that it lived in turbid, muddy waters of lowered visibility. But that is fairly speculative, and the small size of *Haikouella*'s eyes remains a puzzle. Like those of larval lampreys they could have been directional light detectors and secondarily reduced.

25. Lee et al. (2011); Ma et al. (2012).

26. Shu et al. (2010) and Conway Morris and Caron (2012) argue that yunnanozoans are not related to vertebrates. These researchers have questioned the existence of any eyes, brain, or a notochord in yunnanozoans. They interpret yunnanozoans as basal deuterostomes not chordates. But we stand by Chen and Mallatt's (2003) assessment of the yunnanozoan fossils as vertebrate relatives, having found muscle fibers in the body segments (the myomeres, which Shu thought were just segments of a hard cuticle on the outside of the body), and segments along the notochord suggesting the vertebrae of a vertebral column ("Protovertebrae" in figure 5.8B).

27. For the key papers on the earliest fossil vertebrates *Haikouichthys, Metaspriggina*, and their relatives, see Shu et al. (1999, 2003); see also Conway Morris and Caron (2014).

28. Conway Morris and Caron (2014).

29. Shu et al. (2003).

Notes

30. The sensory pathways in mammals, pictured in figures 5.9 through 5.12, are described in various neuroanatomy textbooks such as Brodal (2010), Butler and Hodos (2005), and Kandel et al. (2012).

31. Smell's apparent nonreliance on the thalamus is especially remarkable because the thalamus is an important "gateway" and organizing center for all the other kinds of senses that are headed to the cerebrum at the top of the pathway. Why the olfactory system alone does not need to relay through the thalamus is an interesting question. Jay Gottfried proposed that it is because the olfactory pathways evolved early before the emergence of the other senses, all of which use the thalamus (Gottfried, 2006; see also Shepherd, 2007). Translating Gottfried's smell-first view into our alternate vision-first concept, this would mean the sense of smell elaborated so soon after vision that it was semi-independent and was not as much influenced by the vision blueprint as were the other, later evolving senses of hearing, improved touch, and so on. Or else it could mean that smell does not need the thalamus because some other part of the brain performs the thalamus-like functions for olfaction. That part would be the olfactory cortex of the cerebrum itself, according to table 5.3. The unique aspects of the olfactory pathway have long interested neuroscientists, and they will come up again in chapter 6 (pp. 113–115) and especially in chapter 10 (p. 211).

32. Damasio (2010), pp. 79–80 and chap. 6; Maier et al. (2008); Panagiotaropoulos et al. (2012); Pollen (2011); van Gaal and Lamme (2012); Watanabe et al. (2011).

33. Nevin et al. (2010) describe the circuitry for complex sensory processing within the fish tectum. The hearing and lateral-line pathways to the fish tectum relay through a midbrain region called *torus semicircularis*, accounting for the fourth neuronal level in those pathways (Northmore, 2011).

6 Two-Step Evolution of Sensory Consciousness in Vertebrates

1. Our recent publications are Feinberg and Mallatt (2016b) and Feinberg and Mallatt (2013). While this chapter was in preparation, Fabbro et al. (2015) also came out as advocating consciousness in every vertebrate. By contrast, the investigators who say vertebrate consciousness exists only in mammals and birds include Baars, Franklin, and Ramsoy (2013); Boly et al. (2013); Butler (2008a); Di Perri et al. (2014); Edelman, Gally, and Baars (2011); Min (2010); Ribary (2005); Rose et al. (2014); Seth, Baars, and Edelman (2005); van Gaal and Lamme (2012).

Our theory of consciousness, as developed here and in chapter 5, falls squarely in the category of "first-order representational theories" (Mehta and Mashour, 2013). Such theories say that consciousness consists of sensory representations of the world and body that are directly available to the subject. However, Mehta and Mashour's particular first-order theory is very different from ours because it says the cerebral cortex is necessary for consciousness.

2. Kardong (2012) and Shubin (2008) provide good references on the history of the vertebrates since the Cambrian.

3. Janvier (1996, 2008); Maisey (1996); Sansom et al. (2010).

4. For more information on the lives of hagfish and lampreys, see Hardisty (1979), Jorgensen et al. (1998), and Zintzen et al. (2011).

5. Gelman et al. (2007); Ronan and Northcutt (1987).

6. For more on the origin of jaws in vertebrates, see Mallatt (1996, 2008). See also Kuratani (2012).

7. For information on ancient jawed fish and tetrapods, see Janvier (1996), Maisey (1996), and Kardong (2012). For more on the history of tetrapods and of modern amphibians, see Anderson et al. (2008) and Paton, Smithson, and Clack (1999).

8. Mallatt (1997); Northcutt (1996). For information on the senses and brain of hagfish, see Jorgensen et al. (1998).

9. For studies on why lampreys are said to retain more of the sensory and brain characters of the first vertebrates than do hagfish, and have most of the same neural and brain structures as gnathostomes, see Butler and Hodos (2005); Collin (2009); Glover and Fritzsch (2009); Kardong (2012); Nieuwenhuys and Nicholson (1998); Wicht (1996).

10. For the lamprey brain not being quite as well developed as that of jawed vertebrates, see Northcutt (2002): simpler pallium; Bullock, Moore, and Fields (1984): no myelin; and Glover and Fritzsch (2009): simpler visceral nervous system.

11. The studies that investigated the lamprey nervous system include Cinelli et al. (2013); Grillner, Robertson, and Stephenson-Jones (2013); Lopes and Kampff (2015); Murakami and Kuratani (2008); Nieuwenhuys and Nicholson (1998); Northcutt (1996); Northcutt and Wicht (1997); Ocaña et al. (2015); Repérant et al. (2009); Robertson et al. (2014); Rovainen (1979).

12. For the importance of topographic mapping in sensory hierarchies, mental images, and consciousness, see Damasio (2000); Edelman (1989); Jbabdi, Sotiropoulos, and Behrens (2013); Kaas (1997); Risi and Stanley (2014); Singer (2001); Tinsley (2008); Ursino, Magosso, and Cuppini (2009).

13. For more on oscillatory patterns of neuronal communication, see Boly et al. (2013); Bullock (2002); Bürck et al. (2010); Calabrò et al. (2015); Cardin et al. (2009); Dehaene and Changeux (2011); Feinberg (2012); Feldman (2013); Goddard et al. (2012); Kim et al. (2015); Krauzlis, Lovejoy, and Zénon (2013); Llinás (2001), chap. 6; Magosso (2010); Melloni et al. (2007); Miller and Buschman (2013); Mori et al. (2013); Orpwood (2013); Schiff (2008); Schomburg et al. (2014); Singer (2001); Sridharan, Boahen, and Knudsen (2011); Uhlhaas et al. (2009); Ursino, Magosso, and Cuppini (2009); van Gaal and Lamme (2012), p. 9; Zmigrod and Hommel (2013). For a more skeptical view of the role of these oscillations, see Mazaheri and Van Diepen (2014); Nunez and Srinivasan (2010); Tononi and Koch (2008, 2015).

Notes

An intricate theory of the role of oscillatory communication in consciousness is that of Georg Northoff (2013a, 2013b, in press). He emphasizes the complex networks of neurons in the brain's core and cerebrum, whose oscillatory intercommunication produces the brain's intrinsic activity of brainwaves. This is resting-state activity. Then, according to his model, extrinsic sensory inputs adjust and tune this baseline activity into a complex code of brainwaves of many frequency categories (for example, gamma, alpha, theta, and delta waves). The encoding varies according to the sensory stimulus, both in different brain locations and in the relative timing of the different wave-frequency categories, producing a unique code for each stimulus. Northoff argues that this stimulus-imprinted "spatiotemporal structure" links to or produces sensory consciousness. The extrinsic stimuli are coded into the neural activity directly, in "stimulus-based coding," and, more importantly, the *time-differences* between successive stimuli are encoded as well. He believes this "time-difference based coding" is a key feature of consciousness, going beyond the strictly stimulus-based coding of simple reflexes and of simple, unconscious networks. Northoff's idea that stimuli are translated into a spatiotemporal structure and into difference-based coding may one day prove to be a real neural correlate of consciousness. However, these properties need more exploration, so it seems too early to include them in our universal criteria of consciousness in table 2.1 and table 9.2.

14. For the name "unitary scene," see http://en.wikipedia.org/wiki/Gerald_Edelman.

15. For basic works on the optic tectum see Butler and Hodos (2005); Feinberg and Meister (2015); Guirado and Davila (2009); Northmore (2011); Pérez-Pérez et al. (2003); Saidel (2009); Wullimann and Vernier (2009b).

16. Northmore (2011).

17. Feinberg and Mallatt (2016b). The multisensory tectal mapping is documented by Guirado and Davila (2009); Knudsen (2011); Manger (2009); Saidel (2009); Saitoh, Ménard, and Grillner (2007); Stein and Meredith (1993).

18. Sources on the lamprey tectum include De Arriba and Pombal (2007); Iwahori, Kawawaki, and Baba (1998); Jones, Grillner, and Robertson (2009); Kardamakis et al. (2015); Nieuwenhuys and Nicholson (1998); Pombal, Marín, and González (2001); Repérant et al. (2009); Robertson et al. (2006); Sarvestani et al. (2013).

19. For the tectum's sensory and salience-determining roles, see Ben-Tov et al. (2015); Del Bene et al. (2010); Dudkin and Gruberg (2009); Graham and Northmore (2007); Gruberg et al. (2006); Gutfreund (2012); Nevin et al. (2010); Preuss et al. (2014); Schuelert and Dicke (2005); Temizer et al. (2015).

20. For more on the tectum's motor functions, see Cohen and Castro-Alamancos (2010); Dutta and Gutfreund (2014); Northmore (2011).

21. For the tectum's role in selective attention and isthmus-tectal interactions, see Dutta and Gutfreund (2014); Gruberg et al. (2006); Knudsen (2011); Northmore 2011.

22. On the tectum being for snap choices rather than deep analysis, see Cohen and Castro-Alamancos (2010); Kaas (1997); Krauzlis, Lovejoy, and Zénon (2013); Northmore (2011); Sridharan, Schwarz, and Knudsen (2014).

23. For ideas on why the tectum receives input from senses other than vision, see Bürck et al. (2010); Kaas (1997); Quinn et al. (2014).

24. Kaas (1997), p. 110.

25. Dicke and Roth (2009), p. 1456; italics added.

26. Wullimann and Vernier (2009a), p. 1325.

27. Damasio (2010, pp. 88–91) also had the idea that the vertebrate tectum is involved in consciousness, although he only considered the superior colliculus, especially of humans. He suggested that the human superior colliculus has crude mental images, which could explain the phenomenon of blindsight (see our chapter 10) and why hydranencephalic children born without a cerebral cortex show some visual ability. Our view, by contrast, is that the mammalian superior colliculus is a secondarily simplified tectum that has lost all or almost all of its original conscious role (for this loss, see "Evolutionary Step 2: Consciousness Jumps Ahead in Mammals and Birds," in this chapter). Strehler (1991) and Merker (2005, 2007) go further than Damasio and claim that the superior colliculus, not the cerebral cortex, is the major site of consciousness in mammals.

28. Goddard et al. (2012); Knudsen (2011); Marín et al. (2007); Sridharan, Boahen, and Knudsen (2011); Sridharan, Schwarz, and Knudsen (2014).

29. Caudill et al. (2010); Northmore and Gallagher (2003).

30. De Arriba and Pombal (2007); Pombal, Marín, and González (2001); Robertson et al. (2006).

31. Wullimann and Vernier (2009b).

32. Demski (2013); Hofmann and Northcutt (2012); Repérant et al. (2009); Wullimann and Vernier (2009b).

33. Hofmann and Northcutt (2012); Northcutt and Wicht (1997); Wullimann and Vernier (2009b).

34. Wilczynski (2009), p. 1304.

35. Mueller (2012); Northmore (2011); Wullimann and Vernier (2009b). For a few exceptions, in which some fish have separate pallial areas for the different senses, see Demski (2013) and Prechtl et al. (1998).

36. Brodal (2010); Eisthen (1997); Green et al. (2013); Li et al. (2010); Northcutt and Puzdrowski (1988); Shepherd (2007); Wullimann and Vernier (2009b).

37. De Arriba and Pombal (2007); Saidel (2009).

38. Ocaña et al. (2015); Lopes and Kampff (2015).

Notes 269

39. For fish behaving almost normally after their pallium has been removed, see Northmore (2011). For the memory functions of the anamniote and amniote pallium, see Bingman, Salas, and Rodriquez (2009); Broglio et al. (2005); Fuss, Bleckmann, and Schluessel (2014); Nadel et al. (2012).

40. Butler and Hodos (2005); Northcutt and Wicht (1997).

41. Gerald Edelman discusses the role of memory in Edelman (1989, 1992); Edelman, Gally, and Baars (2011). See also Baars et al. (2013) and Nadel et al. (2012). For the fundamental role of learning and training in object recognition, see Jordan and Mitchell (2015) and LeCun, Bengio, and Hinton (2015).

42. For more on hippocampus function and connections, see Allen and Fortin (2013); Bingman, Salas, and Rodriguez (2009); Brodal (2010); Butler and Hodos (2005); Northcutt and Ronan (1992); Rolls (2012).

43. Fuss, Bleckmann, and Schluessel (2014); O'Connell and Hofmann (2011); Schluessel and Bleckmann (2005).

44. Watanabe, Hirano, and Murakami (2008).

45. For the anatomical relations between olfactory and memory structures in the brain, see Brodal (2010); Butler and Hodos (2005; chaps. 29 and 30). For smells evoking deep memories, see https://www.psychologytoday.com/blog/brain-babble/201501/smells-ring-bells-how-smell-triggers-memories-and-emotions.

46. For more on how general arousal relates to selective attention, see Schiff (2008). For more about how arousal and the level of consciousness are considered basic aspects of consciousness, see Northoff (2013a). For information on the reticular activating system (RAS) and basal forebrain, plus their neuromodulators norepinephrine and acetylcholine, see Butler and Hodos (2005); Calabrò et al. (2015); Fuller et al. (2011); Herculano-Houzel et al. (1999); Kim et al. (2015); Lee and Dan (2012); Manger (2009). Robertson et al. (2007) discuss the RAS and basal forebrain of lampreys.

47. Manger (2009); Schiff (2008).

48. For information on this role of norepinephrine and acetylcholine, see Lee and Dan (2012).

49. Lacalli (2008); Vopalensky et al. (2012); Wicht and Lacalli (2005).

50. Holland and Yu (2002).

51. Candiani et al. (2012).

52. Our first study, in which we found genes associated with consciousness but not with topographic organization, was Feinberg and Mallatt (2013). The study in which we recognized the importance of *Eph/ephrin* genes for topography is Feinberg and Mallatt (2016b). The studies on the *Eph/ephrin* genes are Cramer and Gabriele (2014); Knöll and Drescher (2002); North et al. (2010); Triplett and Feldheim (2012). Bosne (2010) records the apparent absence of *Eph/ephrin* genes in amphioxus.

53. Prediction is at the center of many modern models of consciousness, cognition, learning, and reward. The idea is that the brain builds representations of the world, from which it makes predictions of what will happen next, and which it constantly updates in order to influence decisions and to select behavioral actions. To sample this literature, see Llinás (2001); Gershman et al. (2015); Schultz (2015); Seth (2013). This is related to the idea that conscious systems are difference detectors for integrating novel information (Jonkisz, 2015; Mudrik, Faivre, & Koch, 2014).

Concerning prediction, we do not claim the first fish, lampreys, or other anamniotes predict or imagine the distant future like humans do. The mental prediction of which we speak needs only estimate (roughly) where a swimming prey or approaching predator will be in a second or so. Memories based on past experience would also help in making such short-term predictions.

Llinás's (2001) concept of consciousness as a prediction device is motor-centered whereas ours is sensory-centered. He proposed that consciousness evolved to predict what movements one will make, and the consequences of these movements, so that one can interact smoothly with the outside world. His concept is not completely motor-based, for it has a sensory part. That is, he says exteroceptive sensory inputs are used to build simplified simulations of the outside world that allow fast and efficient decisions about motor actions (see his chaps. 5, 10). By contrast, we put more emphasis on the sensory inputs and say they produce highly detailed simulations and predictions, not simplified ones.

54. Others agree with the main idea of this paragraph, that the main adaptive role of consciousness is to select and guide behavioral actions. Perhaps Jonkisz (2015) emphasizes this most strongly of all. Damasio (2010, pp. 61–62, 284) extends the idea by saying that more highly detailed mental images allow more precise behavioral responses. And the detail lets one choose from a wider range of different responses, in complex environments. This idea is also expressed by Denton (2006, p. 5) and by Keller (2014), who says that it is the "evolutionary function of conscious information processing to guide behaviors in situations in which the organism has to choose between many possible responses."

55. For more on the synapsids and sauropsids, see Kardong (2012), pp. 109–127. McLoughlin (1980) provides a fun introduction to the synapsids. Sidor et al. (2013) and Whiteside et al. (2011) explore why the sauropsid reptiles out-competed the synapsids during the Triassic Period of the Age of Reptiles.

56. For estimated dates of origin of the first mammals, see Bi et al. (2014); of the first birds, see Lee, Cau, et al. (2014).

57. For the brain sizes of reptiles versus birds and mammals, see Alonso et al. (2004); Balanoff et al. (2013); Benton (2014); Jerison (2009); Northcutt (2002, 2012b); Rowe, Macrini, and Luo (2011).

58. Århem et al. (2008); Boly et al. (2013); Butler (2008a); Butler and Cotterill (2006); Butler et al. (2005); Edelman and Seth (2009); Edelman, Gally, and Baars (2011).

Notes 271

59. For the 350 mya age of the first amniotes, see Benton and Donoghue (2007).

60. For the brain sizes of living amphibians versus reptiles, see Northcutt (2002). For comparisons of the brain and behavioral complexity of amphibians versus reptiles, see Århem et al. (2008) and Burghardt (2013).

61. Butler (2008a); Århem et al. (2008).

62. Gerkema et al. (2013), p. 3.

63. Clarke and Pörtner (2010); Grigg, Beard, and Augee (2004); Hopson (2012); Kardong (2012), pp. 117, 123; Nespolo et al. (2011); Newman, Mezentseva, and Badyaev (2013).

64. For evidence that the first mammals relied less on vision and more on smell and touch because they were nocturnal, see Rowe, Macrini, and Luo (2011). See also Aboitiz (1992); Aboitiz and Montiel (2007); Bowmaker (1998); Gerkema et al. (2013); Hall, Kamilar, and Kirk (2012); Jacobs (2012); and the classic book by Walls (1942).

65. Kaas (2011).

66. Benton and Donoghue (2007); Bi et al. (2014); Luo et al. (2011).

67. Lee, Cau et al. (2014); Zelenitsky et al. (2011).

68. For memory storage in the dorsal pallium, see Nadel et al. (2012).

69. Smell memories in mammals: Aboitiz and Montiel (2007); Jacobs (2012). Visual memories in birds: Rattenborg and Martinez-Gonzalez (2011), p. 242.

70. Damasio (2010), chap. 6; Haladjian and Montemayer (2014).

71. Krauzlis, Lovejoy, and Zénon (2013).

72. Seth, Baars, and Edelman (2005), p. 122; Tulving (1985, 1987, 2002a, 2002b).

73. Balanoff et al. (2013); Benton (2014); Kielan-Jaworowska (2013); Kielan-Jaworowska et al. (2013); Lee, Cau et al. (2014). See also Bonaparte (2012); Gill et al. (2014); Hall, Kamilar, and Kirk (2012); Heesy and Hall (2010); Ruta et al. (2013). Finally, see Maynard (2013), commenting on Godefroit et al. (2013).

74. Balanoff et al. (2013); Lee, Cau et al. (2014).

75. Gershman et al. (2015) emphasize that the *speed* of neural calculations must always be considered, as well as the computational costs and the expected value of the resulting improvement in decision quality. Slow, complex systems with long computation times are a handicap in many environments, unable to make decisions fast enough for survival. See their fig. 2.

76. For more on convergent evolution, see http://en.wikipedia.org/wiki/convergent_evolution.

77. On the pallia in mammals, see Kaas (2011); Kandel et al. (2012). On the pallia in birds, see Cunningham et al. (2013); Kalman (2009); Keary et al. (2010); Müller and Leppelsack (1985); Wild, Kubke, and Peña (2008).

78. Butler (2008a); Edelman, Baars, and Seth (2005); Pepperberg (2009).

79. See Wild (2009) for a review article on the bird hyperpallium or Wulst.

80. See Karten (2013) and his earlier papers cited therein; Dugas-Ford, Rowell, and Ragsdale (2012); Shanahan et al. (2013); Butler, Reiner, and Karten (2011). Especially informative is Jarvis et al. (2013).

81. For the current literature on the functions and neural homologies of the reptile pallium, see Dugas-Ford, Rowell, and Ragsdale (2012); Fournier et al. (2015); Nomura et al. (2013, 2015); Northcutt (2012b).

82. The information on the pallium of living amphibians, lungfish, and reptiles comes from Butler and Hodos (2005) and Butler, Reiner, and Karten (2011).

83. The references in endnote 81 document how incompletely the reptile pallium is known. See also Binder et al. (2009).

84. Butler and Hodos (2005); Demski (2013); Hofmann and Northcutt (2012); Huesa et al. (2009); Quintana-Urzainqui et al. (2012); Wullimann and Vernier (2009b); Yopak (2012).

85. Maisey (1996).

86. Northcutt (2002).

87. Braun (1996); von During and Andres (1998).

88. More credit must be given here to the trailblazing work of Bjorn Merker (2005, 2007). Like ours, his theory assigns exteroceptive sensory consciousness to all the vertebrates, emphasizes topographically mapped (isomorphic) images to simulate reality, and says the optic tectum constructs these conscious simulations. Unlike us, however, he says consciousness evolved to block out all the distracting sensations caused by the *movements* of one's own body ("consciousness arose as a comprehensive solution to the logistics problems created by self-motion"), and that the optic tectum/superior colliculus is a major center of consciousness in humans and other mammals.

7 Searching for Sentience: Feelings

1. Balcombe (2009). The dictionary definition of sentience is capacity for experiencing sensations or feelings, but we emphasize the "feelings" aspect.

2. Barrett et al. (2007); Casimir (2009); Damasio and Carvalho (2013); Kittilsen (2013); Kobayashi (2012); Mendl, Paul, and Chittka (2011); Nargeot and Simmers (2011, 2012); Ohira (2010); Wilson-Mendenhall, Barrett, and Barsalou (2013).

3. Duncan and Barrett (2007); Gallese (2013); Kittilsen (2013); Northoff (2012a), p. 12; Solms (2013a).

Notes 273

4. LeDoux (2012); Anderson and Adolphs (2014); Berlin (2013); Millot et al. (2014).

5. Anderson and Adolphs (2014); Berridge and Winkielman (2003); Broom (2001); Critchley and Harrison (2013); Damasio (2010), chap. 5; Denton et al. (2009); Feinstein (2013); Guillory and Bujarski (2014); Plutchik (2001); Rolls (2014, 2015); Seth (2013). See also http://en.wikipedia.org/wiki/Somatic_marker_hypothesis.

6. Many references for this section are given at the end of table 7.1. For more on interoception in general, see Hall (2011).

7. Rolls (2014).

8. Almeida, Roizenblatt, and Tufik (2004); Damasio, Damasio, and Tranel (2012); Farb, Segal, and Anderson (2013); Mouton and Holstege (2000); Saper (2002); Treede et al. (1999); Veinante, Yalcin, and Barrot (2013).

9. Mancini et al. (2012).

10. Craig (2003a, 2003c, 2008).

11. Giordano (2005).

12. The literature on interoceptive and pain pathways is referenced in table 7.2, in mammals and other vertebrates.

13. Saper (2002).

14. Humphries, Gurney, and Prescott (2007), p. 1632; Mouton and Holstege (2000); Veinante, Yalcin, and Barrot (2013).

15. Final loss of interoceptive isomorphism in the anterior cingulate and orbitomedial prefrontal cortex is documented by Craig (2010); Farb, Segal, and Anderson (2013); Treede et al. (1999).

16. Butler and Hodos (2005), p. 553.

17. Ibid., pp. 190–194.

18. For the trigeminal path and spinal trigeminal nucleus occurring in all vertebrates: Butler and Hodos (2005); Butler (2009b); in humans: DaSilva et al. (2002); in lizards: Desfilis, Font, and García-Verdugo (1998); in teleost tilapia: Xue et al. (2006a); in teleost carp: Xue et al. (2006b); in sharks: Anadón et al. (2000); in lampreys: Koyama et al. (1987).

19. For the nucleus of the solitary tract and the reticular formation in various vertebrate groups, see Nieuwenhuys, ten Donkelaar, and Nicholson (1998). For the parabrachial nucleus in various vertebrates, see Mueller, Vernier, and Wullimann (2004); Barreiro-Iglesias, Anadón, and Rodicio (2010); Neary (1995); Wild, Arends, and Ziegler (1990).

20. Nieuwenhuys (1996).

21. Plutchik (2001); Feinstein (2013).

22. The term "internal milieu" is from physiologist Claude Bernard, and refers to the body's interior and to keeping it stable (maintaining homeostasis): see http://en.wikipedia.org/wiki/Milieu_interieur.

23. James (1884); Lange (1885). See also Critchley and Harrison (2013); Damasio, Everitt, and Bishop (1996); Damasio (2010); Norman, Berntson, and Cacioppo (2014); Seth (2013).

24. Cannon (1927). See also Anderson and Adolphs (2014); Denton (2006), p. 112; and Quigley and Barrett (2014). Neuroscientist and philosopher Georg Northoff (2012a) reviewed the evidence, mainly from human studies, and concluded that exteroceptive and interoceptive inputs come together in the brain and balance each other to produce the affects. He called this "relational coding" as opposed to the James–Lange concept of "translational coding" (feelings arising only after translation from physiological change). Thus, Northoff supports View 2 and a more direct role for exteroception in emotional feeling.

25. Critchley and Harrison (2013), p. 632–634; Damasio et al. (2012).

26. For studies that discuss this issue, present the evidence, and agree with us that both views pertain, see Anderson and Adolphs (2014); Dalgleish, Dunn, and Mobbs (2009); Fotopoulou (2013); Montoya and Schandry (1994); Waraczynski (2009).

27. Craig (2010). See also Craig (2003c).

28. The term "material me" originally came from Sherrington (1900).

29. See http://en.wikipedia.org/wiki/Somatic_marker_hypothesis.

30. Butler, Reiner, and Karten (2011).

31. Benton and Donoghue (2007).

32. Damasio, Damasio, and Tranel (2012); Feinstein (2013).

33. Guillory and Bujarski (2014).

34. Vierck et al. (2013).

35. This cortical projection corrects some errors in older, classical ideas that had also claimed SI produces pain affect (see Craig, 2010, p. 564). See also Mancini et al. (2012).

36. Benton and Donoghue (2007).

37. Denton (2006); Denton et al. (2009).

38. Denton et al. (2009), pp. 503–504.

39. Ibid., p. 502.

40. Denton (2006), p. 114.

41. See Panksepp (1998, 2005, 2011, 2013), and his many other publications. Panksepp has a somewhat more motor-based view of affects than we do, claiming that feelings come as much from actions as from sensory inputs (see Panksepp, 2005, p. 39,

and the discussion in Northoff, 2012a, p. 12). In contrast, we have a sensory-centered view of affects as belonging to sensory consciousness, although we certainly admit that affects cause actions, so the two are related.

42. Panksepp (2013), pp. 63–64.

43. Watt (2005).

44. For Panksepp's views on behaviorism see Panksepp (2011). Griffin (2001) also argues against behaviorism. The paper in which Panksepp assigned the date of 560–520 mya to the origin of affective consciousness is Fabbro et al. (2015).

45. Packard and Delafield-Butt (2014).

46. Solms and Panksepp (2012); Solms (2013a).

47. This view is discussed by Barrett et al. (2007); Berlin (2013); Critchley and Harrison (2013); Gallese (2013); Rolls (2014, 2015); Seth (2013); Turnbull (2013).

48. Panksepp (1998).

49. Solms and Panksepp (2012), after Huston and Borbely (1973).

50. Turnbull (2013).

51. Denton et al. (2009).

52. Merker (2007); Damasio (2010, pp. 85–88). Articles by the many authors who accept this evidence and accept Solm's ideas can be found in the journal *Neuropsychoanalysis 15*(1) (2013). Specifically, see Fotopoulou (2013); Friston (2013); Gallese (2013); Kessler (2013); Tsakiris (2013); Turnbull (2013). See also Kittelsen (2013) and Vargas, López, and Portavella (2009).

53. Merker (2007), p. 79.

54. For additional evidence that the cerebral cortex of mammals has only a secondary role of modulating the subcortically generated primal emotions, see Berridge and Kringelbach (2013); Dalgleish, Dunn, and Mobbs (2009); Damasio, Damasio, and Tranel (2012); Quirk and Beer (2006).

55. Cabanac (1996); Cabanac, Cabanac, and Parent (2009).

56. For similar explanations of the adaptive value of affects, see Damasio and Carvalho (2013); Gallese (2013); Giske et al. (2013); Ohira (2010). Schultz (2015, pp. 857–859) similarly emphasizes how rewards function to increase survival and reproductive success.

57. Cabanac (1996).

58. Cabanac, Cabanac, and Parent (2009).

59. For a review of LeDoux's studies on the amygdala, see LeDoux (2007). For his doubts about whether animals have real conscious feelings or emotions, see LeDoux (2012, 2014).

60. LeDoux (2014), p. 2873.

61. LeDoux (2012), p. 656.

62. Eriksson et al. (2012).

63. For the studies that center on the exteroceptive rather than the affective aspect of consciousness, see the dozens presented in chapter 6. See also Crick and Koch (2003); Dehaene et al. (2014); Tononi and Koch (2008, 2015).

8 Finding Sentience

1. Sources that consider pain to be the prototypical and representative emotion include Casimir (2009); Key (2014); Levine (1983); Watt (2005).

2. For more on the definition of pain, see http://www.iasp-pain.org/Taxonomy.

3. See Committee on Recognition and Alleviation of Pain in Laboratory Animals, National Research Council (2009).

4. Sherrington (1906).

5. See Rolls (2000); Winkielman, Berridge, and Wilbarger (2005).

6. Ruppert, Fox, and Barnes (2004).

7. Craig (2003a, 2003b); Vierck et al. (2013).

8. Sivarao et al. (2007).

9. Grau, Barstow, and Joynes (1998); Sherrington (1906).

10. Berridge and Winkielman (2003); Woods (1964).

11. Rose and Woodbury (2008); Rose et al. (2014).

12. Matthies and Franklin (1992).

13. For the definition of operant conditioning, see Brembs (2003a, 2003b); Perry, Barron, and Cheng (2013); https://en.wikipedia.org/wiki/Operant_conditioning.

14. Vierck and Charles (2006) argue that pain studies must use operant tests, not simple-reflex tests; see also Cloninger and Gilligan (1987), p. 466.

15. Rolls (2014).

16. Perry, Barron, and Cheng (2013), table 1.

17. Martin, Kim, and Eisenach (2006).

18. Colpaert et al. (1980); Danbury et al. (2000).

19. Braithwaite (2014); Rose et al. (2014); Sneddon (2011).

20. References that document the criteria for affect that we excluded are given at the end of table 8.1.

Notes

21. The references that support the criteria we did include are listed at the end of table 8.2.

22. Vierck and Charles (2006), p. 181.

23. Mogil (2009); Walters (1994).

24. Anderson and Adolphs (2014); Cloninger and Gilligan (1987); Kittilsen (2013).

25. We realize that this approach has a potential problem, in that it relies on the absence of evidence to draw some of its conclusions. That is, future studies could discover that some of the "lower" animals meet these behavioral criteria after all. Our way of dealing with the problem was to find and include as many studies as possible, and to admit that some of the behaviors we now call "absent" may change to "present" at a later date. The problem is pointed out in http://plato.stanford.edu/entries/consciousness-animal/. (See sec. 6.4 of that essay.)

26. Ruppert, Fox, and Barnes (2004).

27. The definition of behavioral contrast is from Papini (2002).

28. For the structures in tables 8.4 and 8.5 being limbic structures, see Nieuwenhuys (1996); Heimer and Van Hoesen (2006). For those workers who have linked these structures to affects in amniotes and anamniotes, see O'Connell and Hofmann (2011); Panksepp (1998); Watt (2005). See also Damasio (2010).

29. Feinstein (2013); Goodson and Kingsbury (2013); Kender et al. (2008); O'Connell and Hofmann (2011).

30. O'Connell and Hofmann (2011).

31. Solms (2013a, 2013b); Panksepp (1998, 2011); Damasio, Damasio, and Tranel (2012); Denton et al. (2009).

32. For the claim that cerebral cortex modulates emotions but does not generate them, see Berridge and Kringelbach (2013); Dalgleish, Dunn, and Mobbs (2009); Damasio, Damasio, and Tranel (2012); Solms and Panksepp (2012).

33. There may be another way to test for a progressive evolution of the affective structures in vertebrates. All the limbic structures in tables 8.4 and 8.5 contribute to affects, and it has been difficult for researchers to assign positive affects to one particular structure and negative affects to another. But Berridge and Kringelbach (2013) have found "hedonic hotspots" where pleasure, dread, and fear affects are generated in the rat brain. These hotspots are the nucleus accumbens and ventral pallidum for pleasure and dread, and the amygdala for fear and reward (for more on the functions of these structures, see Heimer and Van Hoesen, 2006; Nelson, Lau, & Jarco, 2014; Janek & Tye, 2015). Also especially important are the habenula for coding negative reward, and the ventral tegmental area (or posterior tuberculum), whose neurons release dopamine, which is associated with reward (Barron, Søvik, & Cornish, 2010; Hikosaka, 2010; Kobayashi, 2012; Lammel et al. 2012; Nargeot and Simmers, 2011; Rolls, 2014; Schultz, 2015). Of these five especially important "keystone" structures, tetrapods have all five,

but lampreys and sharks have only two with certainty (ventral tegmental area and habenula) and two "maybes" (amygdala and vental pallidum). Thus, this suggests a progressive increase in important affective structures, from fish to tetrapods.

34. For studies proposing and developing the idea of a social behavioral network, our adaptive behavioral network, see Newman (1999); O'Connell and Hofmann (2011); Goodson and Kingsbury (2013).

35. O'Connell and Hofmann (2011).

36. Like us, Packard and Delefield-Butt (2014) also conclude that core affects evolved in the prevertebrates or very first vertebrates.

37. Almeida, Roizenblatt, and Tufik (2004); McHaffie, Kao, and Stein (1989).

38. See studies cited in Saidel (2009); Northmore (2011).

39. See Rose et al. (2014); Rose (2002). See also Key (2014).

40. Snow, Plenderleith, and Wright (1993) document the absence of C fibers in cartilaginous fish.

41. Sneddon (2002) records the few C fibers in teleost fish.

42. Rose et al. (2014), p. 112.

43. Guo et al. (2003).

44. Rose et al. (2014).

45. Sneddon, Braithwaite, and Gentle (2003); Reilly et al. (2008).

46. Rose et al. (2014); see also Bolles and Faneslow (1980).

47. Crook and Walters (2014).

48. Casimir (2009); Key (2014); Levine (1983); Watt (2005).

49. Damasio (2010), p. 27.

50. Candiani et al. (2012); Reaume and Sokolowski (2011).

51. Packard and Delafield-Butt (2014); Shen (2015).

52. Cabanac (1996); Ohira (2010); Gallese (2013); Giske et al. (2013); Packard and Delafield-Butt (2014); http://en.wikipedia.org/wiki/Somatic_marker_hypothesis.

53. Solms and Panksepp (2012).

54. Schultz (2015); pp. 853–860.

55. After Panksepp, Solms, and Denton: see chapter 7.

Notes 279

9 Does Consciousness Need a Backbone?

1. Many of the references for this chapter are listed in the appendix, but others will be commented upon here in the endnotes.

2. Griffin (2001), chap. 1; http://www.animal-ethics.org/what-beings-are-conscious/.

3. See Ruppert, Fox, and Barnes (2004); Wägele and Bartolomaeus (2014).

4. Ruppert, Fox, and Barnes (2004), chaps. 12, 16–21.

5. See Elwood (2011) for some references on nociceptors and nociception in invertebrates. We are more skeptical about invertebrate pain than he is, however.

6. Brodal (2010); Saidel (2009).

7. Chittka and Niven (2009); Farris (2012); Giurfa (2013); Perry, Barron, and Cheng (2013); Strausfeld (2013). Many of these sources also acknowledge the small size of insect brains and how that may limit cognition.

8. Figure 9.4 is based on the vertebrate graph of Wullimann and Vernier (2009a), with the arthropod values taken from Chittka and Niven (2009), Eberhard and Wcislo (2011), and Martin Heisenberg (personal communication on the fruit fly). Cuttlefish and octopus values are from Packard (1972).

9. In this we agree with Plotnick, Dornbos, and Chen (2010); Sherrington (1947); Trestman (2013).

10. Strausfeld (2013) gives the most complete description of insect and arthropod senses. For more on arthropods' high-resolution vision, see Nilsson (2013) and the references in that paper.

11. Land and Fernald (1992); Nilsson (2013).

12. Strausfeld (2013); Newland et al. (2000); Walters et al. (2001).

13. Seelig and Jayaraman (2013).

14. For more on the central complex, see Strausfeld (2013), chap. 7. New information shows it is also for orienting the body toward visual landmarks, like an internal compass: Seelig and Jayaraman (2015).

15. On arthropod retinotopy: see Harzsch (2002, p. 14) for crustaceans and for the chelicerate horseshoe crab; Strausfeld (2009) for crustaceans; Sombke and Harsch (2015) for a centipede; and Selig and Jayaraman (2013) for insects.

16. Chittka and Niven (2009).

17. Ibid. For studies that found places of convergence of different senses in insect brains, see Strausfeld (2013), pp. 211, 679.

18. Giurfa (2013); Tang and Juusola (2010).

19. See Chittka and Niven (2009); Chittka and Skorupski (2011).

20. Strausfeld (2013), p. 291.

21. Chittka and Niven (2009), p. 2006.

22. Fauria, Colborn, and Collett (2000).

23. Giurfa et al. (2001) and Gould and Gould (1986) have carried out additional studies of bees, with similar implications for mental images.

24. Strausfeld (2013).

25. Cong et al. (2014); Ma et al. (2012, 2014); Ortega-Hernandez (2015).

26. For accounts of arthropod evolution since the Cambrian, see Edgecombe (2010) and Davis, Baldauf, and Mayhew (2010).

27. For information on arthropod molting, see http://evolution.berkeley.edu/evolibrary/article/0_0_0/mantisshrimp_05.

28. For descriptions of the most-studied behaviors of gastropods, see Brembs (2003a, 2003b); Hirayama and Gillette (2012); Hirayama et al. (2012); Jing and Gillette (2000); Noboa and Gillette (2013).

29. For more on the cerebral ganglia of snails and slugs, see http://neuronbank.org/wiki/index.php/Aplysia_cerebral_ganglia.

30. Baxter and Byrne (2006); see also http://neuronbank.org/wiki/index.php/Aplysia_cerebral_ganglia.

31. For more on gastropod eyes, see Nilsson (2013); Ruppert, Fox, and Barnes (2004); Sweeney, Haddock, and Johnsen (2007).

32. Walters et al. (2004); Xin, Weiss, and Kupfermann (1995).

33. For reviews of the complex behaviors of cephalopods, see Mather (2008, 2012), and Mather and Kuba (2013). See also Montgomery (2015). For genetic evidence from the genome of an octopus that the brain and nervous system are remarkably complex, see Albertin et al. (2015).

34. For cephalopod eyes and how they compare to vertebrate eyes, see Nilsson (2013); Ruppert, Fox, and Barnes (2004); Sweeney, Haddock, and Johnsen (2007).

35. For the visual pathway in cephalopods, see Young (1974); Shigeno and Ragsdale (2015); and Williamson and Chrachri (2004). For the tactile pathway see Grasso (2014) and Shigeno and Ragsdale (2015). For the other senses, see the references in table 9.2. Other good studies on cephalopod brains are by Young (1971) and Wollesen, Loesel, and Wanninger (2009).

36. Godfrey-Smith (2013); Grasso (2014); Hochner (2012, 2013); Zullo et al. (2009).

37. For a detailed description of the cephalopod arm-nervous system, see Grasso (2014).

38. Zullo and Hochner (2011); Zullo et al. (2009). Only one study looked for somatotopy in a nonmotor part of the brain: Budelmann and Young (1985).

39. Young (1974).

40. Grasso (2014), p. 111.

41. Gutnick et al. (2011).

42. For figures showing this reciprocal cross-talk, see Williamson and Chrachri (2004).

43. See http://en.wikipedia.org/wiki/Nautilus.

44. Kröger, Vinther, and Fuchs (2011). This tells the dates of first appearance of the cephalopods and the cephalopod subclades, as determined from the fossil record.

45. The classic paper that compares cephalopods with fish and describes the remarkable similarities between these two animal groups is Packard (1972). This paper also compares the adaptive advantages, disadvantages, and evolutionary successes of the two groups.

46. Mather and Kuba (2013).

10 Neurobiological Naturalism: A Consilience

1. These features of neurobiological naturalism were introduced in Feinberg (2012). See also Feinberg and Mallatt (2016a).

2. Thompson (2007), p. 236.

3. For emergent properties in the building of anthills and other engineering in social insects, see Bilchev and Parmee (1995).

4. Jonkisz (2015) has also worked to explain consciousness by identifying its essential ingredients. However, his ingredients, of "individuated information in action," do not require a nervous system and they apply to any living system with actions, including an animal that acts entirely by reflexes, and even a bacterium. In our view, consciousness also requires the nervous elements of complex neurohierarchies.

5. That said, recall that gastropods, whose nervous system may be only slightly more complex than that of the ancestral worms, show some behavioral evidence for affects without exteroceptive consciousness (chapter 9). The references are listed in the appendix under the heading for table 9.1 on p. 242.

6. Kröger, Vinther, and Fuchs (2011).

7. Many of the seven basic emotions proposed by Panksepp (1998, 2011) could have been completed here, or else back in the first amniotes: seeking, rage, lust, fear, care, grief, and play.

8. Eriksson et al. (2012). For a consideration of the leap in consciousness in humans, see LeDoux (2012), pp. 665–666.

9. Gurdasani et al. (2015).

10. Fabbro et al. (2015), table 1.

11. For the many studies by the advocates of cortical consciousness, see chapters 6–8; see also Key (2014) and Tononi and Koch (2008). Concerning humans who lose visual consciousness from cortical lesions, such people can retain some visual abilities even though they do not consciously see, a condition called *blindsight*. If forced to guess, they can detect, locate, and distinguish between some objects with better-than-chance accuracy. One interpretation of blindsight is that it arises from the undamaged tectum/superior colliculus. See http://kin450-neurophysiology.wikispaces.com/Blindsight; Alexander and Cowey (2010); Cowey (2010); Overgaard (2012).

12. The studies of Redouan Bshary and his colleagues demonstrate fish performing advanced behaviors that require sensing many subtle signals, from both social and environmental stimuli. See Abbott (2015) and Bshary and Grutter (2006).

13. For the definitions and descriptions of these phenomena, see Baars (2002); Baars, Franklin, and Ramsoy (2013); Brogaard (2011); Ciaramelli et al. (2010); Kiefer et al. (2011); Kouider and Dehaene (2007); Näätänen et al. (2007); Overgaard (2012); Panagiotaropoulos et al. (2012); van Gaal and Lamme (2012).

14. This point, about the inadequacy of non-conscious perception for survival, is also made in the essay on animal consciousness at http://plato.stanford.edu/entries/consciousness-animal/ (p. 23). That essay also recognizes the diversity of consciousness (sec. 6.6).

15. Kandel et al. (2012); Johnson et al. (2007). These sources also document all the information we present on proprioception.

15. Johnson et al. (2007).

16. Merker (2005).

17. Tononi and Koch (2015), p. 10. See also Baars, Franklin, and Ramsoy (2013), pp. 6, 8, 20.

18. Butler (2008a), p. 442.

19. Gottfried (2010); Mori et al. (2013). But see Courtiol and Wilson (2014) for the possibility that the thalamic relay might participate in smell perception after all.

20. For the gamma oscillations in the olfactory pathway, see Mori et al. (2013).

21. Barrett et al. (2007); Craig (2010), p. 568; Heimer and Van Hoesen (2006); LeDoux (2012).

22. Olfactory to limbic: Butler and Hodos (2005); Heimer and Van Hoesen (2006). Hippocampus to limbic: O'Connell and Hoffman (2011). Another exteroceptive-to-affective link has come to light: Shang et al. (2015) found a pathway from the superior

Notes 283

colliculus through the isthmus nuclei to the amygdala, so that threatening objects in the visual field induce affective fear. This link was only found in mice, however, and it would be interesting if it also exists in fish.

23. Brudzynski (2007, 2014); Lammel et al. (2012).

24. LDN in all vertebrates: Rodríguez-Moldes et al. (2002). LDN in lampreys: Ryczko et al. (2013).

25. Still another common feature of exteroceptive and affective consciousness could be the involvement in both of gamma oscillations. That is, affects might also stem from oscillations. For example, gamma (and theta) synchronization has been measured during emotional memories of fear and safety in rats, as the oscillations run between the amygdala, hippocampus, and prefrontal cortex (Bocchio & Capogna, 2014; Stujenske et al., 2014). Also, Marco-Pallares, Münte, and Rodríguez-Fornells (2015, p. 1) state that "a number of recent studies have revealed an increase of beta-gamma activity (20–35 Hz) after unexpected or relevant positive reward outcomes" (positive affect) in humans and rats. Further study is needed to determine whether all types of affects always have the oscillations, and if they really are similar to the gamma oscillations associated with exteroceptive consciousness (chapter 6). Especially needed are studies on nonmammalian vertebrates.

26. Dopamine genes: Barron, Søvik, and Cornish (2010); Perry and Barron (2013); Waddel (2013). Dopamine function: Rutledge et al. (2015); Schultz (2015).

27. For more discussion of the self in consciousness, see Bekoff and Sherman (2004); Gazzaniga (1997); Griffin (2001); Keenan, Gallup, and Falk (2003); Keenan et al. (2005); Revonsuo (2010); Seth (2013).

28. Damasio (2000).

29. Tulving (1985), p. 3.

30. Tulving (1985, 2002a, 2002b).

31. Allen and Fortin (2013); see also Forster et al. (2005) regarding lizards.

32. Damasio (2010), p. 27; Fabbro et al. (2015), table 1; Pepperberg (2009); Plotnik, De Waal, and Reiss (2006).

33. James (1879); Damasio (2010); Denton (2006); Edelman (1992); Griffin (2001), chap. 1.

34. Block (1995); Bradley (2011); Jackson (1982); Lindahl (1997); Nichols and Grantham (2000); Robinson (2007); Tsou (2013).

35. Nichols and Grantham (2000).

36. Bradley (2011).

37. This correlation is best shown by the fact that damage to specific brain regions interferes with specific perceptions, or relates to "specific psychopathologies": Nichols and Grantham (2000), pp. 650–652. See also Lycan (2009).

38. See http://en.wikipedia.org/wiki/Correlation_does_not_imply_causation.

39. Feinberg and Mallatt (2016a). For more on the idea that mental images must simulate reality well in order to allow survival, see Edwards (2014). Damasio (2010, p. 76) also argues that mental images reflect reality. For a related, information-based argument for why consciousness is adaptive and reflects the environment, see Tononi and Koch (2015), p. 10.

40. For the sharpened nonvisual senses of blind cave animals, see Schlegel et al. (2009).

41. Ruppert, Fox, and Barnes (2004); but see Sombke and Harzsch (2015) for retinotopy in a centipede brain.

42. For a description of the biology and life cycle of rhizocephalans see Ruppert, Fox, and Barnes (2004), p. 685.

43. Before leaving the topic of consciousness as an adaptation, we could consider whether consciousness must be full-blown in order to aid survival. This is the topic of partial consciousness. We have concluded that consciousness evolved relatively fast in vertebrates, and perhaps in arthropods, as part of the Cambrian explosion. But that is only fast in the geological terms of Earth history, and it still could have taken millions of years. This could mean that consciousness evolved slowly enough for the intermediate animals to have been "partially conscious." At first thought, the idea of partial consciousness sounds absurd, like being "half pregnant." But closer consideration says it must exist. In every human life cycle, we start as one fertilized egg cell that must be unconscious, and we are conscious at birth, so there must be a partially conscious stage in between. If partial consciousness is thus possible, what form would it have taken in the adult prevertebrates that were not yet fully conscious? Were their experiences only dimly perceived, or flickering in and out of existence? We propose that these transitional animals were unconscious and "reflexive" most of the time, but they temporarily became conscious whenever it was important to attend to certain stimuli for survival. Upon arousal, both feeling states and mental sensory images were evoked and directed by the brain's attention mechanisms. Then gradually, as the prevertebrate line became more active and more often exposed to danger, consciousness became more continuous (except when the animal was protected and dormant, or in a sleeplike state). At first, consciousness offered its adaptive benefits only when needed, but when not needed it went on idle. For more on the possibility of partial consciousness, see section 4.7 in http://plato.stanford.edu/entries/consciousness-animal/.

44. Globus (1973), p. 1129.

45. Feinberg (2012).

46. Georg Northoff called the brain's inability to reference its own neuronal states as the "autoepistemic limitation of our own brain" (Northoff & Musholt, 2006; Northoff, 2012b). He attributes this inaccessibility to some built-in limitation of the brain's neural code (he defines "neural code" as the way in which the brain's intrinsic activity of brainwaves interacts with the received sensory stimuli to produce conscious qualia).

Notes

By contrast, we just say that the accessibility never evolved because it has no adaptive value and would waste energy. In summary, Northoff proposed an active and mechanical reason, whereas we give a passive and evolutionary reason, for auto-ontological irreducibility.

47. Feigl (1967), n. 16. Even a real sort of autocerebroscope, which is in its earliest stages of testing as a flexible electronic-recording net that is injected apparently without harm into the brain, would not solve the problem of auto-ontological irreducibility because the neuronal firing would be objectively observed but remain subjectively experienced. See Kim and Lee (2015).

48. Merker (2005, pp. 94, 99) makes this same point.

49. Feinberg (2012), p. 27.

50. Ibid., p. 29.

51. Mayr (2004), chap. 3.

52. Chalmers (1996), p. 5.

53. Crick (1995), p. 3.

References

Abbott, A. (2015). Clever fish. *Nature*, *521*, 413–414.

Abitua, P. B., Gainous, T. B., Kaczmarczyk, A. N., Winchell, C. J., Hudson, C., Kamata, K., … Levine, M. (2015). The pre-vertebrate origins of neurogenic placodes. *Nature*, *524*, 462–465.

Abitua, P. B., Wagner, E., Navarrete, I. A., & Levine, M. (2012). Identification of a rudimentary neural crest in a non-vertebrate chordate. *Nature*, *492*(7427), 104–107.

Aboitiz, F. (1992). The evolutionary origin of the mammalian cerebral cortex. *Biological Research*, *25*(1), 41–49.

Aboitiz, F., & Montiel, J. (2007). Origin and evolution of the vertebrate telencephalon, with special reference to the mammalian neocortex. *Advances in Anatomy, Embryology, and Cell Biology*, *193*, 1–116.

Abramson, C. I., & Feinman, R. D. (1990). Lever-press conditioning in the crab. *Physiology & Behavior*, *48*(2), 267–272.

Adan, D. (2005). Scientists say lobsters feel no pain. *Guardian*, Tuesday, Feb. 8. http://www.theguardian.com/world/2005/feb/08/research.highereducation.

Agata, K., Soejima, Y., Kato, K., Kobayashi, C., Umesono, Y., & Watanabe, K. (1998). Structure of the planarian central nervous system (CNS) revealed by neuronal cell markers. *Zoological Science*, *15*(3), 433–440.

Ahl, V., & Allen, T. F. H. (1996). *Hierarchy theory*. New York: Columbia University Press.

Aiello, L. C., & Wheeler, P. (1995). The expensive-tissue hypothesis: The brain and the digestive system in human and primate evolution. *Current Anthropology*, *36*, 199–221.

Albertin, C. B., Simakov, O., Mitros, T., Wang, Z. Y., Pungor, J. R., Edsinger-Gonzales, E., … Rokhsar, D. S. (2015). The octopus genome and the evolution of cephalopod neural and morphological novelties. *Nature*, *524*(7564), 220–224.

Alexander, I., & Cowey, A. (2010). Edges, colour, and awareness in blindsight. *Consciousness and Cognition*, *19*(2), 520–533.

Allen, C., & Bekoff, M. (2010). Animal consciousness. In M. Velmans & S. Schneider (Eds.), *The Blackwell companion to consciousness*. Malden, MA: Blackwell.

Allen, C., Grau, J. W., & Meagher, M. W. (2009). The lower bounds of cognition: What do spinal cords reveal? In J. Bickle (Ed.), *The Oxford handbook of philosophy of neuroscience* (pp. 129–142). Oxford: Oxford University Press.

Allen, T. A., & Fortin, N. J. (2013). The evolution of episodic memory. *Proceedings of the National Academy of Sciences of the United States of America, 110*(Suppl. 2), 10379–10386.

Allen, T. F. H., & Starr, T. B. (1982). *Hierarchy: Perspectives for ecological complexity*. Chicago: University of Chicago Press.

Almeida, T. F., Roizenblatt, S., & Tufik, S. (2004). Afferent pain pathways: A neuroanatomical review. *Brain Research, 1000*(1), 40–56.

Alonso, P. D., Milner, A. C., Ketcham, R. A., Cookson, M. J., & Rowe, T. B. (2004). The avian nature of the brain and inner ear of *Archaeopteryx*. *Nature, 430*(7000), 666–669.

Anadón, R., Molist, P., Rodríguez-Moldes, I., López, J. M., Quintela, I., Cerviño, M. C., ... Gonzalez, A. (2000). Distribution of choline acetyltransferase immunoreactivity in the brain of an elasmobranch, the lesser spotted dogfish (*Scyliorhinus canicula*). *Journal of Comparative Neurology, 420*(2), 139–170.

Anderson, D. J., & Adolphs, R. (2014). A framework for studying emotions across species. *Cell, 157*(1), 187–200.

Anderson, J. S., Reisz, R. R., Scott, D., Fröbisch, N. B., & Sumida, S. S. (2008). A stem batrachian from the Early Permian of Texas and the origin of frogs and salamanders. *Nature, 453*(7194), 515–518.

Anderson, R. C., & Mather, J. A. (2007). The packaging problem: Bivalve prey selection and prey entry techniques of the octopus *Enteroctopus dofleini*. *Journal of Comparative Psychology, 121*(3), 300.

Andrews, P. L., Darmaillacq, A. S., Dennison, N., Gleadall, I. G., Hawkins, P., Messenger, J. B., ... Smith, J. A. (2013). The identification and management of pain, suffering, and distress in cephalopods, including anaesthesia, analgesia, and humane killing. *Journal of Experimental Marine Biology and Ecology, 447*, 46–64.

Anton, S., Evengaard, K., Barrozo, R. B., Anderson, P., & Skals, N. (2011). Brief predator sound exposure elicits behavioral and neuronal long-term sensitization in the olfactory system of an insect. *Proceedings of the National Academy of Sciences of the United States of America, 108*(8), 3401–3405.

Appel, M., & Elwood, R. W. (2009). Motivational trade-offs and potential pain experience in hermit crabs. *Applied Animal Behaviour Science, 119*(1), 120–124.

Århem, P., Lindahl, B. I. B., Manger, P. R., & Butler, A. B. (2008). On the origin of consciousness—some amniote scenarios. In H. Liljenstrom & P. Arhem (Eds.), *Conscious-*

ness transitions: Phylogenetic, ontogenetic, and physiological aspects (pp. 77–96). San Francisco, CA: Elsevier.

Azrin, N. H., Hutchinson, R. R., & Hake, D. F. (1966). Extinction-induced aggression 1. *Journal of the Experimental Analysis of Behavior*, 9(3), 191–204.

Baars, B. J. (1988). *A cognitive theory of consciousness*. New York: Cambridge University Press.

Baars, B. J. (2002). The conscious access hypothesis: Origins and recent evidence. *Trends in Cognitive Sciences*, 6(1), 47–52.

Baars, B. J., Franklin, S., & Ramsoy, T. Z. (2013). Global workspace dynamics: Cortical "binding and propagation" enables conscious contents. *Frontiers in Psychology*, 4(200).

Baars, B. J., & Galge, N. M. (Eds.). (2010). *Cognition, brain, and consciousness: An introduction to cognitive neuroscience* (2nd ed.). San Diego, CA: Academic Press (Elsevier).

Bailey, A. P., Bhattacharyya, S., Bronner-Fraser, M., & Streit, A. (2006). Lens specification is the ground state of all sensory placodes, from which FGF promotes olfactory identity. *Developmental Cell*, 11(4), 505–517.

Bailly, X., Reichert, H., & Hartenstein, V. (2013). The urbilaterian brain revisited: Novel insights into old questions from new flatworm clades. *Development Genes and Evolution*, 223(3), 149–157.

Baker, C. V., Modrell, M. S., & Gillis, J. A. (2013). The evolution and development of vertebrate lateral line electroreceptors. *Journal of Experimental Biology*, 216(13), 2515–2522.

Balaban, P. M., & Maksimova, O. A. (1993). Positive and negative brain zones in the snail. *European Journal of Neuroscience*, 5(6), 768–774.

Balanoff, A. M., Bever, G. S., Rowe, T. B., & Norell, M. A. (2013). Evolutionary origins of the avian brain. *Nature*, 501(7465), 93–96.

Balasko, M., & Cabanac, M. (1998). Behavior of juvenile lizards (*Iguana iguana*) in a conflict between temperature regulation and palatable food. *Brain, Behavior and Evolution*, 52(6), 257–262.

Balcombe, J. (2009). Animal pleasure and its moral significance. *Applied Animal Behaviour Science*, 118(3), 208–216.

Balment, R. J., Lu, W., Weybourne, E., & Warne, J. M. (2006). Arginine vasotocin a key hormone in fish physiology and behaviour: A review with insights from mammalian models. *General and Comparative Endocrinology*, 147(1), 9–16.

Bambach, R. K., Bush, A. M., & Erwin, D. H. (2007). Autecology and the filling of ecospace: Key metazoan radiations. *Palaeontology*, 50(1), 1–22.

Barlow, H. (1995). The neuron doctrine in perception. In M. S. Gazzaniga (Ed.), *The cognitive neurosciences* (pp. 417–435). Cambridge, MA: MIT Press.

Barreiro-Iglesias, A., Anadón, R., & Rodicio, M. C. (2010). The gustatory system of lampreys. *Brain, Behavior and Evolution, 75*(4), 241–250.

Barrett, L. F., Mesquita, B., Ochsner, K. N., & Gross, J. J. (2007). The experience of emotion. *Annual Review of Psychology, 58*, 373.

Barron, A. B., Søvik, E., & Cornish, J. L. (2010). The roles of dopamine and related compounds in reward-seeking behavior across animal phyla. *Frontiers in Behavioral Neuroscience, 4*.

Baxter, D. A., & Byrne, J. H. (2006). Feeding behavior of *Aplysia*: A model system for comparing cellular mechanisms of classical and operant conditioning. *Learning & Memory, 13*(6), 669–680.

Bayne, T. (2010). *The unity of consciousness*. Oxford: Oxford University Press.

Bayne, T., & Chalmers, D. J. (2003). What is the unity of consciousness? In A. Cleeremans (Ed.), *The unity of consciousness: Binding, integration, and dissociation* (pp. 23–58). New York: Oxford University Press.

Beckermann, A., Flohr, H., & Kim, J. (1992). *Emergence or reduction? Prospects for nonreductive physicalism*. Berlin: W. de Gruyter.

Beckers, G., & Zeki, S. (1995). The consequences of inactivating areas V1 and V5 on visual motion perception. *Brain, 118*(1), 49–60.

Bedau, M. A. (1997). Weak emergence. *Noûs, 31*(s11), 375–399.

Bekoff, M., & Sherman, P. W. (2004). Reflections on animal selves. *Trends in Ecology & Evolution, 19*(4), 176–180.

Bennett, A. M., Pereira, D., & Murray, D. L. (2013). Investment into defensive traits by anuran prey (*Lithobates pipiens*) is mediated by the starvation-predation risk trade-off. *PLoS One, 8*(12), e82344.

Benton, M. J. (2014). How birds became birds. *Science, 345*(6196), 508–509.

Benton, M. J., & Donoghue, P. C. (2007). Paleontological evidence to date the tree of life. *Molecular Biology and Evolution, 24*(1), 26–53.

Ben-Tov, M., Donchin, O., Ben-Shahar, O., & Segev, R. (2015). Pop-out in visual search of moving targets in the archer fish. *Nature Communications, 6*(6476), 1–11.

Berlin, H. (2013). The brainstem begs the question: Petitio principii. *Neuro-psychoanalysis, 15*(1), 25–29.

Berridge, K. C., & Kringelbach, M. L. (2013). Neuroscience of affect: Brain mechanisms of pleasure and displeasure. *Current Opinion in Neurobiology, 23*(3), 294–303.

Berridge, K., & Winkielman, P. (2003). What is an unconscious emotion? (The case for unconscious "liking"). *Cognition and Emotion, 17*(2), 181–211.

Berrill, N. J. (1955). *The origin of vertebrates*. Oxford: Clarendon Press.

References

Bi, S., Wang, Y., Guan, J., Sheng, X., & Meng, J. (2014). Three new Jurassic euharamiyidan species reinforce early divergence of mammals. *Nature*, *514*, 579–584.

Bicker, G., Davis, W. J., & Matera, E. M. (1982). Chemoreception and mechanoreception in the gastropod mollusc *Pleurobranchaea californica*. *Journal of Comparative Physiology*, *149*(2), 235–250.

Bilchev, G., & Parmee, I. C. (1995). The ant colony metaphor for searching continuous design spaces. In T. C. Fogerty (Ed.), *Evolutionary computing* (pp. 25–39). Berlin: Springer.

Binder, M. D., Hirokawa, N., & Windhorst, U. (Eds.). (2009). *Encyclopedia of neurosciences*. Berlin: Springer.

Bingman, V. P., Salas, C., & Rodriguez, F. (2009). Evolution of the hippocampus. In M. D. Binder, N. Hirokawa, & U. Windhorst (Eds.), *Encyclopedia of neurosciences* (pp. 1356–1361). Berlin: Springer-Verlag.

Block, N. (1995). How many concepts of consciousness? *Behavioral and Brain Sciences*, *18*(02), 272–287.

Block, N. (2009). Comparing the major theories of consciousness. In M. S. Gazzaniga (Ed.), *The cognitive neurosciences* (Vol. 4, pp. 1111–1122). Cambridge, MA: MIT Press.

Block, N. (2012). The grain of vision and the grain of attention. *Thought: A Journal of Philosophy*, *1*(3), 170–184.

Bocchio, M., & Capogna, M. (2014). Oscillatory substrates of fear and safety. *Neuron*, *83*(4), 753–755.

Bolles, R. C., & Faneslow, M. S. (1980). A perceptual-defensive-recuperative model of fear and pain. *Behavioral and Brain Sciences*, *3*, 291–323.

Boly, M., & Seth, A. K. (2012). Modes and models in disorders of consciousness science. *Archives Italiennes de Biologie*, *150*(2–3), 172–184.

Boly, M., Seth, A. K., Wilke, M., Ingmundson, P., Baars, B., Laureys, S., & Tsuchiya, N. (2013). Consciousness in humans and non-human animals: Recent advances and future directions. *Frontiers in Psychology*, *4*.

Bonaparte, J. F. (2012). Miniaturisation and the origin of mammals. *Historical Biology*, *24*(1), 43–48.

Borszcz, G. S. (2006). Contribution of the ventromedial hypothalamus to generation of the affective dimension of pain. *Pain*, *123*(1), 155–168.

Bosne, S. C. D. M. D. (2010). *The Eph/ephrin gene family in the European amphioxus: An evo-devo approach*. Master's thesis in Biology. University of Lisbon, Department of Animal Biology.

Bowmaker, J. K. (1998). Evolution of colour vision in vertebrates. *Eye*, *12*, 541–547.

Bradley, M. (2011). The causal efficacy of qualia. *Journal of Consciousness Studies, 18*(11–12), 32–44.

Brahic, C. J., & Kelley, D. B. (2003). Vocal circuitry in *Xenopus laevis*: Telencephalon to laryngeal motor neurons. *Journal of Comparative Neurology, 464*(2), 115–130.

Braithwaite, V. A. (2014). Pain perception. In D. H. Evans, J. B. Claiborne, & S. Currie (Eds.), *The physiology of fishes* (pp. 327–344). Boca Raton, FL: CRC Press.

Braun, C. B. (1996). The sensory biology of the living jawless fishes: A phylogenetic assessment. *Brain, Behavior and Evolution, 48*(5), 262–276.

Brembs, B. (2003 a). Operant conditioning in invertebrates. *Current Opinion in Neurobiology, 13*(6), 710–717.

Brembs, B. (2003 b). Operant reward learning in *Aplysia*. *Current Directions in Psychological Science, 12*(6), 218–221.

Brodal, P. (2010). *The central nervous system: Structure and function* (4th ed.). New York: Oxford University Press.

Brogaard, B. (2011). Are there unconscious perceptual processes? *Consciousness and Cognition, 20*(2), 449–463.

Broglio, C., Gomez, A., Duran, E., Ocana, F. M., Jiménez-Moya, F., Rodriguez, F., & Salas, C. (2005). Hallmarks of a common forebrain vertebrate plan: Specialized pallial areas for spatial, temporal, and emotional memory in actinopterygian fish. *Brain Research Bulletin, 66*(4), 277–281.

Broom, D. M. (2001). Evolution of pain. *Vlaams Diergeneeskundig Tijdschrift, 70*(1), 17–21.

Brook, A., & Raymont, P. (2010). The unity of consciousness. In E. N. Zalta (Ed.), *The Stanford encyclopedia of philosophy* (Fall 2010 Edition). http://plato.stanford.edu/archives/fall2010/entries/consciousness-unity.

Brudzynski, S. M. (2007). Ultrasonic calls of rats as indicator variables of negative or positive states: Acetylcholine–dopamine interaction and acoustic coding. *Behavioural Brain Research, 182*(2), 261–273.

Brudzynski, S. M. (2014). The ascending mesolimbic cholinergic system—a specific division of the reticular activating system involved in the initiation of negative emotional states. *Journal of Molecular Neuroscience, 53*(3), 436–445.

Bshary, R., & Grutter, A. S. (2006). Image scoring and cooperation in a cleaner fish mutualism. *Nature, 441*(7096), 975–978.

Budelmann, B. U., & Young, J. Z. (1985). Central pathways of the nerves of the arms and mantle of *Octopus*. *Philosophical Transactions of the Royal Society of London, Series B, Biological Sciences, 310*(1143), 109–122.

Bullock, T. H. (2002). Biology of brain waves: Natural history and evolution of an information-rich sign of activity. In K. Arikan & N. C. Moore (Eds.), *Advances in electrophysiology in clinical practice and research*. Istanbul: Kjellberg.

Bullock, T. H., Moore, J. K., & Fields, R. D. (1984). Evolution of myelin sheaths: Both lamprey and hagfish lack myelin. *Neuroscience Letters, 48*(2), 145–148.

Bürck, M., Friedel, P., Sichert, A. B., Vossen, C., & van Hemmen, J. L. (2010). Optimality in mono-and multisensory map formation. *Biological Cybernetics, 103*(1), 1–20.

Burghardt, G. M. (2005). *The genesis of animal play: Testing the limits*. Cambridge, MA: MIT Press.

Burghardt, G. M. (2013). Environmental enrichment and cognitive complexity in reptiles and amphibians: Concepts, review, and implications for captive populations. *Applied Animal Behaviour Science, 147*(3), 286–298.

Burighel, P., & Cloney, R. A. (1997). Chordata: Ascidiacia. In F. W. Harrison & E. E. Ruppert (Eds.), *Microscopic anatomy of invertebrates, Hemichordata, Chaetognatha, and the invertebrate chordates* (Vol. 15, pp. 221–347). New York: Wiley-Liss.

Buss, L. W. (1987). *The evolution of individuality*. Princeton, NJ: Princeton University Press.

Butler, A. B. (2000). Chordate evolution and the origin of craniates: An old brain in a new head. *Anatomical Record, 261*(3), 111–125.

Butler, A. B. (2006). The serial transformation hypothesis of vertebrate origins: Comment on "The new head hypothesis revisited." *Journal of Experimental Zoology, Part B, Molecular and Developmental Evolution, 306*(5), 419–424.

Butler, A. B. (2008 a). Evolution of brains, cognition, and consciousness. *Brain Research Bulletin, 75*(2), 442–449.

Butler, A. B. (2008 b). Evolution of the thalamus: A morphological and functional review. *Thalamus & Related Systems, 4*(1), 35–58.

Butler, A. B. (2009 a). Evolution of the diencephalon. In M. D. Binder, N. Hirokawa, & U. Windhorst (Eds.), *Encyclopedia of neurosciences* (pp. 1342–1346). Berlin: Springer-Verlag.

Butler, A. B. (2009 b). Evolution of the somatosensory system in nonmammalian vertebrates. In M. D. Binder, N. Hirokawa, & U. Windhorst (Eds.), *Encyclopedia of neurosciences* (pp. 1419–1421). Berlin: Springer-Verlag.

Butler, A. B., & Cotterill, R. M. (2006). Mammalian and avian neuroanatomy and the question of consciousness in birds. *Biological Bulletin, 211*(2), 106–127.

Butler, A. B., & Hodos, W. (2005). *Comparative vertebrate neuroanatomy* (2nd ed.). Hoboken, NJ: Wiley-Interscience.

Butler, A. B., Manger, P. R., Lindahl, B. I. B., & Århem, P. (2005). Evolution of the neural basis of consciousness: A bird–mammal comparison. *BioEssays, 27*(9), 923–936.

Butler, A. B., Reiner, A., & Karten, H. J. (2011). Evolution of the amniote pallium and the origins of mammalian neocortex. *Annals of the New York Academy of Sciences, 1225*(1), 14–27.

Cabanac, M. (1995). Palatability vs. money: Experimental study of a conflict of motivations. *Appetite, 25*(1), 43–49.

Cabanac, M. (1996). On the origin of consciousness, a postulate and its corollary. *Neuroscience and Biobehavioral Reviews, 20*(1), 33–40.

Cabanac, M., Cabanac, A. J., & Parent, A. (2009). The emergence of consciousness in phylogeny. *Behavioural Brain Research, 198*(2), 267–272.

Cabanac, M., & Johnson, K. G. (1983). Analysis of a conflict between palatability and cold exposure in rats. *Physiology & Behavior, 31*(2), 249–253.

Calabrò, R. S., Cacciola, A., Bramanti, P., & Milardi, D. (2015). Neural correlates of consciousness: What we know and what we have to learn! *Neurological Sciences, 36*(4), 505–513.

Camazine, S., Denoubourg, J., Franks, N., Sneyd, J., Theraulaz, G., & Bonabeau, E. (2003). *Self-organization in biological systems*. Princeton, NJ: Princeton University Press.

Cameron, C. B., Garey, J. R., & Swalla, B. J. (2000). Evolution of the chordate body plan: New insights from phylogenetic analyses of deuterostome phyla. *Proceedings of the National Academy of Sciences of the United States of America, 97*, 4469–4474.

Campbell, D. T. (1974). Downward causation in hierarchically organized biological systems. In F. J. Ayala & T. Dobzhansky (Eds.), *Studies in the philosophy of biology* (pp. 179–186). Berkeley: University of California Press.

Campbell, H. J. (1968). Peripheral self-stimulation as a reward. *Nature, 218*, 104.

Campbell, H. J. (1972). Peripheral self-stimulation as a reward in fish, reptile and mammal. *Physiology & Behavior, 8*(4), 637–640.

Candiani, S., Castagnola, P., Oliveri, D., & Pestarino, M. (2002). Cloning and developmental expression of AmphiBrn1/2/4, a POU III gene in amphioxus. *Mechanisms of Development, 116*(1), 231–234.

Candiani, S., Oliveri, D., Parodi, M., Castagnola, P., & Pestarino, M. (2005). AmphiD1/β, a dopamine D1/β-adrenergic receptor from the amphioxus *Branchiostoma floridae*: Evolutionary aspects of the catecholaminergic system during development. *Development Genes and Evolution, 215*(12), 631–638.

Candiani, S., Moronti, L., Ramoino, P., Schubert, M., & Pestarino, M. (2012). A neurochemical map of the developing amphioxus nervous system. *BMC Neuroscience, 13*(1), 59.

Cannon, W. B. (1927). The James–Lange theory of emotions: A critical examination and an alternative theory. *American Journal of Psychology, 39*, 106–124.

Canteras, N. S., Ribeiro-Barbosa, E. R., Goto, M., Cipolla-Neto, J., & Swanson, L. W. (2011). The retinohypothalamic tract: Comparison of axonal projection patterns from four major targets. *Brain Research Reviews, 65*(2), 150–183.

Cardin, J. A., Carlén, M., Meletis, K., Knoblich, U., Zhang, F., Deisseroth, K., ... Moore, C. I. (2009). Driving fast-spiking cells induces gamma rhythm and controls sensory responses. *Nature, 459*(7247), 663–667.

Carr, F. B., & Zachariou, V. (2014). Nociception and pain: Lessons from optogenetics. *Frontiers in Behavioral Neuroscience, 8*.

Carrera, I., Anadón, R., & Rodríguez-Moldes, I. (2012). Development of tyrosine hydroxylase-immunoreactive cell populations and fiber pathways in the brain of the dogfish *Scyliorhinus canicula*: New perspectives on the evolution of the vertebrate catecholaminergic system. *Journal of Comparative Neurology, 520*(16), 3574–3603.

Carruthers, P. (2003). *Phenomenal consciousness: A naturalistic theory*. New York: Cambridge University Press.

Cartron, L., Darmaillacq, A. S., & Dickel, L. (2013). The "prawn-in-the-tube" procedure: What do cuttlefish learn and memorize? *Behavioural Brain Research, 240*, 29–32.

Casimir, M. J. (2009). On the origin and evolution of affective capacities in lower vertebrates. In B. Rottger-Rossler & H. J. Markowitsch (Eds.), *Emotions as bio-cultural processes* (pp. 55–93). New York: Springer.

Castro, L. F. C., Rasmussen, S. L. K., Holland, P. W. H., Holland, N. D., & Holland, L. Z. (2006). A Gbx homeobox gene in amphioxus: Insights into ancestry of the ANTP class and evolution of the midbrain/hindbrain boundary. *Developmental Biology, 295*, 40–51.

Catania, K. C., & Kaas, J. H. (1996). The unusual nose and brain of the star-nosed mole. *Bioscience, 46*(8), 578–586.

Caudill, M. S., Eggebrecht, A. T., Gruberg, E. R., & Wessel, R. (2010). Electrophysiological properties of isthmic neurons in frogs revealed by in vitro and in vivo studies. *Journal of Comparative Physiology. A, Neuroethology, Sensory, Neural, and Behavioral Physiology, 196*(4), 249–262.

Chalmers, D. J. (1995). Facing up to the problem of consciousness. *Journal of Consciousness Studies, 2*, 200–219.

Chalmers, D. J. (1996). *The conscious mind: In search of a fundamental theory*. New York: Oxford University Press.

Chalmers, D. J. (2006). Strong and weak emergence. In P. Clayton & P. Davies (Eds.), *The reemergence of emergence* (pp. 244–256). New York: Oxford University Press.

Chalmers, D. (2010). *The character of consciousness*. New York: Oxford University Press.

Chakraborty, M., & Burmeister, S. S. (2010). Sexually dimorphic androgen and estrogen receptor mRNA expression in the brain of túngara frogs. *Hormones and Behavior*, *58*(4), 619–627.

Chandroo, K. P., Duncan, I. J., & Moccia, R. D. (2004). Can fish suffer? Perspectives on sentience, pain, fear, and stress. *Applied Animal Behaviour Science*, *86*(3), 225–250.

Chen, J. Y. (2012). Evolutionary scenario of the early history of the animal kingdom: Evidence from Precambrian (Ediacaran) Weng'an and Early Cambrian Maotianshan biotas, China. In J. Talent (Ed.), *Earth and life* (pp. 239–379). Dordrecht: Springer Netherlands.

Chen, J. Y., Dzik, J., Edgecombe, G. D., Ramsköld, L., & Zhou, G. Q. (1995). A possible Early Cambrian chordate. *Nature*, *377*(6551), 720–722.

Chen, J. Y., Huang, D. Y., & Li, C. W. (1999). An early Cambrian craniate-like chordate. *Nature*, *402*(6761), 518–522.

Chica, A. B., Lasaponara, S., Lupiáñez, J., Doricchi, F., & Bartolomeo, P. (2010). Exogenous attention can capture perceptual consciousness: ERP and behavioural evidence. *NeuroImage*, *51*(3), 1205–1212.

Chittka, L., & Niven, J. (2009). Are bigger brains better? *Current Biology*, *19*(21), R995–R1008.

Chittka, L., & Skorupski, P. (2011). Information processing in miniature brains. *Proceedings of the Royal Society of London, Series B, Biological Sciences*, *278*(1707), 885–888.

Churchland, P. M. (1985). Reduction, qualia, and the direct introspection of brain states. *Journal of Philosophy*, *82*(1), 8–28.

Churchland, P. M., & Churchland, P. S. (1981). Functionalism, qualia, and intentionality. *Philosophical Topics*, *12*(1), 121–145.

Ciaramelli, E., Rosenbaum, R. S., Solcz, S., Levine, B., & Moscovitch, M. (2010). Mental space travel: Damage to posterior parietal cortex prevents egocentric navigation and reexperiencing of remote spatial memories. *Journal of Experimental Psychology: Learning, Memory, and Cognition*, *36*(3), 619.

Cinelli, E., Robertson, B., Mutolo, D., Grillner, S., Pantaleo, T., & Bongianni, F. (2013). Neuronal mechanisms of respiratory pattern generation are evolutionary conserved. *Journal of Neuroscience*, *33*(21), 9104–9112.

Clarke, A., & Pörtner, H. O. (2010). Temperature, metabolic power, and the evolution of endothermy. *Biological Reviews of the Cambridge Philosophical Society*, *85*(4), 703–727.

Cleeremans, A. (Ed.). (2003). *The unity of consciousness: Binding, integration, and dissociation*. New York: Oxford University Press.

Cloninger, C. R., & Gilligan, S. B. (1987). Neurogenetic mechanisms of learning: A phylogenetic perspective. *Journal of Psychiatric Research*, *21*(4), 457–472.

Cohen, J. D., & Castro-Alamancos, M. A. (2010). Neural correlates of active avoidance behavior in superior colliculus. *Journal of Neuroscience, 30*(25), 8502–8511.

Cohen, M. A., Cavanagh, P., Chun, M. M., & Nakayama, K. (2012). The attentional requirements of consciousness. *Trends in Cognitive Sciences, 16*(8), 411–417.

Collin, S. P. (2009). Evolution of the visual system in fishes. In M. D. Binder, N. Hirokawa, & U. Windhorst (Eds.), *Encyclopedia of neurosciences* (pp. 1459–1466). Berlin: Springer.

Colpaert, F. C., De Witte, P., Maroli, A. N., Awouters, F., Niemegeers, C. J., & Janssen, P. A. (1980). Self-administration of the analgesic suprofen in arthritic rats: Evidence of *Mycobaterium butyricum*-induced arthritis as an experimental model of chronic pain. *Life Sciences, 27*(11), 921–928.

Committee on Recognition and Alleviation of Pain in Laboratory Animals, National Research Council. (2009). *Recognition and Alleviation of Pain in Laboratory Animals*. Institute for Laboratory Animal Research, Division of Earth and Life Sciences, Washington, DC: National Academies Press. http:/www.nap.edu.catalog/12526.html.

Cong, P., Ma, X., Hou, X., Edgecombe, G. D., & Strausfeld, N. J. (2014). Brain structure resolves the segmental affinity of anomalocaridid appendages. *Nature, 513*, 538–542.

Conway Morris, S. (1998). *The crucible of creation: The Burgess Shale and the rise of animals*. Oxford: Oxford University Press.

Conway Morris, S. (2006). Darwin's dilemma: The realities of the Cambrian "explosion." *Philosophical Transactions of the Royal Society of London, Series B, Biological Sciences, 361*(1470), 1069–1083.

Conway Morris, S., & Caron, J. B. (2012). *Pikaia gracilens* Walcott, a stem-group chordate from the Middle Cambrian of British Columbia. *Biological Reviews of the Cambridge Philosophical Society, 87*(2), 480–512.

Conway Morris, S., & Caron, J. B. (2014). A primitive fish from the Cambrian of North America. *Nature, 512*(7515), 419–422.

Courtiol, E., & Wilson, D. A. (2014). Thalamic olfaction: Characterizing odor processing in the mediodorsal thalamus of the rat. *Journal of Neurophysiology, 111*(6), 1274–1285.

Cowey, A. (2010). The blindsight saga. *Experimental Brain Research, 200*(1), 3–24.

Craig, A. D. (2003 a). A new view of pain as a homeostatic emotion. *Trends in Neurosciences, 26*(6), 303–307.

Craig, A. D. (2003 b). Pain mechanisms: Labeled lines versus convergence in central processing. *Annual Review of Neuroscience, 26*(1), 1–30.

Craig, A. D. (2003 c). Interoception: The sense of the physiological condition of the body. *Current Opinion in Neurobiology, 13*(4), 500–505.

Craig, A. D. (2008). Interoception and emotion: A neuroanatomical perspective. *Handbook of Emotions, 3,* 272–288.

Craig, A. D. (2010). The sentient self. *Brain Structure & Function, 214,* 563–577.

Cramer, K. S., & Gabriele, M. L. (2014). Axon guidance in the auditory system: Multiple functions of Eph receptors. *Neuroscience, 277,* 152–162.

Crancher, P., King, M. G., Bennett, A., & Montgomery, R. B. (1972). Conditioning of a free operant in *Octopus cyanus* Grayi. *Journal of the Experimental Analysis of Behavior, 17*(3), 359–362.

Crick, F. (1995). *Astonishing hypothesis: The scientific search for the soul.* New York: Simon & Schuster.

Crick, F., & Koch, C. (2003). A framework for consciousness. *Nature Neuroscience, 6,* 119–126.

Critchley, H. D., & Harrison, N. A. (2013). Visceral influences on brain and behavior. *Neuron, 77*(4), 624–638.

Crook, R. J., Hanlon, R. T., & Walters, E. T. (2013). Squid have nociceptors that display widespread long-term sensitization and spontaneous activity after bodily injury. *Journal of Neuroscience, 33*(24), 10021–10026.

Crook, R. J., & Walters, E. T. (2014). Neuroethology: Self-recognition helps octopuses avoid entanglement. *Current Biology, 24*(11), R520–R521.

Cunningham, S. J., Corfield, J. R., Iwaniuk, A. N., Castro, I., Alley, M. R., Birkhead, T. R., & Parsons, S. (2013). The anatomy of the bill tip of kiwi and associated somatosensory regions of the brain: Comparisons with shorebirds. *PLoS One, 8*(11), e80036.

Dalgleish, T., Dunn, B. D., & Mobbs, D. (2009). Affective neuroscience: Past, present, and future. *Emotion Review, 1*(4), 355–368.

Damasio, A. R. (2000). *The feeling of what happens: Body and emotion in the making of consciousness.* New York: Random House.

Damasio, A. (2010). *Self comes to mind: Constructing the conscious brain.* New York: Vintage.

Damasio, A., & Carvalho, G. B. (2013). The nature of feelings: Evolutionary and neurobiological origins. *Nature Reviews: Neuroscience, 14*(2), 143–152.

Damasio, A., Damasio, H., & Tranel, D. (2012). Persistence of feelings and sentience after bilateral damage of the insula. *Cerebral Cortex, 23,* 833–846.

Damasio, A., Everitt, B. J., & Bishop, D. (1996). The somatic marker hypothesis and the possible functions of the prefrontal cortex [and discussion]. *Philosophical Transactions of the Royal Society of London, Series B, Biological Sciences, 351*(1346), 1413–1420.

Danbury, T. C., Weeks, C. A., Chambers, J. P., Waterman-Pearson, A. E., & Kestin, S. C. (2000). Self-selection of the analgesic drug carprofen by lame broiler chickens. *Veterinary Record, 146*(11), 307–311.

References

Daneri, M. F., Casanave, E., & Muzio, R. N. (2011). Control of spatial orientation in terrestrial toads (*Rhinella arenarum*). *Journal of Comparative Psychology, 125*(3), 296.

Dardis, A. (2008). *Mental causation: The mind–body problem*. New York: Columbia University Press.

Darwin, C. (1859). *On the origins of species by means of natural selection*. London: Murray.

DaSilva, A. F., Becerra, L., Makris, N., Strassman, A. M., Gonzalez, R. G., Geatrakis, N., & Borsook, D. (2002). Somatotopic activation in the human trigeminal pain pathway. *Journal of Neuroscience, 22*(18), 8183–8192.

Davidson, D. (1980). *Essays on actions and events*. Oxford: Clarendon Press.

Davis, R. B., Baldauf, S. L., & Mayhew, P. J. (2010). Many hexapod groups originated earlier and withstood extinction events better than previously realized: Inferences from supertrees. *Proceedings of the Royal Society of London, Series B, Biological Sciences, 277*(1687), 1597–1606.

Deacon, T. W. (2011). *Incomplete nature: How mind emerged from matter*. New York: W. W. Norton.

De Arriba, M. D. C., & Pombal, A. M. (2007). Afferent connections of the optic tectum in lampreys: An experimental study. *Brain, Behavior and Evolution, 69*, 37–68.

De Brigard, F., & Prinz, J. (2010). Attention and consciousness. *Wiley Interdisciplinary Reviews: Cognitive Science, 1*(1), 51–59.

Dehaene, S., & Changeux, J. P. (2005). Ongoing spontaneous activity controls access to consciousness: A neuronal model for inattentional blindness. *PLoS Biology, 3*(5), e141.

Dehaene, S., & Changeux, J. P. (2011). Experimental and theoretical approaches to conscious processing. *Neuron, 70*(2), 200–227.

Dehaene, S., Changeux, J. P., Naccache, L., Sackur, J., & Sergent, C. (2006). Conscious, preconscious, and subliminal processing: A testable taxonomy. *Trends in Cognitive Sciences, 10*(5), 204–211.

Dehaene, S., Charles, L., King, J. R., & Marti, S. (2014). Toward a computational theory of conscious processing. *Current Opinion in Neurobiology, 25*, 76–84.

Dehaene, S., & Naccache, L. (2001). Toward a cognitive science of consciousness: Basic evidence and a workspace framework. *Cognition, 79*, 1–37.

Dehal, P., & Boore, J. L. (2005). Two rounds of whole genome duplication in the ancestral vertebrate. *PLoS Biology, 3*(10), e314.

Del Bene, F., Wyart, C., Robles, E., Tran, A., Looger, L., Scott, E. K., … Baier, H. (2010). Filtering of visual information in the tectum by an identified neural circuit. *Science, 330*(6004), 669–673.

Delius, J. D., & Pellander, K. (1982). Hunger dependence of electrical brain self-stimulation in the pigeon. *Physiology & Behavior, 28*(1), 63–66.

Delsuc, F., Brinkmann, H., Chourrout, D., & Philippe, H. (2006). Tunicates and not cephalochordates are the closest living relatives of vertebrates. *Nature, 439*, 965–968.

Delsuc, F., Tsagogeorga, G., Lartillot, N., & Philippe, H. (2008). Additional molecular support for the new chordate phylogeny. *Genesis, 46*, 592–604.

Demski, L. S. (2013). The pallium and mind/behavior relationships in teleost fishes. *Brain, Behavior and Evolution, 82*(1), 31–44.

Dennett, D. C. (1988). Quining qualia. In A. J. Marcel & E. Bisiac (Eds.), *Consciousness in contemporary science* (pp. 42–77). Oxford: Clarendon Press.

Dennett, D. C. (1991). *Consciousness explained.* Boston: Little, Brown.

Dennett, D. C. (1995). Animal consciousness: What matters and why. *Social Research, 62*, 691–710.

Denton, D. (2006). *The primordial emotions: The dawning of consciousness.* London: Oxford.

Denton, D. A., McKinley, M. J., Farrell, M., & Egan, G. F. (2009). The role of primordial emotions in the evolutionary origin of consciousness. *Consciousness and Cognition, 18*(2), 500–514.

Descartes, R. (1649). *Les passions de l'ame.* Paris: Gallimard.

Desfilis, E., Font, E., & García-Verdugo, J. M. (1998). Trigeminal projections to the dorsal thalamus in a lacertid lizard, *Podarcis hispanica. Brain, Behavior, and Evolution, 52*(2), 99–110.

Dicke, U., & Roth, G. (2009). Evolution of the visual system in amphibians. In M. D. Binder, N. Hirokawa, & U. Windhorst (Eds.), *Encyclopedia of neurosciences* (pp. 1455–1459). Berlin: Springer.

Dickinson, P. S. (2006). Neuromodulation of central pattern generators in invertebrates and vertebrates. *Current Opinion in Neurobiology, 16*(6), 604–614.

Di Perri, C., Stender, J., Laureys, S., & Gosseries, O. (2014). Functional neuroanatomy of disorders of consciousness. *Epilepsy & Behavior, 30*, 28–32.

Dobzhansky, T. (1973). Nothing in biology makes sense except in the light of evolution. *American Biology Teacher, 35*, 125–129.

Domínguez, L., Morona, R., Joven, A., González, A., & López, J. M. (2010). Immunohistochemical localization of orexins (hypocretins) in the brain of reptiles and its relation to monoaminergic systems. *Journal of Chemical Neuroanatomy, 39*(1), 20–34.

Dreborg, S., Sundström, G., Larsson, T. A., & Larhammar, D. (2008). Evolution of vertebrate opioid receptors. *Proceedings of the National Academy of Sciences of the United States of America, 105*(40), 15487–15492.

Dudkin, E., & Gruberg, F. (2009). Evolution of nucleus isthmi. In M. D. Binder, N. Hirokawa, & U. Windhorst (Eds.), *Encyclopedia of neurosciences* (pp. 1258–1262). Berlin: Springer-Verlag.

References

Dugas-Ford, J., Rowell, J. J., & Ragsdale, C. W. (2012). Cell-type homologies and the origins of the neocortex. *Proceedings of the National Academy of Sciences of the United States of America, 109*(42), 16974–16979.

Duncan, S., & Barrett, L. F. (2007). Affect is a form of cognition: A neurobiological analysis. *Cognition and Emotion, 21*(6), 1184–1211.

Dunlop, R., & Laming, P. (2005). Mechano-receptive and nociceptive responses in the central nervous system of goldfish (*Carassius auratus*) and trout (*Oncorhynchus mykiss*). *Journal of Pain, 6*(9), 561–568.

Dunlop, R., Millsopp, S., & Laming, P. (2006). Avoidance learning in goldfish (*Carassius auratus*) and trout (*Oncorhynchus mykiss*) and implications for pain perception. *Applied Animal Behaviour Science, 97*(2), 255–271.

Dutta, A., & Gutfreund, Y. (2014). Saliency mapping in the optic tectum and its relationship to habituation. *Frontiers in Integrative Neuroscience, 8.*

Earp, S. E., & Maney, D. L. (2012). Birdsong: Is it music to their ears? *Frontiers in Evolutionary Neuroscience, 4.*

Eberhard, W. G., & Wcislo, W. T. (2011). Grade changes in brain-body allometry: Morphological and behavioural correlates of brain size in miniature spiders, insects, and other invertebrates. *Advances in Insect Physiology, 40*, 155.

Edelman, G. M. (1989). *The remembered present: A biological theory of consciousness.* New York: Basic Books.

Edelman, G. M. (1992). *Bright air, brilliant fire: On the matter of the mind.* New York: Basic Books.

Edelman, G. M. (1998). Building a picture of the brain. *Daedalus, 127*, 37–69.

Edelman, G. M., Gally, J. A., & Baars, B. J. (2011). Biology of consciousness. *Frontiers in Psychology, 2.*

Edelman, D. B., Baars, B. J., & Seth, A. K. (2005). Identifying hallmarks of consciousness in non-mammalian species. *Consciousness and Cognition, 14*(1), 169–187.

Edelman, D. B., & Seth, A. K. (2009). Animal consciousness: A synthetic approach. *Trends in Neurosciences, 32*(9), 476–484.

Edgecombe, G. D. (2010). Arthropod phylogeny: An overview from the perspectives of morphology, molecular data and the fossil record. *Arthropod Structure & Development, 39*(2), 74–87.

Edwards, C. (2014). What's it like? The science of scientific analogies. *Skeptic, 19*(3), 59–63.

Eisthen, H. L. (1997). Evolution of vertebrate olfactory systems. *Brain, Behavior and Evolution, 50*(4), 222–233.

Ellis, G. F. (2006). On the nature of emergent reality. In P. Clayton & P. Davies (Eds.), *The re-emergence of emergence* (pp. 79–107). Oxford: Oxford University Press.

Elwood, R. W. (2011). Pain and suffering in invertebrates? *ILAR Journal, 52*(2), 175–184.

Elwood, R. W., & Appel, M. (2009). Pain experience in hermit crabs? *Animal Behaviour, 77*(5), 1243–1246.

Endepols, H., Roden, K., & Walkowiak, W. (2005). Hodological characterization of the septum in anuran amphibians: II. Efferent connections. *Journal of Comparative Neurology, 483*(4), 437–457.

Engel, A. K., Fries, P., König, P., Brecht, M., & Singer, W. (1999). Temporal binding, binocular rivalry, and consciousness. *Consciousness and Cognition, 8*(2), 128–151.

Engel, A. K., Fries, P., & Singer, W. (2001). Dynamic predictions: Oscillations and synchrony in top-down processing. *Nature Reviews: Neuroscience, 2*(10), 704–716.

Engel, A. K., & Singer, W. (2001). Temporal binding and the neural correlates of sensory awareness. *Trends in Cognitive Sciences, 5*(1), 16–25.

Ericsson, J., Stephenson-Jones, M., Kardamakis, A., Robertson, B., Silberberg, G., & Grillner, S. (2013). Evolutionarily conserved differences in pallial and thalamic short-term synaptic plasticity in striatum. *Journal of Physiology, 591*(4), 859–874.

Eriksson, A., Betti, L., Friend, A. D., Lycett, S. J., Singarayer, J. S., von Cramon-Taubadel, N., ... Manica, A. (2012). Late Pleistocene climate change and the global expansion of anatomically modern humans. *Proceedings of the National Academy of Sciences of the United States of America, 109*(40), 16089–16094.

Erwin, D. H., & Valentine, J. W. (2013). *The Cambrian explosion*. Greenwood Village, CO: Roberts.

Fabbro, F., Aglioti, S. M., Bergamasco, M., Clarici, A., & Panksepp, J. (2015). Evolutionary aspects of self-and world consciousness in vertebrates. *Frontiers in Human Neuroscience, 9*.

Farb, N. A., Segal, Z. V., & Anderson, A. K. (2013). Attentional modulation of primary interoceptive and exteroceptive cortices. *Cerebral Cortex, 23*(1), 114–126.

Farrell, W. J., & Wilczynski, W. (2006). Aggressive experience alters place preference in green anole lizards, Anolis carolinensis. *Animal Behaviour, 71*(5), 1155–1164.

Farris, S. M. (2005). Evolution of insect mushroom bodies: Old clues, new insights. *Arthropod Structure & Development, 34*(3), 211–234.

Farris, S. M. (2012). Evolution of complex higher brain centers and behaviors: Behavioral correlates of mushroom body elaboration in insects. *Brain, Behavior, and Evolution, 82*(1), 9–18.

Fauria, K., Colborn, M., & Collett, T. S. (2000). The binding of visual patterns in bumblebees. *Current Biology, 10*(15), 935–938.

Feigl, H. (1967). *The "mental" and the "physical."* Minneapolis: University of Minnesota Press.

References

Feinberg, E. H., & Meister, M. (2015). Orientation columns in the mouse superior colliculus. *Nature*. doi:.10.1038/ nature14103

Feinberg, T. E. (1997). The irreducible perspectives of consciousness. *Seminars in Neurology*, *17*(2): 85–93.

Feinberg, T. E. (2000). The nested hierarchy of consciousness: A neurobiological solution to the problem of mental unity. *Neurocase*, *6*(2), 75–81.

Feinberg, T. E. (2001). *Altered egos: How the brain creates the self*. Oxford: Oxford University Press.

Feinberg, T. E. (2009). *From axons to identity: Neurological explorations of the nature of the self*. New York: W. W. Norton.

Feinberg, T. E. (2011). The nested neural hierarchy and the self. *Consciousness and Cognition*, *20*, 4–17.

Feinberg, T. E. (2012). Neuroontology, neurobiological naturalism, and consciousness: A challenge to scientific reduction and a solution. *Physics of Life Reviews*, *9*(1), 13–34.

Feinberg, T. E., & Mallatt, J. (2013). The evolutionary and genetic origins of consciousness in the Cambrian Period over 500 million years ago. *Frontiers in Psychology*, *4*.

Feinberg, T. E., & Mallatt, J. (2016a). Neurobiological naturalism. In R. R. Poznanski, J. Tuszynski, & T. E. Feinberg (Eds.), *Biophysics of consciousness: A foundational approach*. London: World Scientific.

Feinberg, T. E., & Mallatt, J. (2016b). The evolutionary origins of consciousness. In R. R. Poznanski, J. Tuszynski, & T. E. Feinberg (Eds.), *Biophysics of consciousness: A foundational approach*. London: World Scientific.

Feinstein, J. S. (2013). Lesion studies of human emotion and feeling. *Current Opinion in Neurobiology*, *23*(3), 304–309.

Feldman, J. (2013). The neural binding problem(s). *Cognitive Neurodynamics*, *7*(1), 1–11.

Ferreiro-Galve, S., Carrera, I., Candal, E., Villar-Cheda, B., Anadón, R., Mazan, S., & Rodríguez-Moldes, I. (2008). The segmental organization of the developing shark brain based on neurochemical markers, with special attention to the prosencephalon. *Brain Research Bulletin*, *75*(2), 236–240.

Flanagan, O. (1992). *Consciousness reconsidered*. Cambridge, MA: MIT Press.

Fiorito, G., & Scotto, P. (1992). Observational learning in *Octopus vulgaris*. *Science*, *256*(5056), 545–547.

Flood, N. C., Overmier, J. B., & Savage, G. E. (1976). Teleost telencephalon and learning: An interpretive review of data and hypotheses. *Physiology & Behavior*, *16*(6), 783–798.

Forlano, P. M., & Bass, A. H. (2011). Neural and hormonal mechanisms of reproductive-related arousal in fishes. *Hormones and Behavior*, *59*(5), 616–629.

Forlano, P. M., Maruska, K. P., Sower, S. A., King, J. A., & Tricas, T. C. (2000). Differential distribution of gonadotropin-releasing hormone-immunoreactive neurons in the stingray brain: Functional and evolutionary considerations. *General and Comparative Endocrinology, 118*(2), 226–248.

Forster, G. L., Watt, M. J., Korzan, W. J., Renner, K. J., & Summers, C. H. (2005). Opponent recognition in male green anoles, *Anolis carolinensis*. *Animal Behaviour, 69*(3), 733–740.

Fotopoulou, A. (2013). Beyond the reward principle: Consciousness as precision seeking. *Neuro-psychoanalysis, 15*(1), 33–38.

Fournier, J., Müller, C. M., & Laurent, G. (2015). Looking for the roots of cortical sensory computation in three-layered cortices. *Current Opinion in Neurobiology, 31*, 119–126.

Freamat, M., & Sower, S. A. (2013). Integrative neuro-endocrine pathways in the control of reproduction in lamprey: A brief review. *Frontiers in Endocrinology, 4*.

Fried, I., MacDonald, K. A., & Wilson, C. L. (1997). Single neuron activity in human hippocampus and amygdala during recognition of faces and objects. *Neuron, 18*(5), 753–765.

Freidin, E., Cuello, M. I., & Kacelnik, A. (2009). Successive negative contrast in a bird: Starlings' behaviour after unpredictable negative changes in food quality. *Animal Behaviour, 77*(4), 857–865.

Friston, K. (2013). Consciousness and hierarchical inference. *Neuro-psychoanalysis, 15*(1), 38–42.

Fuller, P., Sherman, D., Pedersen, N. P., Saper, C. B., & Lu, J. (2011). Reassessment of the structural basis of the ascending arousal system. *Journal of Comparative Neurology, 519*(5), 933–956.

Fuss, T., Bleckmann, H., & Schluessel, V. (2014). Place learning prior to and after telencephalon ablation in bamboo and coral cat sharks (*Chiloscyllium griseum* and *Atelomycterus marmoratus*). *Journal of Comparative Physiology, A, Neuroethology, Sensory, Neural, and Behavioral Physiology, 200*(1), 37–52.

Gabor, C. R., & Jaeger, R. G. (1995). Resource quality affects the agonistic behaviour of territorial salamanders. *Animal Behaviour, 49*(1), 71–79.

Gallese, V. (2013). Bodily self, affect, consciousness, and the cortex. *Neuro-psychoanalysis, 15*(1), 42–45.

Gans, C., & Northcutt, R. G. (1983). Neural crest and the origin of vertebrates: A new head. *Science, 220*(4594), 268–273.

Ganz, J., Kaslin, J., Freudenreich, D., Machate, A., Geffarth, M., & Brand, M. (2012). Subdivisions of the adult zebrafish subpallium by molecular marker analysis. *Journal of Comparative Neurology, 520*(3), 633–655.

References

García-Bellido, D. C., Lee, M. S., Edgecombe, G. D., Jago, J. B., Gehling, J. G., & Paterson, J. R. (2014). A new vetulicolian from Australia and its bearing on the chordate affinities of an enigmatic Cambrian group. *BMC Evolutionary Biology, 14*(1), 214.

Garstang, W. (1928). The morphology of the tunicata. *Quarterly Journal of Microscopical Science, 72*, 51–187.

Gazzaniga, M. S. (1997). Brain and conscious experience. *Advances in Neurology, 77*, 181–192.

Gelman, S., Ayali, A., Tytell, E. D., & Cohen, A. H. (2007). Larval lampreys possess a functional lateral line system. *Journal of Comparative Physiology, A, Neuroethology, Sensory, Neural, and Behavioral Physiology, 193*(2), 271–277.

Gentle, M. J., Tilston, V., & McKeegan, D. E. F. (2001). Mechanothermal nociceptors in the scaly skin of the chicken leg. *Neuroscience, 106*(3), 643–652.

Gerkema, M. P., Davies, W. I., Foster, R. G., Menaker, M., & Hut, R. A. (2013). The nocturnal bottleneck and the evolution of activity patterns in mammals. *Proceedings of the Royal Society of London, Series B, Biological Sciences, 280*(1765), 1–11.

Gershman, S. J., Horvitz, E. J., & Tenenbaum, J. B. (2015). Computational rationality: A converging paradigm for intelligence in brains, minds, and machines. *Science, 349*(6245), 273–278.

Gill, P. G., Purnell, M. A., Crumpton, N., Brown, K. R., Gostling, N. J., Stampanoni, M., & Rayfield, E. J. (2014). Dietary specializations and diversity in feeding ecology of the earliest stem mammals. *Nature, 512*(7514), 303–305.

Gillette, R., Huang, R. C., Hatcher, N., & Moroz, L. L. (2000). Cost-benefit analysis potential in feeding behavior of a predatory snail by integration of hunger, taste, and pain. *Proceedings of the National Academy of Sciences of the United States of America, 97*(7), 3585–3590.

Gingras, M., Hagadorn, J. W., Seilacher, A., Lalonde, S. V., Pecoits, E., Petrash, D., & Konhauser, K. O. (2011). Possible evolution of mobile animals in association with microbial mats. *Nature Geoscience, 4*(6), 372–375.

Giordano, J. (2005). The neurobiology of nociceptive and anti-nociceptive systems. *Pain Physician, 8*(3), 277–290.

Giske, J., Eliassen, S., Fiksen, Ø., Jakobsen, P. J., Aksnes, D. L., Jørgensen, C., & Mangel, M. (2013). Effects of the emotion system on adaptive behavior. *American Naturalist, 182*(6), 689–703.

Giuliani, A., Minelli, D., Quaglia, A., & Villani, L. (2002). Telencephalo-habenulo-interpeduncular connections in the brain of the shark *Chiloscyllium arabicum*. *Brain Research, 926*(1), 186–190.

Giurfa, M. (2013). Cognition with few neurons: Higher-order learning in insects. *Trends in Neurosciences, 36*(5), 285–294.

Giurfa, M., Zhang, S., Jenett, A., Menzel, R., & Srinivasan, M. V. (2001). The concepts of "sameness" and "difference" in an insect. *Nature, 410*(6831), 930–933.

Globus, G. G. (1973). Unexpected symmetries in the "world knot." *Science, 180,* 1129–1136.

Glover, J. C., & Fritzsch, B. (2009). Brains of primitive chordates. In M. D. Binder, N. Hirokawa, & U. Windhorst (Eds.), *Encyclopedia of neurosciences* (pp. 439–448). Berlin: Springer-Verlag.

Goddard, C. A., Sridharan, D., Huguenard, J. R., & Knudsen, E. I. (2012). Gamma oscillations are generated locally in an attention-related midbrain network. *Neuron, 73*(3), 567–580.

Godefroit, P., Cau, A., Dong-Yu, H., Escuillié, F., Wenhao, W., & Dyke, G. (2013). A Jurassic avialan dinosaur from China resolves the early phylogenetic history of birds. *Nature, 498*(7454), 359–362.

Godfrey-Smith, P. (2013). Cephalopods and the evolution of the mind. *Pacific Conservation Biology, 19,* 4–9.

Gonzalez, A., & Moreno, N. (2009). Evolution of the amygdala, tetrapods. In M. D. Binder, N. Hirokawa, & U. Windhorst (Eds.), *Encyclopedia of neurosciences* (pp. 1282–1286). Berlin: Springer-Verlag.

Goodson, J. L., & Bass, A. H. (2000). Forebrain peptides modulate sexually polymorphic vocal circuitry. *Nature, 403*(6771), 769–772.

Goodson, J. L., & Kingsbury, M. A. (2013). What's in a name? Considerations of homologies and nomenclature for vertebrate social behavior networks. *Hormones and Behavior, 64*(1), 103–112.

Gottfried, J. (2006). Smell: Central nervous processing. *Advances in Otorhinolaryngology, 63*(R), 44–69.

Gottfried, J. A. (2010). Right orbitofrontal cortex mediates conscious olfactory perception. *Psychological Science, 21*(10), 1454–1463.

Gould, J. L., & Gould, C. G. (1986). Invertebrate intelligence. In R. J. Hoage & L. Goldman (Eds.), *Animal intelligence: Insights into the animal mind* (pp. 21–36). Washington, DC: Smithsonian.

Gould, S. J. (1977). *Ontogeny and phylogeny*. Cambridge, MA: Harvard University Press.

Gould, S. J. (1989). *Wonderful life: The Burgess Shale and the nature of history*. New York: W. W. Norton.

Graham, B. J., & Northmore, D. P. (2007). A spiking neural network model of midbrain visuomotor mechanisms that avoids objects by estimating size and distance monocularly. *Neurocomputing, 70*(10), 1983–1987.

Graindorge, N., Alves, C., Darmaillacq, A. S., Chichery, R., Dickel, L., & Bellanger, C. (2006). Effects of dorsal and ventral vertical lobe electrolytic lesions on spatial learning and locomotor activity in *Sepia officinalis*. *Behavioral Neuroscience, 120*(5), 1151.

Grasso, F. W. (2014). The octopus with two brains: How are distributed and central representations integrated in the octopus central nervous system? In A. Darmaillacq, L. Dickel, & J. Mather (Eds.), *Cephalopod cognition* (pp. 94–122). Cambridge, MA: Cambridge University Press.

Grau, J. W., Barstow, D. G., & Joynes, R. L. (1998). Instrumental learning within the spinal cord: I. Behavioral properties. *Behavioral Neuroscience, 112*(6), 1366.

Grau, J. W., Crown, E. D., Ferguson, A. R., Washburn, S. N., Hook, M. A., & Miranda, R. C. (2006). Instrumental learning within the spinal cord: Underlying mechanisms and implications for recovery after injury. *Behavioral and Cognitive Neuroscience Reviews, 5*(4), 191–239.

Green, S. A., Simoes-Costa, M., & Bronner, M. E. (2015). Evolution of the vertebrates as viewed from the crest. *Nature, 520*, 474–482.

Green, W. W., Basilious, A., Dubuc, R., & Zielinski, B. S. (2013). The neuroanatomical organization of projection neurons associated with different olfactory bulb pathways in the sea lamprey, *Petromyzon marinus*. *PLoS One, 8*(7), e69525.

Griffin, D. R. (2001). *Animal minds: Beyond cognition to consciousness*. Chicago: University of Chicago Press.

Grigg, G. C., Beard, L. A., & Augee, M. L. (2004). The evolution of endothermy and its diversity in mammals and birds. *Physiological and Biochemical Zoology, 77*(6), 982–997.

Grillner, S., Hellgren, J., Menard, A., Saitoh, K., & Wikström, M. A. (2005). Mechanisms for selection of basic motor programs—roles for the striatum and pallidum. *Trends in Neurosciences, 28*(7), 364–370.

Grillner, S., Robertson, B., & Stephenson-Jones, M. (2013). The evolutionary origin of the vertebrate basal ganglia and its role in action selection. *Journal of Physiology, 591*(22), 5425–5431.

Gross, C. G. (2002). Genealogy of the "grandmother cell." *Neuroscientist, 8*(5), 512–518.

Gross, C. G. (2008). Single neuron studies of inferior temporal cortex. *Neuropsychologia, 46*(3), 841–852.

Gross, C. G., Bender, D. B., & Rocha-Miranda, C. E. (1969). Visual receptive fields of neurons in inferotemporal cortex of the monkey. *Science, 166*(910), 1303–1306.

Gross, C. G., Rocha-Miranda, C. E., & Bender, D. B. (1972). Visual properties of neurons in inferotemporal cortex of the Macaque. *Journal of Neurophysiology, 35*(1), 96–111.

Gruberg, E., Dudkin, E., Wang, Y., Marín, G., Salas, C., Sentis, E., ... Udin, S. (2006). Influencing and interpreting visual input: The role of a visual feedback system. *Journal of Neuroscience, 26*(41), 10368–10371.

Guillory, S. A., & Bujarski, K. A. (2014). Exploring emotions using invasive methods: Review of 60 years of human intracranial electrophysiology. *Social Cognitive and Affective Neuroscience, 2014.* doi:10.1093/scan/nsu002.

Guirado, S., & Davila, J. C. (2009). Evolution of the optic tectum in amniotes. In M. D. Binder, N. Hirokawa, & U. Windhorst (Eds.), *Encyclopedia of neurosciences* (pp. 1375–1380). Berlin: Springer.

Guirado, S., Davila, J. C., Real, M. A., & Medina, L. (1999). Nucleus accumbens in the lizard *Psammodromus algirus*: Chemoarchitecture and cortical afferent connections. *Journal of Comparative Neurology, 405*(1), 15–31.

Guo, Y. C., Liao, K. K., Soong, B. W., Tsai, C. P., Niu, D. M., Lee, H. Y., & Lin, K. P. (2003). Congenital insensitivity to pain with anhidrosis in Taiwan: A morphometric and genetic study. *European Neurology, 51*(4), 206–214.

Guo, Y., Wang, L., Zhou, Z., Wang, M., Liu, R., Wang, L., ... Song, L. (2013). An opioid growth factor receptor (OGFR) for [Met 5]-enkephalin in *Chlamys farreri*. *Fish & Shellfish Immunology, 34*(5), 1228–1235.

Gurdasani, D., Carstensen, T., Tekola-Ayele, F., Pagani, L., Tachmazidou, I., Hatzikotoulas, K., ... Sandhu, M. S. (2015). The African Genome Variation Project shapes medical genetics in Africa. *Nature, 517*(7534), 327–332.

Gutfreund, Y. (2012). Stimulus-specific adaptation, habituation, and change detection in the gaze control system. *Biological Cybernetics, 106*(11–12), 657–668.

Gutnick, T., Byrne, R. A., Hochner, B., & Kuba, M. (2011). *Octopus vulgaris* uses visual information to determine the location of its arm. *Current Biology, 21*(6), 460–462.

Haladjian, H. H., & Montemayor, C. (2014). On the evolution of conscious attention. *Psychonomic Bulletin & Review, 22*(3), 595–613.

Hall, B. K. (2008). *The neural crest and neural crest cells in vertebrate development and evolution* (Vol. 11). New York: Springer Science & Business Media.

Hall, J. (2011). *Guyton and Hall textbook of medical physiology* (12th ed.). Philadelphia: Saunders.

Hall, M. I., Kamilar, J. M., & Kirk, E. C. (2012). Eye shape and the nocturnal bottleneck of mammals. *Proceedings of the Royal Society B: Biological Sciences*, rspb20122258.

Hamamoto, D. T., & Simone, D. A. (2003). Characterization of cutaneous primary afferent fibers excited by acetic acid in a model of nociception in frogs. *Journal of Neurophysiology, 90*(2), 566–577.

Hameroff, S., & Penrose, R. (2014). Consciousness in the universe: A review of the "Orch OR" theory. *Physics of Life Reviews, 11*(1), 39–78.

Häming, D., Simoes-Costa, M., Uy, B., Valencia, J., Sauka-Spengler, T., & Bronner-Fraser, M. (2011). Expression of sympathetic nervous system genes in lamprey suggests their recruitment for specification of a new vertebrate feature. *PLoS One*, *6*(10), e26543.

Hara, E., Kubikova, L., Hessler, N. A., & Jarvis, E. D. (2007). Role of the midbrain dopaminergic system in modulation of vocal brain activation by social context. *European Journal of Neuroscience*, *25*(11), 3406–3416.

Hardisty, M. W. (1979). *Biology of the cyclostomes*. London: Chapman & Hall.

Harzsch, S. (2002). The phylogenetic significance of crustacean optic neuropils and chiasmata: A re-examination. *Journal of Comparative Neurology*, *453*(1), 10–21.

Heesy, C. P., & Hall, M. I. (2010). The nocturnal bottleneck and the evolution of mammalian vision. *Brain, Behavior and Evolution*, *75*(3), 195–203.

Heil, J., & Mele, A. (Eds.). (1993). *Mental causation*. Oxford: Clarendon Press.

Heim, N. A., Knope, M. L., Schaal, E. K., Wang, S. C., & Payne, J. L. (2015). Cope's rule in the evolution of marine animals. *Science*, *347*(6224), 867–870.

Heimer, L., & Van Hoesen, G. W. (2006). The limbic lobe and its output channels: Implications for emotional functions and adaptive behavior. *Neuroscience and Biobehavioral Reviews*, *30*(2), 126–147.

Herberholz, J., & Marquart, G. D. (2012). Decision making and behavioral choice during predator avoidance. *Frontiers in Neuroscience*, *6*.

Herculano-Houzel, S., Munk, M. H., Neuenschwander, S., & Singer, W. (1999). Precisely synchronized oscillatory firing patterns require electroencephalographic activation. *Journal of Neuroscience*, *19*(10), 3992–4010.

Hikosaka, O. (2010). The habenula: From stress evasion to value-based decision-making. *Nature Reviews: Neuroscience*, *11*(7), 503–513.

Hirayama, K., Catanho, M., Brown, J. W., & Gillette, R. (2012). A core circuit module for cost/benefit decision. *Frontiers in Neuroscience*, *6*.

Hirayama, K., & Gillette, R. (2012). A neuronal network switch for approach/avoidance toggled by appetitive state. *Current Biology*, *22*(2), 118–123.

Hisajima, T., Kishida, R., Atobe, Y., Nakano, M., Goris, R. C., & Funakoshi, K. (2002). Distribution of myelinated and unmyelinated nerve fibers and their possible role in blood flow control in crotaline snake infrared receptor organs. *Journal of Comparative Neurology*, *449*(4), 319–329.

Hochner, B. (2012). An embodied view of octopus neurobiology. *Current Biology*, *22*(20), R887–R892.

Hochner, B. (2013). How nervous systems evolve in relation to their embodiment: What we can learn from octopuses and other molluscs. *Brain, Behavior and Evolution*, *82*(1), 19–30.

Hochner, B., Shomrat, T., & Fiorito, G. (2006). The octopus: A model for a comparative analysis of the evolution of learning and memory mechanisms. *Biological Bulletin*, *210*(3), 308–317.

Hofmann, M. H., & Northcutt, R. G. (2012). Forebrain organization in elasmobranchs. *Brain, Behavior, and Evolution*, *80*(2), 142–151.

Hoke, K. L., Ryan, M. J., & Wilczynski, W. (2007). Integration of sensory and motor processing underlying social behaviour in túngara frogs. *Proceedings of the Royal Society B: Biological Sciences*, *274*(1610), 641–649.

Holland, L. Z. (2013). Evolution of new characters after whole genome duplications: Insights from amphioxus. *Seminars in Cell & Developmental Biology*, *24*(2), 101–109.

Holland, L. Z. (2014). Genomics, evolution, and development of amphioxus and tunicates: The Goldilocks principle. *Journal of Experimental Zoology, Part B*, *9999B*, 1–11.

Holland, L. Z., Carvalho, J. E., Escriva, H., Laudet, V., Schubert, M., Shimeld, S. M., & Yu, J.-K. (2013). Evolution of bilaterian central nervous systems: A single origin? *EvoDevo*, *4*, 27.

Holland, N. D., Holland, L. Z., & Holland, P. W. H. (2015). Scenarios for the making of vertebrates. *Nature*, *520*, 450–455.

Holland, N. D., & Yu, K.-Jr. (2002). Epidermal receptor development and sensory pathways in vitally stained amphioxus (*Branchiostoma floridae*). *Acta Zoologica*, *83*(4), 309–319.

Holland, P. W. (2015). Did homeobox gene duplications contribute to the Cambrian explosion? *Zoological Letters*, *1*(1).

Hopson, J. A. (2012). The role of foraging mode in the origin of therapsids: Implications for the origin of mammalian endothermy. *Fieldiana: Life and Earth Sciences*, *5*, 126–148.

Hou, X. G., Ramsköld, L., & Bergström, J. (1991). Composition and preservation of the Chengjiang fauna—a Lower Cambrian soft-bodied biota. *Zoologica Scripta*, *20*(4), 395–411.

Huber, R. (2005). Amines and motivated behaviors: A simpler systems approach to complex behavioral phenomena. *Journal of Comparative Physiology, A, Neuroethology, Sensory, Neural, and Behavioral Physiology*, *191*(3), 231–239.

Huber, R., Panksepp, J. B., Nathaniel, T., Alcaro, A., & Panksepp, J. (2011). Drug-sensitive reward in crayfish: An invertebrate model system for the study of seeking, reward, addiction, and withdrawal. *Neuroscience and Biobehavioral Reviews*, *35*(9), 1847–1853.

Huesa, G., Anadon, R., Folgueira, M., & Yanez, J. (2009). Evolution of the pallium in fishes. In M. D. Binder, N. Hirokawa, & U. Windhorst (Eds.), *Encyclopedia of neurosciences* (pp. 1400–1404). Berlin: Springer.

Humphries, M. D., Gurney, K., & Prescott, T. J. (2007). Is there a brainstem substrate for action selection? *Philosophical Transactions of the Royal Society of London, Series B, Biological Sciences, 362*(1485), 1627–1639.

Hurtado-Parrado, C. (2010). Mecanismos neuronales del aprendizaje en los peces teleósteos. *Universitas Psychologica, 9*(3), 663–678.

Huston, J. P., & Borbely, A. A. (1973). Operant conditioning in forebrain ablated rats by use of rewarding hypothalamic stimulation. *Brain Research, 50*(2), 467–472.

Huston, J. P., Silva, M. A., Topic, B., & Müller, C. P. (2013). What's conditioned in conditioned place preference? *Trends in Pharmacological Sciences, 34*(3), 162–166.

Im, S. H., & Galko, M. J. (2012). Pokes, sunburn, and hot sauce: *Drosophila* as an emerging model for the biology of nociception. *Developmental Dynamics, 241*(1), 16–26.

Ivashkin, E., & Adameyko, I. (2013). Progenitors of the protochordate ocellus as an evolutionary origin of the neural crest. *EvoDevo, 4*(1), 12.

Iwahori, N., Kawawaki, T., & Baba, J. (1998). Neuronal organization of the optic tectum in the river lamprey, *Lampetra japonica*: A Golgi study. *Journal für Hirnforschung, 39*(3), 409–424.

Jabr, F. (2012). Know your neurons: How to classify different types of neurons in the brain's forest. http://blogs.scientificamerican.com/brainwaves/2012/05/16/know-your-neurons-classifying-the-many-types-of-cells-in-the-neuron-forest/.

Jackson, F. (1982). Epiphenomenal qualia. *Philosophical Quarterly, 32*, 127–136.

Jackson, R. R., & Wilcox, R. S. (1993). Observations in nature of detouring behaviour by *Portia fimbriata*, a web-invading aggressive mimic jumping spider from Queensland. *Journal of Zoology, 230*(1), 135–139.

Jackson, R. R., & Wilcox, R. S. (1998). Spider-eating spiders: Despite the small size of their brain, jumping spiders in the genus *Portia* outwit other spiders with hunting techniques that include trial and error. *American Scientist, 86*, 350–357.

Jacobs, L. F. (2012). From chemotaxis to the cognitive map: The function of olfaction. *Proceedings of the National Academy of Sciences of the United States of America, 109*(Suppl. 1), 10693–10700.

Jacobson, G. A., & Friedrich, R. W. (2013). Neural circuits: Random design of a higher-order olfactory projection. *Current Biology, 23*(10), R448–R451.

James, W. (1879). I.—Are we automata? *Mind, 13*, 1–22.

James, W. (1884). What is an emotion? *Mind, 19*, 188–205.

James, W. (1890). *The principles of psychology*. New York: Holt.

James, W. (1904). Does consciousness exist? *Journal of Philosophy, Psychology, and Scientific Methods, 1*(18), 477–491.

Janek, P. H., & Tye, K. M. (2015). From circuits to behavior in the amygdala. *Nature*, *517*, 284–292.

Janvier, P. (1996). *Early vertebrates* (Vol. 33). Oxford: Clarendon Press.

Janvier, P. (2008). Early jawless vertebrates and cyclostome origins. *Zoological Science*, *25*(10), 1045–1056.

Jarvis, E. D., Yu, J., Rivas, M. V., Horita, H., Feenders, G., Whitney, O., ... Wada, K. (2013). Global view of the functional molecular organization of the avian cerebrum: Mirror images and functional columns. *Journal of Comparative Neurology*, *521*(16), 3614–3665.

Jbabdi, S., Sotiropoulos, S. N., & Behrens, T. E. (2013). The topographic connectome. *Current Opinion in Neurobiology*, *23*(2), 207–215.

Jerison, H. J. (2009). Evolution and brain-body allometry. In M. D. Binder, N. Hirokawa, & U. Windhorst (Eds.), *Encyclopedia of neurosciences* (pp. 1161–1165). Berlin: Springer-Verlag.

Jing, J., & Gillette, R. (2000). Escape swim network interneurons have diverse roles in behavioral switching and putative arousal in *Pleurobranchaea*. *Journal of Neurophysiology*, *83*(3), 1346–1355.

Johnson, E. O., Babis, G. C., Soultanis, K. C., & Soucacos, P. N. (2007). Functional neuroanatomy of proprioception. *Journal of Surgical Orthopaedic Advances*, *17*(3), 159–164.

Jones, M. R., Grillner, S., & Robertson, B. (2009). Selective projection patterns from subtypes of retinal ganglion cells to tectum and pretectum: Distribution and relation to behavior. *Journal of Comparative Neurology*, *517*(3), 257–275.

Jonkisz, J. (2015). Consciousness: Individuated information in action. *Frontiers in psychology*, *6*.

Jordan, M. I., & Mitchell, T. M. (2015). Machine learning: Trends, perspectives, and prospects. *Science*, *349*(6245), 255–260.

Jorgensen, J., Lomholt, J., Weber, R., & Malte, H. (Eds.). (1998). *The biology of hagfishes*. London: Chapman & Hall.

Kaas, J. H. (1997). Topographic maps are fundamental to sensory processing. *Brain Research Bulletin*, *44*(2), 107–112.

Kaas, J. H. (2011). Neocortex in early mammals and its subsequent variations. *Annals of the New York Academy of Sciences*, *1225*(1), 28–36.

Kalman, M. (2009). Evolution of the brain at reptile-bird transition. In M. D. Binder, N. Hirokawa, & U. Windhorst (Eds.), *Encyclopedia of neurosciences* (pp. 1305–1312). Berlin: Springer.

Kalueff, A. V., Stewart, A. M., & Gerlai, R. (2014). Zebrafish as an emerging model for studying complex brain disorders. *Trends in Pharmacological Sciences*, *35*(2), 63–75.

Kamikouchi, A. (2013). Auditory neuroscience in fruit flies. *Neuroscience Research*, *76*(3), 113–118.

Kandel, E. R. (2007). *In search of memory: The emergence of a new science of mind.* New York: W. W. Norton.

Kandel, E. R. (2009). The biology of memory: A forty-year perspective. *Journal of Neuroscience*, *29*(41), 12748–12756.

Kandel, E. R., Schwartz, J. H., Jessell, T. M., Siegelbaum, S. A., & Hudspeth, A. J. (2012). *Principles of neural science* (5th ed.). New York: McGraw Hill.

Kardamakis, A. A., Saitoh, K., & Grillner, S. (2015). Tectal microcircuit generating visual selection commands on gaze-controlling neurons. *Proceedings of the National Academy of Sciences*, *112*(15), E1956–E1965.

Kardong, K. (2012). *Vertebrates: Comparative anatomy, function, evolution* (6th ed.). Dubuque, IA: McGraw-Hill Higher Education.

Karten, H. J. (2013). Neocortical evolution: Neuronal circuits arise independently of lamination. *Current Biology*, *23*(1), R12–R15.

Kavanaugh, S. I., Nozaki, M., & Sower, S. A. (2008). Origins of gonadotropin-releasing hormone (GnRH) in vertebrates: Identification of a novel GnRH in a basal vertebrate, the sea lamprey. *Endocrinology*, *149*(8), 3860–3869.

Kawai, N., Kono, R., & Sugimoto, S. (2004). Avoidance learning in the crayfish (*Procambarus clarkii*) depends on the predatory imminence of the unconditioned stimulus: A behavior systems approach to learning in invertebrates. *Behavioural Brain Research*, *150*(1), 229–237.

Keary, N., Voss, J., Lehmann, K., Bischof, H. J., & Löwel, S. (2010). Optical imaging of retinotopic maps in a small songbird, the zebra finch. *PLoS One*, *5*(8), e11912.

Keenan, J. P., Gallup, G. C., & Falk, D. (2003). *The face in the mirror: The search for the origins of consciousness.* New York: HarperCollins.

Keenan, J. P., Rubio, J., Racioppi, C., Johnson, A., & Barnacz, A. (2005). The right hemisphere and the dark side of consciousness. *Cortex*, *41*(5), 695–704.

Keller, A. (2014). The evolutionary function of conscious information processing is revealed by its task-dependency in the olfactory system. *Frontiers in Psychology*, *5*.

Kender, R. G., Harte, S. E., Munn, E. M., & Borszcz, G. S. (2008). Affective analgesia following muscarinic activation of the ventral tegmental area in rats. *Journal of Pain*, *9*(7), 597–605.

Kessler, L. (2013). Conscious Id or unconscious Id or both: An attempt at "self"-help. *Neuro-psychoanalysis*, *15*(1), 48–51.

Key, B. (2014). Fish do not feel pain and its implications for understanding phenomenal consciousness. *Biology & Philosophy, 30*, 149–165.

Kiefer, M., Ansorge, U., Haynes, J. D., Hamker, F., Mattler, U., Verleger, R., & Niedeggen, M. (2011). Neuro-cognitive mechanisms of conscious and unconscious visual perception: From a plethora of phenomena to general principles. *Advances in Cognitive Psychology, 7*, 55–67.

Kielan-Jaworowska, Z. (2013). *In pursuit of early mammals*. Bloomington: Indiana University Press.

Kielan-Jaworowska, Z., Cifelli, R. L., Cifelli, R., & Luo, Z. X. (2013). *Mammals from the age of dinosaurs: Origins, evolution, and structure*. New York: Columbia University Press.

Kily, L. J., Cowe, Y. C., Hussain, O., Patel, S., McElwaine, S., Cotter, F. E., & Brennan, C. H. (2008). Gene expression changes in a zebrafish model of drug dependency suggest conservation of neuro-adaptation pathways. *Journal of Experimental Biology, 211*(10), 1623–1634.

Kim, D. H., & Lee, Y. (2015). Bioelectronics: Injection and unfolding. *Nature Nanotechnology, 10*(7), 570–571.

Kim, J. (1992). "Downward causation" in emergentism and nonreductive physicalism. In A. Beckermann, H. Flohr, & J. Kim (Eds.), *Emergence or reduction? Essays on the prospects of nonreductive physicalism* (pp. 119–138). New York: de Gruyter.

Kim, J. (1995). The non-reductivist's troubles with mental causation. In J. Heil & A. Mele (Eds.), *Mental causation* (pp. 189–210). Oxford: Clarendon Press.

Kim, J. (1998). *Mind in a physical world: An essay on the mind–body problem and mental causation*. Cambridge, MA: MIT Press.

Kim, T., Thankachan, S., McKenna, J. T., McNally, J. M., Yang, C., Choi, J. H., ... McCarley, R. W. (2015). Cortically projecting basal forebrain parvalbumin neurons regulate cortical gamma band oscillations. *Proceedings of the National Academy of Sciences of the United States of America, 112*(11), 3535–3540.

Kingsbury, M. A., Kelly, A. M., Schrock, S. E., & Goodson, J. L. (2011). Mammal-like organization of the avian midbrain central gray and a reappraisal of the intercollicular nucleus. *PLoS One, 6*(6), e20720.

Kisch, J., & Erber, J. (1999). Operant conditioning of antennal movements in the honey bee. *Behavioural Brain Research, 99*(1), 93–102.

Kirk, R. (1994). *Raw feeling*. Cambridge, MA: MIT Press.

Kirsch, J. A., Güntürkün, O., & Rose, J. (2008). Insight without cortex: Lessons from the avian brain. *Consciousness and Cognition, 17*(2), 475–483.

Kittelberger, J. M., Land, B. R., & Bass, A. H. (2006). Midbrain periaqueductal gray and vocal patterning in a teleost fish. *Journal of Neurophysiology, 96*(1), 71–85.

Kittilsen, S. (2013). Functional aspects of emotions in fish. *Behavioural Processes, 100*, 153–159.

Klee, E. W., Ebbert, J. O., Schneider, H., Hurt, R. D., & Ekker, S. C. (2011). Zebrafish for the study of the biological effects of nicotine. *Nicotine & Tobacco Research, 13*(5), 301–312.

Knoll, A. H., Javaux, E. J., Hewitt, D., & Cohen, P. (2006). Eukaryotic organisms in Proterozoic oceans. *Philosophical Transactions of the Royal Society of London, Series B, Biological Sciences, 361*(1470), 1023–1038.

Knoll, A. H., & Sperling, E. A. (2014). Oxygen and animals in Earth history. *Proceedings of the National Academy of Sciences, 111*(11), 3907–3908.

Knöll, B., & Drescher, U. (2002). Ephrin-As as receptors in topographic projections. *Trends in Neurosciences, 25*(3), 145–149.

Knudsen, E. I. (2011). Control from below: The role of a midbrain network in spatial attention. *European Journal of Neuroscience, 33*(11), 1961–1972.

Kobayashi, S. (2012). Organization of neural systems for aversive information processing: Pain, error, and punishment. *Frontiers in Neuroscience, 6*(136).

Koch, C. (2004). *The quest for consciousness. A neurobiological approach.* Englewood, CO: Roberts.

Kohl, J. V. (2013). Nutrient-dependent/pheromone-controlled adaptive evolution: A model. *Socioaffective Neuroscience & Psychology, 3*, 20553.

Kotrschal, A., Rogell, B., Bundsen, A., Svensson, B., Zajitschek, S., Brännström, I., ... Kolm, N. (2013). Artificial selection on relative brain size in the guppy reveals costs and benefits of evolving a larger brain. *Current Biology, 23*(2), 168–171.

Kouider, S., & Dehaene, S. (2007). Levels of processing during non-conscious perception: A critical review of visual masking. *Philosophical Transactions of the Royal Society of London, Series B: Biological Sciences, 362*(1481), 857–875.

Kouider, S., Stahlhut, C., Gelskov, S. V., Barbosa, L. S., Dutat, M., De Gardelle, V., ... Dehaene-Lambertz, G. (2013). A neural marker of perceptual consciousness in infants. *Science, 340*(6130), 376–380.

Koyama, H., Kishida, R., Goris, R. C., & Kusunoki, T. (1987). Organization of sensory and motor nuclei of the trigeminal nerve in lampreys. *Journal of Comparative Neurology, 264*(4), 437–448.

Krauzlis, R. J., Lovejoy, L. P., & Zénon, A. (2013). Superior colliculus and visual spatial attention. *Annual Review of Neuroscience, 36*, 165–182.

Kreiman, G., Koch, C., & Fried, I. (2000). Category-specific visual responses of single neurons in the human medial temporal lobe. *Nature Neuroscience, 3*(9), 946–953.

Kröger, B., Vinther, J., & Fuchs, D. (2011). Cephalopod origin and evolution: A congruent picture emerging from fossils, development, and molecules. *BioEssays, 33*(8), 602–613.

Kuba, M. J., Byrne, R. A., & Burghardt, G. M. (2010). A new method for studying problem solving and tool use in stingrays (*Potamotrygon castexi*). *Animal Cognition, 13*(3), 507–513.

Kuba, M. J., Byrne, R. A., Meisel, D. V., & Mather, J. A. (2006). When do octopuses play? Effects of repeated testing, object type, age, and food deprivation on object play in *Octopus vulgaris*. *Journal of Comparative Psychology, 120*(3), 184–190.

Kubokawa, K., Tando, Y., & Roy, S. (2010). Evolution of the reproductive endocrine system in chordates. *Integrative and Comparative Biology, 50*(1), 53–62.

Kuraku, S., Meyer, A., & Kuratani, S. (2009). Timing of genome duplications relative to the origin of the vertebrates: Did cyclostomes diverge before or after? *Molecular Biology and Evolution, 26*(1), 47–59.

Kuratani, S. (2012). Evolution of the vertebrate jaw from developmental perspectives. *Evolution & Development, 14*(1), 76–92.

Kusayama, T., & Watanabe, S. (2000). Reinforcing effects of methamphetamine in planarians. *Neuroreport, 11*(11), 2511–2513.

Lacalli, T. C. (2008). Basic features of the ancestral chordate brain: A protochordate perspective. *Brain Research Bulletin, 75*, 319–323.

Lacalli, T. C. (2013). Looking into eye evolution: Amphioxus points the way. *Pigment Cell & Melanoma Research, 26*, 162–164.

Lacalli, T. C. (2015). The origin of vertebrate neural organization. In A. Schmidt-Rhaesa, S. Harszch, & G. Purschke (Eds.), *Structure and evolution of invertebrate nervous systems*. Oxford: Oxford University Press.

Lacalli, T. C., & Holland, L. Z. (1998). The developing dorsal ganglion of the salp *Thalia democratica*, and the nature of the ancestral chordate brain. *Philosophical Transactions of the Royal Society of London, Series B, Biological Sciences, 353*, 1943–1967.

Lacalli, T. C., & Kelly, S. J. (2003). Sensory pathways in amphioxus larvae. II. Dorsal tracts and translumenal cells. *Acta Zoologica Stockholm, 84*, 1–13.

Lacalli, T. C., & Stach, T. (2015). Acrania (Cephalochordata). In A. Schmidt-Rhaesa, S. Harszch, & G. Purschke (Eds.), *Structure and evolution of invertebrate nervous systems*. Oxford: Oxford University Press.

Laland, K., Uller, T., Feldman, M., Sterelny, K., Müller, G. B., Moczek, A., ... Strassmann, J. E. (2014). Does evolutionary theory need a rethink? *Nature, 514*(7521), 161.

Lamb, T. D. (2013). Evolution of phototransduction, vertebrate photoreceptors, and retina. *Progress in Retinal and Eye Research, 36*, 52–119.

References

Lammel, S., Lim, B. K., Ran, C., Huang, K. W., Betley, M. J., Tye, K. M., … Malenka, R. C. (2012). Input-specific control of reward and aversion in the ventral tegmental area. *Nature, 491*(7423), 212–217.

Land, M. F., & Fernald, R. D. (1992). The evolution of eyes. *Annual Review of Neuroscience, 15*(1), 1–29.

Lane, N., & Martin, W. (2010). The energetics of genome complexity. *Nature, 467*(7318), 929–934.

Lange, C. G. (1885). *The mechanism of the emotions. The emotions* (pp. 33–92). Baltimore, MD: Williams & Wilkins.

Lanuza, E., & Martinez-Garcia, F. (2009). Evolution of septal nuclei. In M. D. Binder, N. Hirokawa, & U. Windhorst (Eds.), *Encyclopedia of neurosciences* (pp. 1270–1278). Berlin: Springer.

LeCun, Y., Bengio, Y., & Hinton, G. (2015). Deep learning. *Nature, 521*(7553), 436–444.

LeDoux, J. (2007). The amygdala. *Current Biology, 17*(20), R868–R874.

LeDoux, J. (2012). Rethinking the emotional brain. *Neuron, 73*(4), 653–676.

LeDoux, J. E. (2014). Coming to terms with fear. *Proceedings of the National Academy of Sciences of the United States of America, 111*(8), 2871–2878.

Lee, A. W., & Brown, R. E. (2007). Comparison of medial preoptic, amygdala, and nucleus accumbens lesions on parental behavior in California mice (*Peromyscus californicus*). *Physiology & Behavior, 92*(4), 617–628.

Lee, H., Kim, D. W., Remedios, R., Anthony, T. E., Chang, A., Madisen, L., … Anderson, D. J. (2014). Scalable control of mounting and attack by Esr1+ neurons in the ventromedial hypothalamus. *Nature, 509*(7502), 627–632.

Lee, M. S., Cau, A., Naish, D., & Dyke, G. J. (2014). Sustained miniaturization and anatomical innovation in the dinosaurian ancestors of birds. *Science, 345*(6196), 562–566.

Lee, M. S., Jago, J. B., García-Bellido, D. C., Edgecombe, G. D., Gehling, J. G., & Paterson, J. R. (2011). Modern optics in exceptionally preserved eyes of Early Cambrian arthropods from Australia. *Nature, 474*(7353), 631–634.

Lee, S. H., & Dan, Y. (2012). Neuromodulation of brain states. *Neuron, 76*(1), 209–222.

Lett, B. T., & Grant, V. L. (1989). The hedonic effects of amphetamine and pentobarbital in goldfish. *Pharmacology, Biochemistry, and Behavior, 32*(1), 355–356.

Levens, N., & Akins, C. K. (2001). Cocaine induces conditioned place preference and increases locomotor activity in male Japanese quail. *Pharmacology, Biochemistry, and Behavior, 68*(1), 71–80.

Levine, J. (1983). Materialism and qualia: The explanatory gap. *Pacific Philosophical Quarterly, 64*(4), 354–361.

Li, W., Lopez, L., Osher, J., Howard, J. D., Parrish, T. B., & Gottfried, J. A. (2010). Right orbitofrontal cortex mediates conscious olfactory perception. *Psychological Science*, *21*(10), 1454–1463.

Liang, Y. F., & Terashima, S. I. (1993). Physiological properties and morphological characteristics of cutaneous and mucosal mechanical nociceptive neurons with A-δ peripheral axons in the trigeminal ganglia of crotaline snakes. *Journal of Comparative Neurology*, *328*(1), 88–102.

Lillvis, J. L., Gunaratne, C. A., & Katz, P. S. (2012). Neurochemical and neuroanatomical identification of central pattern generator neuron homologues in Nudipleura molluscs. *PLoS One*, *7*(2), e31737.

Lindahl, B. I. B. (1997). Consciousness and biological evolution. *Journal of Theoretical Biology*, *187*(4), 613–629.

Liu, L., Wolf, R., Ernst, R., & Heisenberg, M. (1999). Context generalization in *Drosophila* visual learning requires the mushroom bodies. *Nature*, *400*(6746), 753–756.

Llinás, R. R. (2002). *I of the vortex: From neurons to self*. Cambridge, MA: MIT Press.

Lopes, G., & Kampff, A. R. (2015). Cortical control: Learning from the lamprey. *Current Biology*, *25*(5), R203–R205.

López, J. C., Vargas, J. P., Gómez, Y., & Salas, C. (2003). Spatial and non-spatial learning in turtles: The role of medial cortex. *Behavioural Brain Research*, *143*(2), 109–120.

Lowe, C. J., Clarke, D. N., Medeiros, D. M., Rockhsar, D. S., & Gerhart, J. (2015). The deuterostome context of chordate origins. *Nature*, *520*, 456–465.

Luisi, P. L. (1998). About various definitions of life. *Origins of Life and Evolution of the Biosphere*, *28*(4–6), 613–622.

Luo, Z. X., Yuan, C. X., Meng, Q. J., & Ji, Q. (2011). A Jurassic eutherian mammal and divergence of marsupials and placentals. *Nature*, *476*(7361), 442–445.

Lycan, W. G. (2009). Giving dualism its due. *Australasian Journal of Philosophy*, *87*(4), 551–563.

Ma, X. Y., Hou, X. G., Edgecombe, G. D., & Strausfeld, N. J. (2012). Complex brain and optic lobes in an early Cambrian arthropod. *Nature*, *490*, 258–261.

Ma, X., Cong, P., Hou, X., Edgecombe, G. D., & Strausfeld, N. J. (2014). An exceptionally preserved arthropod cardiovascular system from the early Cambrian. *Nature Communications*, *5*(3560), 1–7.

Machin, K. L. (1999). Amphibian pain and analgesia. *Journal of Zoo and Wildlife Medicine*, *30*(1), 2–10.

Mackie, G. O., & Burighel, P. (2005). The nervous system in adult tunicates: Current research directions. *Canadian Journal of Zoology*, *83*, 151–183.

MacLean, P. D. (1975). Sensory and perceptive factors in emotional functions of the triune brain. *Biological Foundations of Psychiatry*, *1*, 177–198.

Magosso, E. (2010). Integrating information from vision and touch: A neural network modeling study. *Information Technology in Biomedicine. IEEE Transactions on*, *14*(3), 598–612.

Maier, A., Wilke, M., Aura, C., Zhu, C., Frank, Q. Y., & Leopold, D. A. (2008). Divergence of fMRI and neural signals in V1 during perceptual suppression in the awake monkey. *Nature Neuroscience*, *11*(10), 1193–1200.

Maisey, J. G. (1996). *Discovering fossil fishes*. New York: Holt.

Mallatt, J. (1982). Pumping rates and particle retention efficiencies of the larval lamprey, an unusual suspension feeder. *Biological Bulletin*, *163*(1), 197–210.

Mallatt, J. (1996). Ventilation and the origin of jawed vertebrates: A new mouth. *Zoological Journal of the Linnean Society*, *117*(4), 329–404.

Mallatt, J. (1997). Hagfish do not resemble ancestral vertebrates. *Journal of Morphology*, *232*, 293.

Mallatt, J. (2008). The origin of the vertebrate jaw: Neoclassical ideas versus newer, development-based ideas. *Zoological Science*, *25*(10), 990–998.

Mallatt, J. (2009). Evolution and phylogeny of chordates. In M. D. Binder, N. Hirokawa, & U. Windhorst (Eds.), *Encyclopedia of neurosciences* (pp. 1201–1208). Berlin: Springer-Verlag.

Mallatt, J., & Chen, J. Y. (2003). Fossil sister group of craniates: Predicted and found. *Journal of Morphology*, *258*(1), 1–31.

Mallatt, J., & Holland, N. D. (2013). *Pikaia gracilens* Walcott: Stem chordate, or already specialized in the Cambrian? *Journal of Experimental Zoology*, *320B*, 247–271.

Maloof, A. C., Rose, C. V., Beach, R., Samuels, B. M., Calmet, C. C., Erwin, D. H., … Simons, F. J. (2010). Possible animal-body fossils in pre-Marinoan limestones from South Australia. *Nature Geoscience*, *3*(9), 653–659.

Mancini, F., Haggard, P., Iannetti, G. D., Longo, M. R., & Sereno, M. I. (2012). Fine-grained nociceptive maps in primary somatosensory cortex. *Journal of Neuroscience*, *32*(48), 17155–17162.

Manger, P. R. (2009). Evolution of the reticular formation. In M. D. Binder, N. Hirokawa, & U. Windhorst (Eds.), *Encyclopedia of neurosciences* (pp. 1413–1416). Berlin: Springer.

Marco-Pallarés, J., Münte, T. F., & Rodríguez-Fornells, A. (2015). The role of high-frequency oscillatory activity in reward processing and learning. *Neuroscience and Biobehavioral Reviews*, *49*, 1–7.

Marieb, E., Wilhelm, P., & Mallatt, J. (2014). *Human anatomy* (7th ed.). San Francisco, CA: Pearson.

Marín, G., Salas, C., Sentis, E., Rojas, X., Letelier, J. C., & Mpodozis, J. (2007). A cholinergic gating mechanism controlled by competitive interactions in the optic tectum of the pigeon. *Journal of Neuroscience, 27*(30), 8112–8121.

Martin, T. J., Kim, S. A., & Eisenach, J. C. (2006). Clonidine maintains intrathecal self-administration in rats following spinal nerve ligation. *Pain, 125*(3), 257–263.

Martínez-de-la-Torre, M., Pombal, M. A., & Puelles, L. (2011). Distal-less-like protein distribution in the larval lamprey forebrain. *Neuroscience, 178*, 270–284.

Mather, J. A. (2008). Cephalopod consciousness: Behavioural evidence. *Consciousness and Cognition, 17*(1), 37–48.

Mather, J. (2012). Cephalopod intelligence. In J. Vonk & T. K. Shackelford (Eds.), *The Oxford handbook of comparative evolutionary psychology* (pp. 118–128). Oxford: Oxford University Press.

Mather, J. A., & Kuba, M. J. (2013). The cephalopod specialties: Complex nervous system, learning, and cognition 1. *Canadian Journal of Zoology, 91*(6), 431–449.

Matheson, T. (2002). *Invertebrate nervous systems: Encyclopedia of life sciences* (pp. 1–5). London: Nature Publishing Group.

Mathur, P., Berberoglu, M. A., & Guo, S. (2011). Preference for ethanol in zebrafish following a single exposure. *Behavioural Brain Research, 217*(1), 128–133.

Matthews, G., & Wickelgren, W. O. (1978). Trigeminal sensory neurons of the sea lamprey. *Journal of Comparative Physiology, 123*(4), 329–333.

Matthies, B. K., & Franklin, K. B. (1992). Formalin pain is expressed in decerebrate rats but not attenuated by morphine. *Pain, 51*(2), 199–206.

Maximino, C., Lima, M. G., Oliveira, K. R. M., Batista, E. D. J. O., & Herculano, A. M. (2013). "Limbic associative" and "autonomic" amygdala in teleosts: A review of the evidence. *Journal of Chemical Neuroanatomy, 48*, 1–13.

Maynard, J. (2013). Archaeopteryx is a bird after all—just not the first. *iTech Post*, May 29. http://www.itechpost.com/articles/9940/20130529/archaeopteryx-bird-first.htm.

Mayr, E. (1982). *The growth of biological thought: Diversity, evolution, and inheritance.* Cambridge, MA: Harvard University Press.

Mayr, E. (2004). *What makes biology unique? Considerations on the autonomy of a scientific discipline.* Cambridge: Cambridge University Press.

Mazaheri, A., & Van Diepen, R. (2014). Gamma oscillations in a bind? *Cerebral Cortex.* doi:10.1093/cercor/bhu136.

Mazet, F., & Shimeld, S. M. (2002). The evolution of chordate neural segmentation. *Developmental Biology, 251*, 258–270.

McHaffie, J. G., Kao, C. Q., & Stein, B. E. (1989). Nociceptive neurons in rat superior colliculus: Response properties, topography, and functional implications. *Journal of Neurophysiology*, *62*(2), 510–525.

McClendon, J., Lecaude, S., Dores, A. R., & Dores, R. M. (2010). Evolution of the opioid/ORL-1 receptor gene family. *Annals of the New York Academy of Sciences*, *1200*(1), 85–94.

McLoughlin, J. C. (1980). *Synapsida: A new look into the origins of mammals*. New York: Viking Press.

Medina, L. (2009). Evolution and embryological development of the forebrain. In M. D. Binder, N. Hirokawa, & U. Windhorst (Eds.), *Encyclopedia of neurosciences* (pp. 1172–1192). Berlin: Springer.

Meehl, P. (1966). The complete autocerebroscopist: A thought experiment on Professor Feigl's mind/body identify thesis. In P. K. Feyerabend & G. Maxwell (Eds.), *Mind, matter, and method* (p. 103–180). Minneapolis: University of Minnesota Press.

Mehta, N., & Mashour, G. A. (2013). General and specific consciousness: A first-order representationalist approach. *Frontiers in psychology*, *4*.

Melloni, L., Molina, C., Pena, M., Torres, D., Singer, W., & Rodriguez, E. (2007). Synchronization of neural activity across cortical areas correlates with conscious perception. *Journal of Neuroscience*, *27*(11), 2858–2865.

Ménard, A., & Grillner, S. (2008). Diencephalic locomotor region in the lamprey—afferents and efferent control. *Journal of Neurophysiology*, *100*(3), 1343–1353.

Mendl, M., Paul, E. S., & Chittka, L. (2011). Animal behaviour: Emotion in invertebrates? *Current Biology*, *21*(12), R463–R465.

Merker, B. (2005). The liabilities of mobility: A selection pressure for the transition to consciousness in animal evolution. *Consciousness and Cognition*, *14*(1), 89–114.

Merker, B. (2007). Consciousness without a cerebral cortex: A challenge for neuroscience and medicine. *Behavioral and Brain Sciences*, *30*(01), 63–81.

Mersha, M., Formisano, R., McDonald, R., Pandey, P., Tavernarakis, N., & Harbinder, S. (2013). GPA-14, a Gαi subunit mediates dopaminergic behavioral plasticity in *C. elegans*. *Behavioral and Brain Functions*, *9*(1), 16.

Mescher, A. L. (2013). *Junquiera's basic histology: Text and atlas* (13th ed.). New York: McGraw Hill.

Mesquita, F. D. O. (2011). *Coping styles and learning in fish: Developing behavioural tools for welfare-friendly aquaculture*. Doctoral dissertation, University of Glasgow.

Messenger, J. B. (1979). The nervous system of *Loligo*: IV. The peduncle and olfactory lobes. *Philosophical Transactions of the Royal Society of London, Series B, Biological Sciences*, *285*(1008), 275–309.

Metzinger, T. (2003). *Being no one: The self-model theory of subjectivity.* Cambridge, MA: MIT Press.

Meysman, F. J., Middelburg, J. J., & Heip, C. H. (2006). Bioturbation: A fresh look at Darwin's last idea. *Trends in Ecology & Evolution, 21*(12), 688–695.

Miller, E. K., & Buschman, T. J. (2013). Cortical circuits for the control of attention. *Current Opinion in Neurobiology, 23*(2), 216–222.

Millot, S., Cerqueira, M., Castanheira, M. F., Øverli, Ø., Martins, C. I., & Oliveira, R. F. (2014). Use of conditioned place preference/avoidance tests to assess affective states in fish. *Applied Animal Behaviour Science, 154,* 104–111.

Millsopp, S., & Laming, P. (2008). Trade-offs between feeding and shock avoidance in goldfish (*Carassius auratus*). *Applied Animal Behaviour Science, 113*(1), 247–254.

Min, B. K. (2010). A thalamic reticular networking model of consciousness. *Theoretical Biology & Medical Modelling, 7*(10).

Minakata, H. (2010). Oxytocin/vasopressin and gonadotropin-releasing hormone from cephalopods to vertebrates. *Annals of the New York Academy of Sciences, 1200*(1), 3.

Mitchell, M. (2009). *Complexity: A guided tour.* Oxford: Oxford University Press.

Mobley, A. S., Michel, W. C., & Lucero, M. T. (2008). Odorant responsiveness of squid olfactory receptor neurons. *Anatomical Record, 291*(7), 763–774.

Mogil, J. S. (2009). Animal models of pain: Progress and challenges. *Nature Reviews: Neuroscience, 10*(4), 283–294.

Mole, C. (2008). Attention and consciousness. *Journal of Consciousness Studies, 15*(4), 86–104.

Monod, J. (1971). *Chance and necessity: An essay on the natural philosophy of modern biology.* New York: Alfred A. Knopf.

Montgomery, S. (2015). *The soul of an octopus: A surprising exploration into the world of consciousness.* New York: Atria Books.

Montoya, P., & Schandry, R. (1994). Emotional experience and heartbeat perception in patients with spinal cord injury and control subjects. *Journal of Psychophysiology, 8,* 289–296.

Moreno, N., & González, A. (2007). Regionalization of the telencephalon in urodele amphibians and its bearing on the identification of the amygdaloid complex. *Frontiers in Neuroanatomy, 1.*

Moret, F., Christiaen, L., Deyts, C., Blin, M., Joly, J. S., & Vernier, P. (2005). The dopamine-synthesizing cells in the swimming larva of the tunicate *Ciona intestinalis* are located only in the hypothalamus-related domain of the sensory vesicle. *European Journal of Neuroscience, 21*(11), 3043–3055.

Moret, F., Guilland, J. C., Coudouel, S., Rochette, L., & Vernier, P. (2004). Distribution of tyrosine hydroxylase, dopamine, and serotonin in the central nervous system of amphioxus (*Branchiostoma lanceolatum*): Implications for the evolution of catecholamine systems in vertebrates. *Journal of Comparative Neurology, 468*(1), 135–150.

Mori, I. (1999). Genetics of chemotaxis and thermotaxis in the nematode *Caenorhabditis elegans*. *Annual Review of Genetics, 33*(1), 399–422.

Mori, K., Manabe, H., Narikiyo, K., & Onisawa, N. (2013). Olfactory consciousness and gamma oscillation couplings across the olfactory bulb, olfactory cortex, and orbitofrontal cortex. *Frontiers in Psychology, 4*.

Moroz, L. L., Kocot, K. M., Citarella, M. R., Dosung, S., Norekian, T. P., Povolotskaya, I. S., … Kohn, A. B. (2014). The ctenophore genome and the evolutionary origins of neural systems. *Nature, 510*(7503), 109–114.

Mouton, L. J., & Holstege, G. (2000). Segmental and laminar organization of the spinal neurons projecting to the periaqueductal gray (PAG) in the cat suggests the existence of at least five separate clusters of spino-PAG neurons. *Journal of Comparative Neurology, 428*(3), 389–410.

Mudrik, L., Faivre, N., & Koch, C. (2014). Information integration without awareness. *Trends in Cognitive Sciences, 18*(9), 488–496.

Mueller, T. (2012). What is the thalamus in zebrafish? *Frontiers in Neuroscience, 6*.

Mueller, T., Vernier, P., & Wullimann, M. F. (2004). The adult central nervous cholinergic system of a neurogenetic model animal, the zebrafish *Danio rerio*. *Brain Research, 1011*(2), 156–169.

Müller, C. M., & Leppelsack, H. J. (1985). Feature extraction and tonotopic organization in the avian auditory forebrain. *Experimental Brain Research, 59*(3), 587–599.

Murakami, Y., & Kuratani, S. (2008). Brain segmentation and trigeminal projections in the lamprey; with reference to vertebrate brain evolution. *Brain Research Bulletin, 75*(2), 218–224.

Muzio, R. N., Segura, E. T., & Papini, M. R. (1994). Learning under partial reinforcement in the toad (Bufo arenarum): Effects of lesions in the medial pallium. *Behavioral and Neural Biology, 61*(1), 36–46.

Näätänen, R., Paavilainen, P., Rinne, T., & Alho, K. (2007). The mismatch negativity (MMN) in basic research of central auditory processing: A review. *Clinical Neurophysiology, 118*(12), 2544–2590.

Nadel, L., Hupbach, A., Gomez, R., & Newman-Smith, K. (2012). Memory formation, consolidation, and transformation. *Neuroscience and Biobehavioral Reviews, 36*(7), 1640–1645.

Nagel, T. (1974). What is it like to be a bat? *Philosophical Review, 83*(4), 435–450.

Nagel, T. (1989). *The view from nowhere*. New York: Oxford University Press.

Namburi, P., Beyeler, A., Yorozu, S., Calhoo, G. G., Halbert, S. A., Wichmann, R., ... Tye, K. M. (2015). A circuit mechanism for differentiating positive and negative mechanisms. *Nature*, *520*, 675–678.

Nargeot, R., & Simmers, J. (2011). Neural mechanisms of operant conditioning and learning-induced behavioral plasticity in Aplysia. *Cellular and Molecular Life Sciences*, *68*(5), 803–816.

Nargeot, R., & Simmers, J. (2012). Functional organization and adaptability of a decision-making network in Aplysia. *Frontiers in Neuroscience*, *6*(113).

Navratilova, E., Xie, J. Y., Okun, A., Qu, C., Eyde, N., Ci, S., ... Porreca, F. (2012). Pain relief produces negative reinforcement through activation of mesolimbic reward-valuation circuitry. *Proceedings of the National Academy of Sciences of the United States of America*, *109*(50), 20709–20713.

Neary, T. J. (1995). Afferent projections to the hypothalamus in ranid frogs. *Brain, Behavior and Evolution*, *46*(1), 1–13.

Nebeling, B. (2000). Morphology and physiology of auditory and vibratory ascending interneurones in bushcrickets. *Journal of Experimental Zoology*, *286*(3), 219–230.

Nelson, E. E., Lau, J. Y., & Jarcho, J. M. (2014). Growing pains and pleasures: How emotional learning guides development. *Trends in Cognitive Sciences*, *18*(2), 99–108.

Nespolo, R. F., Bacigalupe, L. D., Figueroa, C. C., Koteja, P., & Opazo, J. C. (2011). Using new tools to solve an old problem: The evolution of endothermy in vertebrates. *Trends in Ecology & Evolution*, *26*(8), 414–423.

Nevin, L. M., Robles, E., Baier, H., & Scott, E. K. (2010). Focusing on optic tectum circuitry through the lens of genetics. *BMC Biology*, *8*(1), 126.

Newland, P. L., Rogers, S. M., Gaaboub, I., & Matheson, T. (2000). Parallel somatotopic maps of gustatory and mechanosensory neurons in the central nervous system of an insect. *Journal of Comparative Neurology*, *425*(1), 82–96.

Newman, S. A., Mezentseva, N. V., & Badyaev, A. V. (2013). Gene loss, thermogenesis, and the origin of birds. *Annals of the New York Academy of Sciences*, *1289*(1), 36–47.

Newman, S. W. (1999). The medial extended amygdala in male reproductive behavior a node in the mammalian social behavior network. *Annals of the New York Academy of Sciences*, *877*(1), 242–257.

Nichols, S., & Grantham, T. (2000). Adaptive complexity and phenomenal consciousness. *Philosophy of Science*, *67*, 648–670.

Nieuwenhuys, R. (1996). The greater limbic system, the emotional motor system, and the brain. *Progress in Brain Research*, *107*, 551–580.

Nieuwenhuys, R., & Nicholson, C. (1998). Lampreys, petromyzontoidea. In R. Nieuwenhuys, H. J. ten Donkelaar, & C. Nicholson (Eds.), *The central nervous system of vertebrates* (Vol. 1, pp. 397–495). Heidelberg: Springer Berlin.

Nieuwenhuys, R., ten Donkelaar, H. J., & Nicholson, C. (1998). *The central nervous system of vertebrates*. New York: Springer.

Nieuwenhuys, R., Veening, J. G., & Van Domburg, P. (1987). Cores and paracores: Some new chemoarchitectural entities in the mammalian neuraxis. *Acta Morphologica Neerlando-Scandinavica, 26*, 131.

Niimura, Y. (2009). On the origin and evolution of vertebrate olfactory receptor genes: Comparative genome analysis among 23 chordate species. *Genome Biology and Evolution, 1*, 34–44.

Nilsson, D.-E. (2009). The evolution of eyes and visually guided behaviour. *Philosophical Transactions of the Royal Society of London, Series B, Biological Sciences, 364*(1531), 2833–2847.

Nilsson, D.-E. (2013). Eye evolution and its functional basis. *Visual Neuroscience, 30*(1–2), 5–20.

Nilsson, S. (2011). Comparative anatomy of the autonomic nervous system. *Autonomic Neuroscience, 165*(1), 3–9.

Ninkovic, J., & Bally-Cuif, L. (2006). The zebrafish as a model system for assessing the reinforcing properties of drugs of abuse. *Methods (San Diego, Calif.), 39*(3), 262–274.

Noboa, V., & Gillette, R. (2013). Selective prey avoidance learning in the predatory sea slug *Pleurobranchaea californica*. *Journal of Experimental Biology, 216*(17), 3231–3236.

Nomura, T., Kawaguchi, M., Ono, K., & Murakami, Y. (2013). Reptiles: A new model for brain evo-devo research. *Journal of Experimental Zoology, Part B, Molecular and Developmental Evolution, 320*(2), 57–73.

Nomura, T., Murakami, Y., Gotoh, H., & Ono, K. (2014). Reconstruction of ancestral brains: Exploring the evolutionary process of encephalization in amniotes. *Neuroscience Research, 86*, 25–36.

Nordström, K. J., Fredriksson, R., & Schiöth, H. B. (2008). The amphioxus (*Branchiostoma floridae*) genome contains a highly diversified set of G protein-coupled receptors. *BMC Evolutionary Biology, 8*(1), 9.

Norman, G. J., Berntson, G. G., & Cacioppo, J. T. (2014). Emotion, somatovisceral afference, and autonomic regulation. *Emotion Review, 6*(2), 113–123.

North, H. A., Karim, A., Jacquin, M. F., & Donoghue, M. J. (2010). EphA4 is necessary for spatially selective peripheral somatosensory topography. *Developmental Dynamics, 239*(2), 630–638.

Northcutt, R. G. (1995). The forebrain of gnathostomes: In search of a morphotype. *Brain, Behavior and Evolution, 46*(4–5), 304–318.

Northcutt, R. G. (1996). The agnathan ark: The origin of craniate brains. *Brain, Behavior and Evolution, 48*(5), 237–247.

Northcutt, R. G. (2002). Understanding vertebrate brain evolution. *Integrative and Comparative Biology, 42*(4), 743–756.

Northcutt, R. G. (2005). The new head revisited. *Journal of Experimental Zoology, 304B*, 274–297.

Northcutt, R. G. (2012 a). Evolution of centralized nervous systems: Two schools of evolutionary thought. *Proceedings of the National Academy of Sciences of the United States of America, 109*(Suppl. 1), 10626–10633.

Northcutt, R. G. (2012 b). Variation in reptilian brains and cognition. *Brain, Behavior, and Evolution, 82*(1), 45–54.

Northcutt, R. G., & Puzdrowski, R. L. (1988). Projections of the olfactory bulb and nervus terminalis in the silver lamprey. *Brain, Behavior and Evolution, 32*(2), 96–107.

Northcutt, R. G., & Ronan, M. (1992). Afferent and efferent connections of the bullfrog medial pallium. *Brain, Behavior and Evolution, 40*(1), 1–16.

Northcutt, R. G., & Wicht, H. (1997). Afferent and efferent connections of the lateral and medial pallia of the silver lamprey. *Brain, Behavior and Evolution, 49*(1), 1–19.

Northmore, D. (2011). The optic tectum. In A. Farrell (Ed.), *The encyclopedia of fish physiology: From genome to environment* (pp. 131–142). San Diego, CA: Academic Press.

Northmore, D. P., & Gallagher, S. P. (2003). Functional relationship between nucleus isthmi and tectum in teleosts: Synchrony but no topography. *Visual Neuroscience, 20*(03), 335–348.

Northoff, G. (2012 a). From emotions to consciousness—a neuro-phenomenal and neuro-relational approach. *Frontiers in Psychology, 3*.

Northoff, G. (2012 b). Autoepistemic limitation and the brain's neural code: Comment on "Neuroontology, neurobiological naturalism, and consciousness: A challenge to scientific reduction and a solution" by Todd E. Feinberg. *Physics of Life Reviews, 9*(1), 38–39.

Northoff, G. (2013 a). What the brain's intrinsic activity can tell us about consciousness? A tri-dimensional view. *Neuroscience and Biobehavioral Reviews, 37*(4), 726–738.

Northoff, G. (2013 b). *Consciousness: Vol. 2. Unlocking the brain*. Oxford: Oxford University Press.

Northoff, G. (in press). Slow cortical potentials and "inner time consciousness"—A neuro-phenomenal hypothesis about the "width of present." *International Journal of Psychophysiology*.

Northoff, G., & Musholt, K. (2006). How can Searle avoid property dualism? Epistemic-ontological inference and autoepistemic limitation. *Philosophical Psychology, 19*(5), 589–605.

Nunez, P. L., & Srinivasan, R. (2010). Scale and frequency chauvinism in brain dynamics: Too much emphasis on gamma band oscillations. *Brain Structure & Function*, *215*(2), 67–71.

Ocaña, F. M., Suryanarayana, S. M., Saitoh, K., Kardamakis, A. A., Capantini, L., Robertson, B., & Grillner, S. (2015). The lamprey pallium provides a blueprint of the mammalian motor projections from cortex. *Current Biology*, *25*(4), 413–423.

O'Connell, L. A., & Hofmann, H. A. (2011). The vertebrate mesolimbic reward system and social behavior network: A comparative synthesis. *Journal of Comparative Neurology*, *519*(18), 3599–3639.

O'Connell, L. A., & Hofmann, H. A. (2012). Evolution of a vertebrate social decision-making network. *Science*, *336*(6085), 1154–1157.

Ohira, H. (2010). The somatic marker revisited: Brain and body in emotional decision making. *Emotion Review*, *2*, 245.

Ohyama, T., Schneider-Mizell, C., Fetter, R. D., Aleman, J. V., Franconville, R., Rivera-Alba, M., … Zlatic, M. (2015). A multilevel multimodal circuit enhances action selection in *Drosophila*. *Nature*, *520*, 633–639.

Okamoto, H., Agetsuma, M., & Aizawa, H. (2012). Genetic dissection of the zebrafish habenula, a possible switching board for selection of behavioral strategy to cope with fear and anxiety. *Developmental Neurobiology*, *72*(3), 386–394.

Orpwood, R. (2013). Qualia could arise from information processing in local cortical networks. *Frontiers in Psychology*, *4*.

Ortega-Hernandez, J. (2015). Homology of head sclerites in Burgess Shale euarthropods. *Current Biology*, *25*, 1–7.

Ostrowski, T. D., & Stumpner, A. (2010). Frequency processing at consecutive levels in the auditory system of bush crickets (Tettigoniidae). *Journal of Comparative Neurology*, *518*(15), 3101–3116.

Overgaard, M. (2012). Blindsight: Recent and historical controversies on the blindness of blindsight. *Wiley Interdisciplinary Reviews: Cognitive Science*, *3*(6), 607–614.

Packard, A. (1972). Cephalopods and fish: The limits of convergence. *Biological Reviews of the Cambridge Philosophical Society*, *47*(2), 241–307.

Packard, A., & Delafield-Butt, J. T. (2014). Feelings as agents of selection: Putting Charles Darwin back into (extended neo-) Darwinism. *Biological Journal of the Linnean Society of London*, *112*(2), 332–353.

Pagán, O. R., Deats, S., Baker, D., Montgomery, E., Wilk, G., Tenaglia, M., & Semon, J. (2013). Planarians require an intact brain to behaviorally react to cocaine, but not to react to nicotine. *Neuroscience*, *246*, 265–270.

Pain, S. P. (2009). Signs of anger: Representation of agonistic behaviour in invertebrate cognition. *Biosemiotics*, *2*(2), 181–191.

Panagiotaropoulos, T. I., Deco, G., Kapoor, V., & Logothetis, N. K. (2012). Neuronal discharges and gamma oscillations explicitly reflect visual consciousness in the lateral prefrontal cortex. *Neuron, 74*(5), 924–935.

Pani, A. M., Mullarkey, E. E., Aronowicz, J., Assimacopoulos, S., Grove, E. A., & Lowe, C. J. (2012). Ancient deuterostome origins of vertebrate brain signalling centres. *Nature, 483*, 289–294.

Panksepp, J. (1998). *Affective neuroscience: The foundations of human and animal emotions.* Oxford: Oxford University Press.

Panksepp, J. (2005). Affective consciousness: Core emotional feelings in animals and humans. *Consciousness and Cognition, 14*(1), 30–80.

Panksepp, J. (2011). Cross-species affective neuroscience decoding of the primal affective experiences of humans and related animals. *PLoS One, 6*(9), e21236.

Panksepp, J. (2013). Toward an understanding of the constitution of consciousness through the laws of affect. *Neuropsychoanalysis: An Interdisciplinary Journal for Psychoanalysis and the Neurosciences, 15*(1), 62–65.

Papini, M. R. (2002). Pattern and process in the evolution of learning. *Psychological Review, 109*(1), 186.

Papini, M. R., & Bitterman, M. E. (1991). Appetitive conditioning in *Octopus cyanea*. *Journal of Comparative Psychology, 105*(2), 107.

Papini, M. R., & Dudley, R. T. (1997). Consequences of surprising reward omissions. *Review of General Psychology, 1*(2), 175.

Papini, M. R., Muzio, R. N., & Segura, E. T. (1995). Instrumental learning in toads (*Bufo arenarum*): Reinforcer magnitude and the medial pallium. *Brain, Behavior and Evolution, 46*(2), 61–71.

Parker, A. (2009). *In the blink of an eye: How vision sparked the big bang of evolution.* New York: Basic Books.

Paterson, J. R., García-Bellido, D. C., Lee, M. S., Brock, G. A., Jago, J. B., & Edgecombe, G. D. (2011). Acute vision in the giant Cambrian predator *Anomalocaris* and the origin of compound eyes. *Nature, 480*(7376), 237–240.

Paton, R. L., Smithson, T. R., & Clack, J. A. (1999). An amniote-like skeleton from the early Carboniferous of Scotland. *Nature, 398*(6727), 508–513.

Pattee, H. H. (1970). The problem of biological hierarchy. In C. H. Waddington (Ed.), *Towards a theoretical biology* (Vol. 3, pp. 117–136). Chicago: Aldine.

Patthey, C., Schlosser, G., & Shimeld, S. M. (2014). The evolutionary history of vertebrate cranial placodes—I: Cell type evolution. *Developmental Biology, 389*, 82–97.

Pecoits, E., Konhauser, K. O., Aubet, N. R., Heaman, L. M., Veroslavsky, G., Stern, R. A., & Gingras, M. K. (2012). Bilaterian burrows and grazing behavior at >585 million years ago. *Science, 336*(6089), 1693–1696.

Penrose, R. (1994). *Shadows of the mind: A search for the missing science of consciousness.* Oxford: Oxford University Press.

Pepperberg, I. (2009). *Alex and me.* New York: HarperCollins.

Pérez-Pérez, M. P., Luque, M. A., Herrero, L., Nunez-Abades, P. A., & Torres, B. (2003). Afferent connectivity to different functional zones of the optic tectum in goldfish. *Visual Neuroscience, 20*(04), 397–410.

Perry, C. J., & Barron, A. B. (2013). Neural mechanisms of reward in insects. *Annual Review of Entomology, 58,* 543–562.

Perry, C. J., Barron, A. B., & Cheng, K. (2013). Invertebrate learning and cognition: Relating phenomena to neural substrate. *Wiley Interdisciplinary Reviews: Cognitive Science, 4*(5), 561–582.

Petersen, C. L., Timothy, M., Kim, D. S., Bhandiwad, A. A., Mohr, R. A., Sisneros, J. A., & Forlano, P. M. (2013). Exposure to advertisement calls of reproductive competitors activates vocal-acoustic and catecholaminergic neurons in the plainfin midshipman fish, *Porichthys notatus. PLoS One, 8*(8), e70474.

Philippe, H., Brinkmann, H., Lavrov, D. V., Littlewood, D. T. J., Manuel, M., Wörheide, G., & Baurain, D. (2011). Resolving difficult phylogenetic questions: Why more sequences are not enough. *PLoS Biology, 9*(3), e1000602.

Pierre, J., Mahouche, M., Suderevskaya, E. I., Reperant, J., & Ward, R. (1997). Immunocytochemical localization of dopamine and its synthetic enzymes in the central nervous system of the lamprey *Lampetra fluviatilis. Journal of Comparative Neurology, 380*(1), 119–135.

Pilastro, A., Griggio, M., Biddau, L., & Mingozzi, T. (2002). Extrapair paternity as a cost of polygyny in the rock sparrow: Behavioural and genetic evidence of the 'trade-off' hypothesis. *Animal Behaviour, 63*(5), 967–974.

Planavsky, N. J., Reinhard, C. T., Wang, X., Thomson, D., McGoldrick, P., Rainbird, R. H., ... Lyons, T. W. (2014). Low Mid-Proterozoic atmospheric oxygen levels and the delayed rise of animals. *Science, 346*(6209), 635–638.

Plotnick, R. E., Dornbos, S. Q., & Chen, J. (2010). Information landscapes and sensory ecology of the Cambrian Radiation. *Paleobiology, 36,* 303–317.

Plotnik, J. M., De Waal, F. B., & Reiss, D. (2006). Self-recognition in an Asian elephant. *Proceedings of the National Academy of Sciences of the United States of America, 103*(45), 17053–17057.

Plutchik, R. (2001). The nature of emotions. *American Scientist, 89*(4), 344–350.

Pollen, D. A. (2011). On the emergence of primary visual perception. *Cerebral Cortex, 21*(9), 1941–1953.

Pombal, M. A., Manira, A. E., & Grillner, S. (1997). Afferents of the lamprey striatum with special reference to the dopaminergic system: A combined tracing and immunohistochemical study. *Journal of Comparative Neurology, 386*(1), 71–91.

Pombal, M. A., Marín, O., & González, A. (2001). Distribution of choline acetyltransferase-immunoreactive structures in the lamprey brain. *Journal of Comparative Neurology*, *431*(1), 105–126.

Popper, K. R., & Eccles, J. C. (1977). *The self and its brain*. New York: Springer.

Portavella, M., Vargas, J. P., Torres, B., & Salas, C. (2002). The effects of telencephalic pallial lesions on spatial, temporal, and emotional learning in goldfish. *Brain Research Bulletin*, *57*(3), 397–399.

Prechtl, J. C., von der Emde, G., Wolfart, J., Karamürsel, S., Akoev, G. N., Andrianov, Y. N., & Bullock, T. H. (1998). Sensory processing in the pallium of a mormyrid fish. *Journal of Neuroscience*, *18*(18), 7381–7393.

Presley, G. M., Lonergan, W., & Chu, J. (2010). Effects of amphetamine on conditioned place preference and locomotion in the male green tree frog, *Hyla cinerea*. *Brain, Behavior and Evolution*, *75*(4), 262–270.

Preuss, S. J., Trivedi, C. A., vom Berg-Maurer, C. M., Ryu, S., & Bollmann, J. H. (2014). Classification of object size in retinotectal microcircuits. *Current Biology*, *24*(20), 2376–2385.

Putnam, N. H., Butts, T., Ferrier, D. E. K., Furlong, R. F., Hellsten, U., Kawashima, T., … Rokhsar, D. S. (2008). The amphioxus genome and the evolution of the chordate karyotype. *Nature*, *453*, 1064–1071.

Puzdrowski, R. L. (1988). Afferent projections of the trigeminal nerve in the goldfish, *Carassius auratus*. *Journal of Morphology*, *198*(2), 131–147.

Qin, J., & Wheeler, A. R. (2007). Maze exploration and learning in *C. elegans*. *Lab on a Chip*, *7*(2), 186–192.

Quigley, K. S., & Barrett, L. F. (2014). Is there consistency and specificity of autonomic changes during emotional episodes? Guidance from the Conceptual Act Theory and psychophysiology. *Biological Psychology*, *98*, 82–94.

Quinn, B. T., Carlson, C., Doyle, W., Cash, S. S., Devinsky, O., Spence, C., … Thesen, T. (2014). Intracranial cortical responses during visual-tactile integration in humans. *Journal of Neuroscience*, *34*(1), 171–181.

Quintana-Urzainqui, I., Sueiro, C., Carrera, I., Ferreiro-Galve, S., Santos-Durán, G., Mazan, S., … Rodríguez-Moldez, I. (2012). Contributions of developmental studies in the dogfish *Scyliorhinus canicula* to the brain anatomy of elasmobranchs: Insights on the basal ganglia. *Brain, Behavior and Evolution*, *80*(2), 127–141.

Quintino-dos-Santos, J. W., Müller, C. J. T., Bernabé, C. S., Rosa, C. A., Tufik, S., & Schenberg, L. C. (2014). Evidence that the periaqueductal gray matter mediates the facilitation of panic-like reactions in neonatally-isolated adult rats. *PLoS One*, *9*(3), e90726.

Quirk, G. J., & Beer, J. S. (2006). Prefrontal involvement in the regulation of emotion: Convergence of rat and human studies. *Current Opinion in Neurobiology*, *16*(6), 723–727.

Quiroga, R. Q., Kreiman, G., Koch, C., & Fried, I. (2008). Sparse but not "grandmother-cell"coding in the medial temporal lobe. *Trends in Cognitive Sciences, 12*(3), 87–91.

Quiroga, R. Q., Reddy, L., Kreiman, G., Koch, C., & Fried, I. (2005). Invariant visual representation by single neurons in the human brain. *Nature, 435*(7045), 1102–1107.

Raffa, R. B., Baron, S., Bhandal, J. S., Brown, T., Song, K., Tallarida, C. S., & Rawls, S. M. (2013). Opioid receptor types involved in the development of nicotine physical dependence in an invertebrate (*Planaria*) model. *Pharmacology, Biochemistry, and Behavior, 112*, 9–14.

Rattenborg, N. C., & Martinez-Gonzalez, D. (2011). A bird-brain view of episodic memory. *Behavioural Brain Research, 222*(1), 236–245.

Reaume, C. J., & Sokolowski, M. B. (2011). Conservation of gene function in behaviour. *Philosophical Transactions of the Royal Society of London, Series B, Biological Sciences, 366*(1574), 2100–2110.

Reilly, S. C., Quinn, J. P., Cossins, A. R., & Sneddon, L. U. (2008). Behavioural analysis of a nociceptive event in fish: Comparisons between three species demonstrate specific responses. *Applied Animal Behaviour Science, 114*(1), 248–259.

Repérant, J., Ward, R., Médina, M., Kenigfest, N. B., Rio, J. P., Miceli, D., & Jay, B. (2009). Synaptic circuitry in the retinorecipient layers of the optic tectum of the lamprey (*Lampetra fluviatilis*): A combined hodological, GABA, and glutamate immunocytochemical study. *Brain Structure & Function, 213*(4–5), 395–422.

Revonsuo, A. (2006). *Inner presence: Consciousness as a biological phenomenon*. Cambridge, MA: MIT Press.

Revonsuo, A. (2010). *Consciousness: The science of subjectivity*. Hove: Psychology Press.

Ribary, U. (2005). Dynamics of thalamo-cortical network oscillations and human perception. *Progress in Brain Research, 150*, 127–142.

Rink, E., & Wullimann, M. F. (1998). Some forebrain connections of the gustatory system in the goldfish *Carassius auratus* visualized by separate DiI application to the hypothalamic inferior lobe and the torus lateralis. *Journal of Comparative Neurology, 394*(2), 152–170.

Rink, E., & Wullimann, M. F. (2001). The teleostean (zebrafish) dopaminergic system ascending to the subpallium (striatum) is located in the basal diencephalon (posterior tuberculum). *Brain Research, 889*(1), 316–330.

Risi, S., & Stanley, K. O. (2014). Guided self-organization in indirectly encoded and evolving topographic maps. In *Proceedings of the 2014 Conference on Genetic and Evolutionary Computation* (pp. 713–720). New York: ACM Press.

Robertson, B., Auclair, F., Ménard, A., Grillner, S., & Dubuc, R. (2007). GABA distribution in lamprey is phylogenetically conserved. *Journal of Comparative Neurology, 503*(1), 47–63.

Robertson, B., Huerta-Ocampo, I., Ericsson, J., Stephenson-Jones, M., Pérez-Fernández, J., Bolam, J. P., ... Grillner, S. (2012). The dopamine D2 receptor gene in lamprey, its expression in the striatum, and cellular effects of D2 receptor activation. *PLoS One*, *7*(4), e35642.

Robertson, B., Kardamakis, A., Capantini, L., Perez-Fernandez, J., Suryanarayana, S. M., Wallen, P., ... Grillner, S. (2014). The lamprey blueprint of the mammalian nervous system. *Progress in Brain Research*, *212*, 337–349.

Robertson, B., Saitoh, K., Ménard, A., & Grillner, S. (2006). Afferents of the lamprey optic tectum with special reference to the GABA input: Combined tracing and immunohistochemical study. *Journal of Comparative Neurology*, *499*(1), 106–119.

Robinson, W. (2007). Evolution and epiphenomenalism. *Journal of Consciousness Studies*, *14*(11), 27–42.

Roch, G. J., Tello, J. A., & Sherwood, N. M. (2014). At the transition from invertebrates to vertebrates, a novel GnRH-like peptide emerges in amphioxus. *Molecular Biology and Evolution*, *31*(4), 765–778.

Rodríguez-Moldes, I., Molist, P., Adrio, F., Pombal, M. A., Yáñez, S. E. P., Mandado, M., ... Anadón, R. (2002). Organization of cholinergic systems in the brain of different fish groups: A comparative analysis. *Brain Research Bulletin*, *57*(3), 331–334.

Rolls, E. T. (2000). On the brain and emotion. *Behavioral and Brain Sciences*, *23*(02), 219–228.

Rolls, E. T. (2012). Multisensory neuronal convergence of taste, somatosensory, visual, olfactory, and auditory inputs. In B. E. Stein (Ed.), *New handbook of multisensory processes* (pp. 311–331). Cambridge, MA: MIT Press.

Rolls, E. T. (2014). Emotion and decision-making explained: Précis: Synopsis of book published by Oxford University Press 2014. *Cortex*, *59*, 185–193.

Rolls, E. T. (2015). Limbic systems for emotion and for memory, but no single limbic system. *Cortex*, *62*, 119–157.

Romer, A. S. (1970). *The vertebrate body* (4th ed.). Philadelphia: Saunders.

Ronan, M., & Northcutt, R. G. (1987). Primary projections of the lateral line nerves in adult lampreys. *Brain, Behavior and Evolution*, *30*(1–2), 62–81.

Ronan, M., & Northcutt, R. G. (1998). The central nervous system of hagfishes. In J. Jorgensen, J. Lomholt, R. Weber, & H. Malte (Eds.), *The biology of hagfishes* (pp. 452–479). Dordrecht: Springer Netherlands.

Roozendaal, B. (2009). Emotional learning/memory. In M. D. Binder, N. Hirokawa, & U. Windhorst (Eds.), *Encyclopedia of neurosciences* (pp. 1095–1100). Berlin: Springer.

Roques, J. A., Abbink, W., Geurds, F., van de Vis, H., & Flik, G. (2010). Tailfin clipping, a painful procedure: Studies on *Nile tilapia* and common carp. *Physiology & Behavior*, *101*(4), 533–540.

References

Rose, J. D. (2002). The neurobehavioral nature of fishes and the question of awareness and pain. *Reviews in Fisheries Science*, *10*(1), 1–38.

Rose, J. D., Arlinghaus, R., Cooke, S. J., Diggles, B. K., Sawynok, W., Stevens, E. D., & Wynne, C. D. L. (2014). Can fish really feel pain? *Fish and Fisheries*, *15*(1), 97–133.

Rose, J. D., & Flynn, F. W. (1993). Lordosis response components can be elicited in decerebrate rats by combined flank and cervix stimulation. *Physiology & Behavior*, *54*(2), 357–361.

Rose, J. D., & Woodbury, C. J. (2008). Animal models of nociception and pain. In P. M. Conn (Ed.), *Sourcebook of models for biomedical research* (pp. 333–339). New York: Humana Press.

Roskies, A. L. (1999). The binding problem. *Neuron*, *24*(1), 7–9.

Rovainen, C. M. (1979). Neurobiology of lampreys. *Physiological Reviews*, *59*(4), 1007–1077.

Rovainen, C. M., & Yan, Q. (1985). Sensory responses of dorsal cells in the lamprey brain. *Journal of Comparative Physiology, A, Neuroethology, Sensory, Neural, and Behavioral Physiology*, *156*(2), 181–183.

Rowe, T. B., Macrini, T. E., & Luo, Z. X. (2011). Fossil evidence on origin of the mammalian brain. *Science*, *332*(6032), 955–957.

Ruppert, E. E. (1997). Cephalochordata (Acrania). In F. W. Harrison & E. E. Ruppert (Eds.), *Microscopic anatomy of invertebrates, Hemichordata, Chaetognatha, and the invertebrate chordates* (Vol. 15, pp. 349–504). New York: Wiley-Liss.

Ruppert, E., Fox, R., & Barnes, R. (2004). *Invertebrate zoology, a functional evolutionary approach* (7th ed.). Belmont, CA: Thomson: Brooks/Cole.

Ruse, M. (2005). Reductionism. In T. Honderich (Ed.), *The Oxford companion to philosophy* (2nd ed.), p. 793. Oxford: Oxford University Press.

Ruta, M., Botha-Brink, J., Mitchell, S. A., & Benton, M. J. (2013). The radiation of cynodonts and the ground plan of mammalian morphological diversity. *Proceedings of the Royal Society B, Biological Sciences*, *280*(1769), 20131865.

Rutledge, R. B., Skandali, N., Dayan, P., & Dolan, R. J. (2015). Dopaminergic modulation of decision making and subjective well-being. *Journal of Neuroscience*, *35*(27), 9811–9822.

Rutten, K., Van Der Kam, E. L., De Vry, J., Bruckmann, W., & Tzschentke, T. M. (2011). The mGluR5 antagonist 2-methyl-6-(phenylethynyl)-pyridine (MPEP) potentiates conditioned place preference induced by various addictive and non-addictive drugs in rats. *Addiction Biology*, *16*(1), 108–115.

Ryczko, D., Grätsch, S., Auclair, F., Dubé, C., Bergeron, S., Alpert, M. H., ... Dubuc, R. (2013). Forebrain dopamine neurons project down to a brainstem region controlling locomotion. *Proceedings of the National Academy of Sciences of the United States of America*, *110*(34), E3235–E3242.

Sada, N., Lee, S., Katsu, T., Otsuki, T., & Inoue, T. (2015). Targeting LDH enzymes with a stiripentol analog to treat epilepsy. *Science, 347*(6228), 1362–1367.

Saidel, W. M. (2009). Evolution of the optic tectum in anamniotes. In M. D. Binder, N. Hirokawa, & U. Windhorst (Eds.), *Encyclopedia of neurosciences* (pp. 1380–1387). Berlin: Springer.

Saitoh, K., Ménard, A., & Grillner, S. (2007). Tectal control of locomotion, steering, and eye movements in lamprey. *Journal of Neurophysiology, 97*(4), 3093–3108.

Salthe, S. N. (1985). *Evolving hierarchical systems: Their structure and representation*. New York: Columbia University Press.

Sánchez-Camacho, C., López, J. M., & González, A. (2006). Basal forebrain cholinergic system of the anuran amphibian *Rana perezi*: Evidence for a shared organization pattern with amniotes. *Journal of Comparative Neurology, 494*(6), 961–975.

Sansom, R. S., Freedman, K. I. M., Gabbott, S. E., Aldridge, R. J., & Purnell, M. A. (2010). Taphonomy and affinity of an enigmatic Silurian vertebrate, *Jamoytius kerwoodi* White. *Palaeontology, 53*(6), 1393–1409.

Saper, C. B. (2002). The central autonomic nervous system: Conscious visceral perception and autonomic pattern generation. *Annual Review of Neuroscience, 25*(1), 433–469.

Sarvestani, I. K., Kozlov, A., Harischandra, N., Grillner, S., & Ekeberg, Ö. (2013). A computational model of visually guided locomotion in lamprey. *Biological Cybernetics, 107*(5), 497–512.

Satoh, G. (2005). Characterization of novel GPCR gene coding locus in amphioxus genome: Gene structure, expression, and phylogenetic analysis with implications for its involvement in chemoreception. *Genesis, 41*, 47–57.

Schiff, N. D. (2008). Central thalamic contributions to arousal regulation and neurological disorders of consciousness. *Annals of the New York Academy of Sciences, 1129*(1), 105–118.

Schifirneţ, E., Bowen, S. E., & Borszcz, G. S. (2014). Separating analgesia from reward within the ventral tegmental area. *Neuroscience, 263*, 72–87.

Schlegel, P. A., Steinfartz, S., & Bulog, B. (2009). Non-visual sensory physiology and magnetic orientation in the Blind Cave Salamander, *Proteus anguinus* (and some other cave-dwelling urodele species). Review and new results on light-sensitivity and non-visual orientation in subterranean urodeles (Amphibia). *Animal Biology, 59*(3), 351–384.

Schlosser, G. (2014). Development and evolution of vertebrate cranial placodes. *Developmental Biology, 389*, 1.

Schmajuk, N. A., Segura, E. T., & Reboreda, J. C. (1980). Appetitive conditioning and discriminatory learning in toads. *Behavioral and Neural Biology, 28*(4), 392–397.

Schmalbruch, H. (1986). Fiber composition of the rat sciatic nerve. *Anatomical Record, 215*(1), 71–81.

Schluessel, V., & Bleckmann, H. (2005). Spatial memory and orientation strategies in the elasmobranch *Potamotrygon motoro*. *Journal of Comparative Physiology, A, Neuroethology, Sensory, Neural, and Behavioral Physiology*, *191*(8), 695–706.

Schmidt, M., & Ache, B. W. (1996). Processing of antennular input in the brain of the spiny lobster, *Panulirus argus*. I. Non-olfactory chemosensory and mechanosensory pathway of the lateral and median antennular neuropils. *Journal of Comparative Physiology, A, Neuroethology, Sensory, Neural, and Behavioral Physiology*, *178*, 579–604.

Schmidt, M., Van Ekeris, L., & Ache, B. W. (1992). Antennular projections to the midbrain of the spiny lobster. I. Sensory innervation of the lateral and medial antennular neuropils. *Journal of Comparative Neurology*, *318*(3), 277–290.

Schomburg, E. W., Fernández-Ruiz, A., Mizuseki, K., Berényi, A., Anastassiou, C. A., Koch, C., & Buzsáki, G. (2014). Theta phase segregation of input-specific gamma patterns in entorhinal-hippocampal networks. *Neuron*, *84*(2), 470–485.

Schopenhauer, A. (1813). *On the fourfold root of the principle of sufficient reason*. London: G. Bell.

Schopf, J. W., & Kudryavtsev, A. B. (2012). Biogenicity of Earth's earliest fossils: A resolution of the controversy. *Gondwana Research*, *22*(3), 761–771.

Schrödel, T., Prevedel, R., Aumayr, K., Zimmer, M., & Vaziri, A. (2013). Brain-wide 3D imaging of neuronal activity in *Caenorhabditis elegans* with sculpted light. *Nature Methods*, *10*(10), 1013–1020.

Schrödinger, E. (1967). *What is life? The physical aspects of living cell with mind and matter and autobiographical sketches*. Cambridge: Cambridge University Press.

Schuelert, N., & Dicke, U. (2005). Dynamic response properties of visual neurons and context-dependent surround effects on receptive fields in the tectum of the salamander *Plethodon shermani*. *Neuroscience*, *134*(2), 617–632.

Schultz, W. (2015). Neuronal reward and decision signals: From theories to data. *Physiological Reviews*, *95*(3), 853–951.

Searle, J. (1992). *The rediscovery of the mind*. Cambridge, MA: MIT Press.

Searle, J. (1997). *The mystery of consciousness*. New York: New York Review of Books.

Searle, J. R. (2002). Why I am not a property dualist. *Journal of Consciousness Studies*, *9*(12), 57–64.

Seelig, J. D., & Jayaraman, V. (2013). Feature detection and orientation tuning in the *Drosophila* central complex. *Nature*, *503*, 262–266.

Seelig, J. D., & Jayaraman, V. (2015). Neural dynamics for landmark orientation and angular path integration. *Nature*, *521*(7551), 186–191.

Sellars, W. (1963). *Science, perception, and reality*. London: Routledge & Kegan Paul.

Sellars, W. (1965). The identity approach to the mind–body problem. *Review of Metaphysics*, *18*(3), 430–451.

Šestak, M. S., Božičević, V., Bakarić, R., Dunjko, V., & Domazet-Lošo, T. (2013). Phylostratigraphic profiles reveal a deep evolutionary history of the vertebrate head sensory systems. *Frontiers in Zoology*, *10*(1).

Šestak, M. S., & Domazet-Lošo, T. (2015). Phylostratigraphic profiles in zebrafish uncover chordate origins of the vertebrate brain. *Molecular Biology and Evolution*, *32*(2), 299–312.

Seth, A. K. (2013). Interoceptive inference, emotion, and the embodied self. *Trends in Cognitive Sciences*, *17*(11), 565–573.

Seth, A. K., Baars, B. J., & Edelman, D. B. (2005). Criteria for consciousness in humans and other mammals. *Consciousness and Cognition*, *14*(1), 119–139.

Sevush, S. (2006). Single-neuron theory of consciousness. *Journal of Theoretical Biology*, *238*(3), 704–725.

Sewards, T. V., & Sewards, M. A. (2002). The medial pain system: Neural representations of the motivational aspect of pain. *Brain Research Bulletin*, *59*(3), 163–180.

Sha, A., Sun, H., & Wang, Y. (2013). Immunohistochemical observations of methionine-enkephalin and delta opioid receptor in the digestive system of *Octopus ocellatus*. *Tissue & Cell*, *45*(1), 83–87.

Shanahan, M., Bingman, V. P., Shimizu, T., Wild, M., & Güntürkün, O. (2013). Large-scale network organization in the avian forebrain: A connectivity matrix and theoretical analysis. *Frontiers in Computational Neuroscience*, *7*.

Shang, C., Liu, Z., Chen, Z., Shi, Y., Wang, Q., Liu, S., … Cao, P. (2015). A parvalbumin-positive excitatory visual pathway to trigger fear responses in mice. *Science*, *348*(6242), 1472–1477.

Shen, H. (2015). Neuroscience: The hard science of oxytocin. *Nature*, *522*(7557), 410–412.

Shepherd, G. M. (2007). Perspectives on olfactory processing, conscious perception, and orbitofrontal cortex. *Annals of the New York Academy of Sciences*, *1121*(1), 87–101.

Shepherd, G. M. (2012). *Neurogastronomy*. New York: Columbia University Press.

Sherrington, C. S. (1900). Cutaneous sensations. *Textbook of Physiology*, *2*, 920–1001.

Sherrington, C. S. (1906). *The integrative action of the nervous system*. Oxford: Oxford University Press.

Sherrington, C. S. (1947). *The integrative action of the nervous system* (with a new preface and full bibliography). New Haven, CT: Yale University Press.

Sherwood, N. M., & Lovejoy, D. A. (1993). Gonadotropin-releasing hormone in cartilaginous fishes: Structure, location, and transport. *Environmental Biology of Fishes*, *38*(1–3), 197–208.

Shibasaki, M., & Ishida, M. (2012). Effects of overtraining on extinction in newts (*Cynops pyrrhogaster*). *Journal of Comparative Psychology, 126*(4), 368.

Shigeno, S., & Ragsdale, C. W. (2015). The gyri of the octopus vertical lobe have distinct neurochemical identities. *Journal of Comparative Neurology, 523*(9), 1297–1317.

Shohat-Ophir, G., Kaun, K. R., Azanchi, R., Mohammed, H., & Heberlein, U. (2012). Sexual deprivation increases ethanol intake in *Drosophila*. *Science, 335*(6074), 1351–1355.

Shomrat, T., Graindorge, N., Bellanger, C., Fiorito, G., Loewenstein, Y., & Hochner, B. (2011). Alternative sites of synaptic plasticity in two homologous "fan-out fan-in" learning and memory networks. *Current Biology, 21*(21), 1773–1782.

Shomrat, T., Turchetti-Maia, A. L., Stern-Mentch, N., Basil, J. A., & Hochner, B. (2015). The vertical lobe of cephalopods: An attractive brain structure for understanding the evolution of advanced learning and memory systems. *Journal of Comparative Physiology A, 201*(9), 947–956.

Shu, D. G., Conway Morris, S., Han, J., Zhang, Z. F., Yasui, K., Janvier, P., ... Liu, H.-Q. (2003). Head and backbone of the Early Cambrian vertebrate *Haikouichthys*. *Nature, 421*(6922), 526–529.

Shu, D. G., Conway Morris, S., Zhang, Z. F., & Han, J. (2010). The earliest history of the deuterostomes: The importance of the Chengjiang Fossil-Lagerstätte. *Proceedings of the Royal Society of London, Series B, Biological Sciences, 277*(1679), 165–174.

Shu, D. G., Luo, H. L., Conway Morris, S., Zhang, X. L., Hu, S. X., Chen, L., ... Chen, L.-Z. (1999). Lower Cambrian vertebrates from south China. *Nature, 402*(6757), 42–46.

Shubin, N. (2008). *Your inner fish: A journey into the 3.5-billion-year history of the human body*. New York: Vintage.

Sidor, C. A., Vilhena, D. A., Angielczyk, K. D., Huttenlocker, A. K., Nesbitt, S. J., Peecook, B. R., ... Tsuji, L. A. (2013). Provincialization of terrestrial faunas following the end-Permian mass extinction. *Proceedings of the National Academy of Sciences of the United States of America, 110*(20), 8129–8133.

Simon, H. A. (1962). The architecture of complexity. *Proceedings of the American Philosophical Society, 106*(6), 467–482.

Simon, H. A. (1973). The organization of complex systems. In H. H. Pattee (Ed.), *Hierarchy theory: The challenge of complex systems* (pp. 1–27). New York: George Braziller.

Singer, W. (1999). Neuronal synchrony: A versatile code for the definition of relations? *Neuron, 24*(1), 49–65.

Singer, W. (2001). Consciousness and the binding problem. *Annals of the New York Academy of Sciences, 929*(1), 123–146.

Singewald, G. M., Rjabokon, A., Singewald, N., & Ebner, K. (2010). The modulatory role of the lateral septum on neuroendocrine and behavioral stress responses. *Neuropsychopharmacology, 36*(4), 793–804.

Sivarao, D. V., Langdon, S., Bernard, C., & Lodge, N. (2007). Colorectal distension-induced pseudoaffective changes as indices of nociception in the anesthetized female rat: Morphine and strain effects on visceral sensitivity. *Journal of Pharmacological and Toxicological Methods, 56*(1), 43–50.

Smith, C. U. (2006). The "hard problem" and the quantum physicists. Part 1: The first generation. *Brain and Cognition, 61*(2), 181–188.

Smith, C. U. (2009). The "hard problem" and the quantum physicists. Part 2: Modern times. *Brain and Cognition, 71*(2), 54–63.

Smith, E. S. J., & Lewin, G. R. (2009). Nociceptors: A phylogenetic view. *Journal of Comparative Physiology, A, Neuroethology, Sensory, Neural, and Behavioral Physiology, 195*(12), 1089–1106.

Sneddon, L. U. (2002). Anatomical and electrophysiological analysis of the trigeminal nerve in a teleost fish, *Oncorhynchus mykiss*. *Neuroscience Letters, 319*(3), 167–171.

Sneddon, L. U. (2004). Evolution of nociception in vertebrates: Comparative analysis of lower vertebrates. *Brain Research Reviews, 46*(2), 123–130.

Sneddon, L. (2011). Pain perception in fish. *Journal of Consciousness Studies, 18*(9–10), 209–229.

Sneddon, L. U., Braithwaite, V. A., & Gentle, M. J. (2003). Do fishes have nociceptors? Evidence for the evolution of a vertebrate sensory system. *Proceedings of the Royal Society of London, Series B, Biological Sciences, 270*(1520), 1115–1121.

Snow, P. J., Plenderleith, M. B., & Wright, L. L. (1993). Quantitative study of primary sensory neurone populations of three species of elasmobranch fish. *Journal of Comparative Neurology, 334*(1), 97–103.

Solinas, M., Panlilio, L. V., Justinova, Z., Yasar, S., & Goldberg, S. R. (2006). Using drug-discrimination techniques to study the abuse-related effects of psychoactive drugs in rats. *Nature Protocols, 1*(3), 1194–1206.

Solms, M. (2013 a). The conscious id. *Neuro-psychoanalysis, 15*(1), 5–19.

Solms, M. (2013 b). Response to commentaries. *Neuro-psychoanalysis, 15*(1), 79–85.

Solms, M., & Panksepp, J. (2012). The "id" knows more than the "ego" admits: Neuro-psychoanalytic and primal consciousness perspectives on the interface between affective and cognitive neuroscience. *Brain Sciences, 2*(2), 147–175.

Sombke, A., & Harzsch, S. (2015). Immunolocalization of histamine in the optic neuropils of *Scutigera coleoptrata* (Myriapoda, Chilopoda) reveals the basal organization of visual systems in the Mandibulata. *Neuroscience Letters, 594*, 111–116.

Sommer, L. (2013). Specification of neural crest- and placode-derived neurons. In J. Rubenstein & P. Rakic (Eds.), *Patterning and cell type specification in the developing CNS and PNS* (pp. 385–400). New York: Academic Press.

Søvik, E., & Barron, A. B. (2013). Invertebrate models in addiction research. *Brain, Behavior and Evolution, 82*(3), 153–165.

Spaethe, J., Tautz, J., & Chittka, L. (2006). Do honeybees detect colour targets using serial or parallel visual search? *Journal of Experimental Biology, 209*(6), 987–993.

Spang, A., Saw, J. H., Jørgensen, S. L., Zaremba-Niedzwiedzka, K., Martijn, J., Lind, A. E. … Ettema, T. J. G. (2015). Complex archaea that bridge the gap between prokaryotes and eukaryotes. *Nataure, 521*, 173–179.

Sperling, E. A., Frieder, C. A., Raman, A. V., Girguis, P. R., Levin, L. A., & Knoll, A. H. (2013). Oxygen, ecology, and the Cambrian radiation of animals. *Proceedings of the National Academy of Sciences, 110*(33), 13446–13451.

Sperling, E. A., Wolock, C. J., Morgan, A. S., Gill, B. C., Kunzmann, M., Halverson, G. P., … Johnston, D. T. (2015). Statistical analysis of iron geochemical data suggests limited late Proterozoic oxygenation. *Nature, 523*(7561), 451–454.

Sperling, E. A., Peterson, K. J., & Laflamme, M. (2011). Rangeomorphs, *Thectardis* (Porifera?) and dissolved organic carbon in the Ediacaran oceans. *Geobiology, 9*(1), 24–33.

Sperry, R. W. (1977). Forebrain commissurotomy and conscious awareness. *Journal of Medicine and Philosophy, 2*(2), 101–126.

Sridharan, D., Boahen, K., & Knudsen, E. I. (2011). Space coding by gamma oscillations in the barn owl optic tectum. *Journal of Neurophysiology, 105*(5), 2005–2017.

Sridharan, D., Schwarz, J. S., & Knudsen, E. I. (2014). Selective attention in birds. *Current Biology, 24*(11), R510–R513.

Stach, T., Winter, J., Bouquet, J. M., Chourrout, D., & Schnabel, R. (2008). Embryology of a planktonic tunicate reveals traces of sessility. *Proceedings of the National Academy of Sciences of the United States of America, 105*(20), 7229–7234.

Stein, B. E., & Meredith, M. A. (1993). *The merging of the senses*. Cambridge, MA: MIT Press.

Steinfartz, S., & Bulog, B. (2009). Non-visual sensory physiology and magnetic orientation in the Blind Cave Salamander, *Proteus anguinus* (and some other cave-dwelling urodele species). Review and new results on light-sensitivity and non-visual orientation in subterranean urodeles (Amphibia). *Animal Biology, 59*(3), 351–384.

Stephenson-Jones, M., Floros, O., Robertson, B., & Grillner, S. (2012). Evolutionary conservation of the habenular nuclei and their circuitry controlling the dopamine and 5-hydroxytryptophan (5-HT) systems. *Proceedings of the National Academy of Sciences of the United States of America, 109*(3), E164–E173.

Stevens, C. W. (2004). Opioid research in amphibians: An alternative pain model yielding insights on the evolution of opioid receptors. *Brain Research Reviews, 46*(2), 204–215.

Stevens, C. W. (2011). Analgesia in amphibians: Preclinical studies and clinical applications. *Veterinary Clinics of North America: Exotic Animal Practice, 14*(1), 33–44.

Stevenson, P. A., & Schildberger, K. (2013). Mechanisms of experience dependent control of aggression in crickets. *Current Opinion in Neurobiology, 23*(3), 318–323.

Strausfeld, N. J. (2009). Brain organization and the origin of insects: An assessment. *Proceedings of the Royal Society of London, Series B, Biological Sciences, 276*(1664), 1929–1937.

Strausfeld, N. J. (2013). *Arthropod brains: Evolution, functional elegance, and historical significance.* Cambridge, MA: Harvard University Press.

Strehler, B. L. (1991). Where is the self? A neuroanatomical theory of consciousness. *Synapse, 7*(1), 44–91.

Stujenske, J. M., Likhtik, E., Topiwala, M. A., & Gordon, J. A. (2014). Fear and safety engage competing patterns of theta-gamma coupling in the basolateral amygdala. *Neuron, 83*(4), 919–933.

Sundström, G., Dreborg, S., & Larhammar, D. (2010). Concomitant duplications of opioid peptide and receptor genes before the origin of jawed vertebrates. *PLoS One, 5*(5), e10512.

Suzuki, D. G., Murakami, Y., Escriva, H., & Wada, H. (2015a). A comparative examination of neural circuit and brain patterning between the lamprey and amphioxus reveals the evolutionary origin of the vertebrate visual center. *Journal of Comparative Neurology, 523*(2), 251–261.

Suzuki, D. G., Murakami, Y., Yamazaki, Y., & Wada, H. (2015b). Expression patterns of *Eph* genes in the "dual visual development" of the lamprey and their significance in the evolution of vision in vertebrates. *Evolution & Development, 17*(2), 139–147.

Sweeney, A. M., Haddock, S. H., & Johnsen, S. (2007). Comparative visual acuity of coleoid cephalopods. *Integrative and Comparative Biology, 47*(6), 808–814.

Tallon-Baudry, C. (2011). On the neural mechanisms subserving consciousness and attention. *Frontiers in Psychology, 2.*

Talsma, D., Senkowski, D., Soto-Faraco, S., & Woldorff, M. G. (2010). The multifaceted interplay between attention and multisensory integration. *Trends in Cognitive Sciences, 14*(9), 400–410.

Tang, S., & Juusola, M. (2010). Intrinsic activity in the fly brain gates visual information during behavioral choices. *PLoS One, 5*(12), e14455.

Teller, P. (1992). Subjectivity and knowing what it's like. In A. Berckermann, H. Flohr, & J. Kim (Eds.), *Emergence or reduction? Essays on the prospects of nonreductive physicalism* (pp. 180–200). Berlin: de Gruyter.

Temizer, I., Donovan, J. C., Baier, H., & Semmelhack, J. L. (2015). A visual pathway for looming-evoked escape in larval zebrafish. *Current Biology, 25*(14), 1823–1834.

ten Donkelaar, H. J., & de Boer-van Huizen, R. (1987). A possible pain control system in a non-mammalian vertebrate (a lizard, *Gekko gecko*). *Neuroscience Letters, 83*(1), 65–70.

Thompson, E. (2007). *Mind in life: Biology, phenomenology, and the sciences of mind*. Cambridge, MA: Harvard University Press.

Tinsley, C. J. (2008). Using topographic networks to build a representation of consciousness. *Biosystems, 92*(1), 29–41.

Tobet, S. A., Chickering, T. W., & Sower, S. A. (1996). Relationship of gonadotropin-releasing hormone (GnRH) neurons to the olfactory system in developing lamprey (*Petromyzon marinus*). *Journal of Comparative Neurology, 376*(1), 97–111.

Todd, A. J. (2010). Neuronal circuitry for pain processing in the dorsal horn. *Nature Reviews: Neuroscience, 11*(12), 823–836.

Tomina, Y., & Takahata, M. (2010). A behavioral analysis of force-controlled operant tasks in American lobster. *Physiology & Behavior, 101*(1), 108–116.

Tononi, G. (2004). An information integration theory of consciousness. *BMC Neuroscience, 5*, 42–72.

Tononi, G. (2008). Consciousness as integrated information: A provisional manifesto. *Biological Bulletin, 215*(3), 216–242.

Tononi, G., & Koch, C. (2008). The neural correlates of consciousness. *Annals of the New York Academy of Sciences, 1124*(1), 239–261.

Tononi, G., & Koch, C. (2015). Consciousness: Here, there, and everywhere? *Philosophical Transactions of the Royal Society of London, Series B, Biological Sciences, 370*(1668), 1–18.

Treede, R. D., Kenshalo, D. R., Gracely, R. H., & Jones, A. K. (1999). The cortical representation of pain. *Pain, 79*(2), 105–111.

Trent, N. L., & Menard, J. L. (2013). Lateral septal infusions of the neuropeptide Y Y2 receptor agonist, NPY 13–36 differentially affect different defensive behaviors in male, Long Evans rats. *Physiology & Behavior, 110*, 20–29.

Trestman, M. (2013). The Cambrian explosion and the origins of embodied cognition. *Biological Theory, 8*(1), 80–92.

Triplett, J. W., & Feldheim, D. A. (2012). Eph and ephrin signaling in the formation of topographic maps. In D. B. Nikolov (Ed.), *Seminars in cell and developmental biology* (Vol. 23, No. 1) (pp. 7–15). Waltham, MA: Academic Press.

Tsakiris, M. (2013). Self-specificity of the external body. *Neuro-psychoanalysis, 15*(1), 66–69.

Tsou, J. Y. (2013). Origins of the qualitative aspects of consciousness: Evolutionary answers to Chalmers' hard problem. In L. Swan (Ed.), *Origins of mind* (pp. 259–269). Dordrecht: Springer Netherlands.

Tsuchiya, N., & van Boxtel, J. (2013). Introduction to research topic: Attention and consciousness in different senses. *Frontiers in Psychology, 4*.

Tulving, E. (1985). Memory and consciousness. *Canadian Psychology, 26*(1), 1.

Tulving, E. (1987). Multiple memory systems and consciousness. *Human Neurobiology, 6*(2), 67–80.

Tulving, E. (2002 a). Episodic memory: From mind to brain. *Annual Review of Psychology, 53*(1), 1–25.

Tulving, E. (2002 b). Chronesthesia: Conscious awareness of subjective time. In D. T. Stuss & R. C. Knight (Eds.), *Principles of frontal lobe function* (pp. 311–325). Oxford: Oxford University Press.

Turnbull, O. H. (2013). Facing inconvenient truths. *Neuro-psychoanalysis, 15*(1), 69–72.

Tye, M. (2000). *Consciousness, color, and content.* Cambridge, MA: MIT Press.

Uhlhaas, P., Pipa, G., Lima, B., Melloni, L., Neuenschwander, S., Nikolić, D., & Singer, W. (2009). Neural synchrony in cortical networks: History, concept, and current status. *Frontiers in Integrative Neuroscience, 3*.

Underwood, E. (2015). The brain's identity crisis. *Science, 349*, 575–577.

Ursino, M., Magosso, E., & Cuppini, C. (2009). Recognition of abstract objects via neural oscillators: Interaction among topological organization, associative memory, and gamma band synchronization. *IEEE Transactions on Neural Networks, 20*(2), 316–335.

Uylings, H., Groenewegen, H. J., & Kolb, B. (2003). Do rats have a prefrontal cortex? *Behavioural Brain Research, 146*(1–2), 3–17.

van Boxtel, J. J., Tsuchiya, N., & Koch, C. (2010). Consciousness and attention: On sufficiency and necessity. *Frontiers in Psychology, 1*.

van Gaal, S., & Lamme, V. A. (2012). Unconscious high-level information processing implication for neurobiological theories of consciousness. *Neuroscientist, 18*(3), 287–301.

Vannier, J., Liu, J., Lerosey-Aubril, R., Vinther, J., & Daley, A. C. (2014). Sophisticated digestive systems in early arthropods. *Nature communications, 5*.

Van Roy, P., Daley, A. C., & Briggs, D. E. (2015). Anomalocaridid trunk limb homology revealed by a giant filter-feeder with paired flaps. *Nature, 522*, 77–80.

van Swinderen, B., & Andretic, R. (2011). Dopamine in *Drosophila*: Setting arousal thresholds in a miniature brain. *Proceedings of the Royal Society of London, Series B, Biological Sciences*, rspb20102564.

van Swinderen, B., & Greenspan, R. J. (2003). Salience modulates 20–30 Hz brain activity in *Drosophila*. *Nature Neuroscience, 6*(6), 579–586.

References

Vargas, J. P., López, J. C., & Portavella, M. (2009). What are the functions of fish brain pallium? *Brain Research Bulletin*, *79*(6), 436–440.

Vargas, J. P., López, J. C., & Portavella, M. (2012). Amygdala and emotional learning in vertebrates—a comparative perspective. *InTech*. doi:.10.5772/51552

Veinante, P., Yalcin, I., & Barrot, M. (2013). The amygdala between sensation and affect: A role in pain. *Journal of Molecular Psychiatry*, *1*(1), 9.

Veit, L., Hartmann, K., & Nieder, A. (2014). Neuronal correlates of visual working memory in the corvid endbrain. *Journal of Neuroscience*, *34*(23), 7778–7786.

Velmans, M. (Ed.). (2000). *Investigating phenomenal consciousness: New methodologies and maps* (Vol. 13). Amsterdam: John Benjamins.

Velmans, M. (2009). *Understanding consciousness*. London: Routledge.

Vierck, C. J., & Charles, J. (2006). Animal models of pain. *Wall and Melzack's textbook of pain*, *5*, 175–185.

Vierck, C. J., Whitsel, B. L., Favorov, O. V., Brown, A. W., & Tommerdahl, M. (2013). Role of primary somatosensory cortex in the coding of pain. *Pain*, *154*(3), 334–344.

Vindas, M. A., Johansen, I. B., Vela-Avitua, S., Nørstrud, K. S., Aalgaard, M., Braastad, B. O., … Øverli, Ø. (2014). Frustrative reward omission increases aggressive behaviour of inferior fighters. *Proceedings of the Royal Society of London, Series B, Biological Sciences*, *281*(1784), 1–8.

von Bertalanffy, L. (1974). *Perspectives on general system theory*. E. Taschdjian (Ed.). New York: George Braziller.

von der Malsburg, C. (1995). Binding in models of perception and brain function. *Current Opinion in Neurobiology*, *5*(4), 520–526.

von Düring, M., & Andres, K. H. (1998). Skin sensory organs in the Atlantic hagfish *Myxine glutinosa*. In J. Jorgensen, J. Lomholt, R. Weber, & H. Malte (Eds.), *The biology of hagfishes* (pp. 499–511). Dordrecht: Springer Netherlands.

Vopalensky, P., Pergner, J., Liegertova, M., Benito-Gutierrez, E., Arendt, D., & Kozmik, Z. (2012). Molecular analysis of the amphioxus frontal eye unravels the evolutionary origin of the retina and pigment cells of the vertebrate eye. *Proceedings of the National Academy of Sciences of the United States of America*, *109*(38), 15383–15388.

Wada, H. (1998). Evolutionary history of free-swimming and sessile lifestyles in urochordates as deduced from 18S rDNA molecular phylogeny. *Molecular Biology and Evolution*, *15*(9), 1189–1194.

Waddell, S. (2013). Reinforcement signalling in *Drosophila*; dopamine does it all after all. *Current Opinion in Neurobiology*, *23*(3), 324–329.

Wägele, J. W., & Bartolomaeus, T. (Eds.). (2014). *Deep metazoan phylogeny: The backbone of the tree of life: New insights from analyses of molecules, morphology, and theory of data analysis*. Berlin: de Gruyter.

Walls, G. L. (1942). *The vertebrate eye and its adaptive radiation*. Bloomfield Hills, MI: Cranbrook Institute of Science.

Walter, S., & Heckmann, H.-D. (Eds.). (2003). *Physicalism and mental causation*. Exeter: Imprint Academic.

Walters, E. T. (1994). Injury-related behavior and neuronal plasticity: An evolutionary perspective on sensitization, hyperalgesia, and analgesia. *International Review of Neurobiology, 36*, 325–427.

Walters, E. T., Bodnarova, M., Billy, A. J., Dulin, M. F., Díaz-Ríos, M., Miller, M. W., & Moroz, L. L. (2004). Somatotopic organization and functional properties of mechanosensory neurons expressing sensorin-A mRNA in Aplysia californica. *Journal of Comparative Neurology, 471*(2), 219–240.

Walters, E. T., Illich, P., Weeks, J., & Lewin, M. (2001). Defensive responses of larval Manduca sexta and their sensitization by noxious stimuli in the laboratory and field. *Journal of Experimental Biology, 204*(3), 457–469.

Wang, J., Wu, X., Li, C., Wei, J., Jiang, H., Liu, C., … Ma, Y. (2012). Effect of morphine on conditioned place preference in rhesus monkeys. *Addiction Biology, 17*(3), 539–546.

Waraczynski, M. (2009). Emotion. In M. D. Binder, N. Hirokawa, & U. Windhorst (Eds.), *Encyclopedia of neurosciences* (pp. 1088–1092). Berlin: Springer.

Watanabe, A., Hirano, S., & Murakami, Y. (2008). Development of the lamprey central nervous system, with reference to vertebrate evolution. *Zoological Science, 25*(10), 1020–1027.

Watanabe, M., Cheng, K., Murayama, Y., Ueno, K., Asamizuya, T., Tanaka, K., & Logothetis, N. (2011). Attention but not awareness modulates the BOLD signal in the human V1 during binocular suppression. *Science, 334*(6057), 829–831.

Watt, D. F. (2005). Panksepp's common sense view of affective neuroscience is not the common sense view in large areas of neuroscience. *Consciousness and Cognition, 14*(1), 81–88.

Webb, J. E. (1969). On the feeding and behavior of the larva of *Branchiostoma lanceolatum*. *Marine Biology, 3*, 58–72.

Webb, K. J., Norton, W. H., Trumbach, D., Meijer, A. H., Ninkovic, J., Topp, S., … Bally-Cuif, L. (2009). Zebrafish reward mutants reveal novel transcripts mediating the behavioral effects of amphetamine. *Genome Biology, 10*(7), R81.

Weiss, E., & Wilson, S. (2003). The use of classical and operant conditioning in training Aldabra tortoises (*Geochelone gigantea*) for venipuncture and other husbandry issues. *Journal of Applied Animal Welfare Science, 6*(1), 33–38.

Weissman, A. (1976). The discriminability of aspirin in arthritic and nonarthritic rats. *Pharmacology, Biochemistry, and Behavior, 5*(5), 583–586.

Wertz, A., Rössler, W., Obermayer, M., & Bickmeyer, U. (2006). Functional neuroanatomy of the rhinophore of *Aplysia punctata*. *Frontiers in Zoology*, 3(6).

Whiteside, J. H., Grogan, D. S., Olsen, P. E., & Kent, D. V. (2011). Climatically driven biogeographic provinces of Late Triassic tropical Pangea. *Proceedings of the National Academy of Sciences of the United States of America*, 108(22), 8972–8977.

Wicht, H. (1996). The brains of lampreys and hagfishes: Characteristics, characters, and comparisons. *Brain, Behavior and Evolution*, 48(5), 248–261.

Wicht, H., & Lacalli, T. C. (2005). The nervous system of amphioxus: Structure, development, and evolutionary significance. *Canadian Journal of Zoology*, 83, 122–150.

Wicht, H., Laedtke, E., Korf, H.-W., & Schomerus, C. (2013). Spatial and temporal expression patterns of Bmal delineate a circadian clock in the nervous system of *Branchiostoma lanceolatum*. *Journal of Comparative Neurology*, 518, 1837–1846.

Wilczynski, W. (2009). Evolution of the brain in amphibians. In M. D. Binder, N. Hirokawa, & U. Windhorst (Eds.), *Encyclopedia of neurosciences* (pp. 1301–1305). Berlin: Springer.

Wild, J. M. (2009). Evolution of the Wulst. In M. D. Binder, N. Hirokawa, & U. Windhorst (Eds.), *Encyclopedia of neurosciences* (pp. 1475–1478). Berlin: Springer.

Wild, J. M., Arends, J. J., & Ziegler, H. P. (1990). Projections of the parabrachial nucleus in the pigeon (*Columbia livia*). *Journal of Comparative Neurology*, 293(4), 499–523.

Wild, J. M., Kubke, M. F., & Peña, J. L. (2008). A pathway for predation in the brain of the barn owl (Tyto alba): Projections of the gracile nucleus to the "claw area" of the rostral wulst via the dorsal thalamus. *Journal of Comparative Neurology*, 509(2), 156–166.

Williamson, R., & Chrachri, A. (2004). Cephalopod neural networks. *Neuro-Signals*, 13(1–2), 87–98.

Wilson-Mendenhall, C. D., Barrett, L. F., & Barsalou, L. W. (2013). Neural evidence that human emotions share core affective properties. *Psychological Science*, 24(6), 947–956.

Winkielman, P., Berridge, K. C., & Wilbarger, J. L. (2005). Emotion and consciousness. In P. Winkielman (Ed.), *Emotion, behavior, and conscious experience: Once more without feeling* (pp. 335–362). New York: Guilford Press.

Wolf, H. (2008). The pectine organs of the scorpion, *Vaejovis spinigerus*: Structure and (glomerular) central projections. *Arthropod Structure & Development*, 37(1), 67–80.

Wollesen, T., Loesel, R., & Wanninger, A. (2009). Pygmy squids and giant brains: Mapping the complex cephalopod CNS by phalloidin staining of vibratome sections and whole-mount preparations. *Journal of Neuroscience Methods*, 179(1), 63–67.

Woods, J. W. (1964). Behavior of chronic decerebrate rats. *Journal of Neurophysiology*, 27, 635–644.

Wullimann, M. F. (2014). Ancestry of basal ganglia circuits: New evidence in teleosts. *Journal of Comparative Neurology, 522*(9), 2013–2018.

Wullimann, M. F., & Vernier, P. (2009 a). Evolution of the brain in fishes. In M. D. Binder, N. Hirokawa, & U. Windhorst (Eds.), *Encyclopedia of neurosciences* (pp. 1318–1326). Berlin: Springer.

Wullimann, M. F., & Vernier, P. (2009 b). Evolution of the telencephalon in anamniotes. In M. D. Binder, N. Hirokawa, & U. Windhorst (Eds.), *Encyclopedia of neurosciences* (pp. 1423–1431). Berlin: Springer.

Xin, Y., Weiss, K. R., & Kupfermann, I. (1995). Distribution in the central nervous system of *Aplysia* of afferent fibers arising from cell bodies located in the periphery. *Journal of Comparative Neurology, 359*(4), 627–643.

Xue, H. G., Yamamoto, N., Yang, C. Y., Kerem, G., Yoshimoto, M., Sawai, N., … Ozawa, H. (2006 a). Projections of the sensory trigeminal nucleus in a percomorph teleost, tilapia (*Oreochromis niloticus*). *Journal of Comparative Neurology, 495*(3), 279–298.

Xue, H. G., Yang, C. Y., Ito, H., Yamamoto, N., & Ozawa, H. (2006 b). Primary and secondary sensory trigeminal projections in a cyprinid teleost, carp (*Cyprinus carpio*). *Journal of Comparative Neurology, 499*(4), 626–644.

Yafremava, L. S., & Gillette, R. (2011). Putative lateral inhibition in sensory processing for directional turns. *Journal of Neurophysiology, 105*(6), 2885–2890.

Yáñez, J., Busch, J., Anadón, R., & Meissl, H. (2009). Pineal projections in the zebrafish (*Danio rerio*): Overlap with retinal and cerebellar projections. *Neuroscience, 164*(4), 1712–1720.

Yang, L. M., Yu, L., Jin, H. J., & Zhao, H. (2014). Substance P receptor antagonist in lateral habenula improves rat depression-like behavior. *Brain Research Bulletin, 100*, 22–28.

Yang, P. F., Chen, D. Y., Hu, J. W., Chen, J. H., & Yen, C. T. (2011). Functional tracing of medial nociceptive pathways using activity-dependent manganese-enhanced MRI. *Pain, 152*(1), 194–203.

Yin, Z., Zhu, M., Davidson, E. H., Bottjer, D. J., Zhao, F., & Tafforeau, P. (2015). Sponge grade body fossil with cellular resolution dating 60 Myr before the Cambrian. *Proceedings of the National Academy of Sciences of the United States of America, 112*(12), E1453–E1460.

Yopak, K. E. (2012). Neuroecology of cartilaginous fishes: The functional implications of brain scaling. *Journal of Fish Biology, 80*(5), 1968–2023.

Young, J. Z. (1971). *The anatomy of the nervous system of Octopus vulgaris*. Oxford: Clarendon Press.

Young, J. Z. (1974). The central nervous system of *Loligo*. I. The optic lobe. *Philosophical Transactions of the Royal Society of London, Series B, Biological Sciences, 267*(885), 263–302.

References

Zeisel, A., Muñoz-Manchado, A. B., Codeluppi, S., Lönnerberg, P., La Manno, G., Juréus, A., ... Linnarsson, S. (2015). Cell types in the mouse cortex and hippocampus revealed by single-cell RNA-seq. *Science*, *347*(6226), 1138–1142.

Zeki, S., & Marini, L. (1998). Three cortical stages of colour processing in the human brain. *Brain*, *121*, 1669–1685.

Zelenitsky, D. K., Therrien, F., Ridgely, R. C., McGee, A. R., & Witmer, L. M. (2011). Evolution of olfaction in non-avian theropod dinosaurs and birds. *Proceedings of the Royal Society of London, Series B, Biological Sciences*, *278*(1725), 3625–3634.

Zhang, Y., Lu, H., & Bargmann, C. I. (2005). Pathogenic bacteria induce aversive olfactory learning in *Caenorhabditis elegans*. *Nature*, *438*(7065), 179–184.

Zhukov, V. V., & Tuchina, O. P. (2008). Structure of visual pathways in the nervous system of freshwater pulmonate molluscs. *Journal of Evolutionary Biochemistry and Physiology*, *44*(3), 341–353.

Zintzen, V., Roberts, C. D., Anderson, M. J., Stewart, A. L., Struthers, C. D., & Harvey, E. S. (2011). Hagfish predatory behaviour and slime defence mechanism. *Scientific Reports*, *1*.

Zmigrod, S., & Hommel, B. (2013). Feature integration across multimodal perception and action: A review. *Multisensory Research*, *26*(1–2), 143–157.

Zullo, L., & Hochner, B. (2011). A new perspective on the organization of an invertebrate brain. *Communicative & Integrative Biology*, *4*(1), 26–29.

Zullo, L., Sumbre, G., Agnisola, C., Flash, T., & Hochner, B. (2009). Nonsomatotopic organization of the higher motor centers in octopus. *Current Biology*, *19*(19), 1632–1636.

Index

Note: Page numbers followed by "f" or "t" refer to figures and tables, respectively.

Aboitiz, Francisco, 122
Access consciousness, 2, 251n4
Acetylcholine, 115, 116, 168
Active, mobile animals and consciousness. *See* Locomotion
Adaptation, 24, 117–118, 217–220, 255n21, 270n54, 284n39, 284n43
Adaptive behavior network (social behavior network), 155, 164–165t, 166
Aδ nerve fibers, 133–137, 158–159t, 167
Affective consciousness, 6, 32f, 129–130, 131t, 138–170
  adaptive function of, 145, 169, 218
  advances through vertebrate evolution, 154, 155, 166, 168, 277n33
  behavioral criteria for and against, 149, 152–154, 152t, 153t, 277n25
  cross-species behavioral evidence for, 154, 156–157t, 174, 175t
  defined, viii–ix, 129, 138
  exteroceptive consciousness in relation to, 131t, 138, 140, 148, 187, 211–212, 213f, 283n25
  genes and, 168, 214
  interoceptive consciousness in relation to, 130, 131t, 138, 140, 211–212, 213f
  invertebrates and, 154, 155–156t, 170, 175t, 187
  limbic system and, 138–140, 154–155, 213f, 277n33, 283n25
  as mental states (affective states), 18t, 30, 34, 129–131, 139, 145, 152, 222
  neuroanatomic structures, 131t, 149, 154–166
  origin, date, and basis of, 142–148, 147t, 166, 168–170, 174–175, 191–192, 199f, 211, 283n25
  pain pathways, 133–138
  subcortical, not cortical, 144, 155, 166, 170, 211–214
  theories of, 142–148, 167–168
  vertebrate, 154, 155, 168–170
Affective neurostructures, 135f, 155, 158–165t, 213f, 277n33
Affects
  as conscious always, 130
  defined, 129–130
  negative and positive, 138, 156–157
Algae, 20f, 55
Allo-ontological irreducibility, 221f, 221–222, 224
Amniotes
  brains of, 72f, 120–121f
  consciousness possessed by, 118–128, 158–165t
  defined, 104
  extinct, 102f, 119f

Amoebas, 13, 19
Amphibians. *See* Anamniotes
Amphioxus
  affective behaviors unknown, 154
  brain of, 39, 44–45, 46f, 47–49, 79, 87–88, 87f, 116, 258n17, 258n20, 259n26
  central nervous system of, 77t
  consciousness lacking in, 48, 116, 117
  eye of (frontal eye), 45, 46f, 80f, 82, 84, 116
  genes and, 45, 117, 259–260n26
  overview of, 41f, 43–44
  primary motor center of, 45, 46f, 49
  senses of, 40, 42, 45, 79, 86, 87
  similar to early prevertebrates, 43, 67
  swimming mechanism of, 67
Amygdala, 74–76t, 115, 133, 139f, 146, 158t, 162t, 164t, 213f, 277n33
Analgesics (and pain), 151, 153, 153t
Anamniotes (fish and amphibians)
  brains of, 72f, 106f, 111–114, 112f
  consciousness possessed by, 104–108, 111, 115, 154–155, 208–211
  defined, 104
Animal intelligence, 12
Animals. *See also* Living animals; Paleobiological domain
  of Cambrian Period, 53, 54f, 57–59f, 60, 66, 201f
  of Carboniferous Period, 202–203f
  of Ediacaran Period, 55, 56f, 61–62, 61f
  pain in, 149–151, 166–168, 174
  phylogenetic relations of conscious, 206, 207f
  phylogenetic tree of, 54f
  of Triassic Period, 204–205f, 207–208
Anoetic consciousness (Tulving), 216
Anomalocarids, 57f, 59f, 60, 185
Ant colony (as nonconscious system), 197
Anterior cingulate cortex, 134–135f, 139f, 147t, 162t, 212, 213f
Appendicularians. *See* Larvaceans
Aplysia (gastropod), 171, 185, 186f

Apple falling, represented in brain, 30, 224
Approaching reinforcing drugs, 153, 156–157t, 174t
Archaeans, 18–20, 254n5
Arms of octopus, cephalopods, 189–190
Arousal, 115, 198, 214
Arthropods. *See also* Protostome invertebrates
  and affective consciousness, 174
  appendages as manipulative limbs, 63, 66, 191
  brains of, 181–185
  of Cambrian Period, 57f, 58f, 60, 64–66, 185, 192
  caveats to consciousness, 184, 192
  consciousness possessed by, ix, 174, 184–185, 192–193, 199f, 200, 206, 207
  and exteroceptive consciousness, 178–185
  humans' and vertebrates' relations to, 60, 207f
  nervous systems of, 179–180f
  overview of, 171
  relation to other animal clades with consciousness, 207f
  vs. vertebrates, 64–67, 104, 185, 192–193
  vision in, 66, 81, 175–176t, 181
Ascidians, 41f, 42, 44f, 258n11
Attention, 35, 257n47. *See also* Selective attention
Auditory. *See* Hearing
Autobiographical self (Damasio), 216
Autocerebroscope, 220, 221, 284n47
Autonoetic consciousness (Tulving), 216–217
Autonomic nervous system, 75t, 77t, 158–159t
Auto-ontological irreducibility, 220–224, 221f, 284n46, 284n47
Axons, 25f, 39

Baars, Bernard, 118
Bacteria, 19–20, 55

Index

Balanoff, Amy, 125
Basic motor programs, 26, 47, 62, 74t, 152, 255n23
Bees, 14, 179f, 183, 184, 200
Behavioral negative contrast, 153, 154, 156–157t, 175t, 200
Behavioral trade-offs, 152, 153t, 156t, 175t
Behaviorism, 143
Benton, Michael, 124
Bernard, Claude, 274n22
Bilaterian animals, 20f, 37–38, 38f, 54f, 55, 57, 60, 61, 61f
  complex ancestor, 61
  simple ancestor (worm), 20f, 54f, 61–62, 79, 82, 187, 222
Biological naturalism (Searle), 14
Birds
  affective consciousness, 151, 154, 156–165t
  brain of, vs. mammal brain, 72f, 118, 120–121f, 122, 125–127
  cognitive functions, 126, 217
  consciousness advancing in, 118, 122–125, 155, 200
  environments of first, 124–125
  optic tectum of, 108, 111
  vision vs. smell in, 123–124
Bivalves, 59f, 173
Blind men and elephant parable, 3, 217, 226
Blindsight, 209, 268n27, 282n11
Boly, Melanie, 118
Bony fish. *See* Osteichthyes
Brachiopods, 54f, 57f, 59f, 60, 64
Bradley, Mark, 218
Brains
  amniote, 72f, 104, 120–121f
  amphioxus, 39, 44–45, 46f, 47–49, 77t, 79, 87f, 87–88, 116, 258n17, 258n20, 259n26
  anamniote, 72f, 104, 112–114
  arthropod, 179–180f, 181–185
  bird, 118, 120–121f, 122, 125–127
  cephalopod, 187–190
  chordate, 37, 39, 46–47f, 48
  core region of, 47–48, 78, 212, 213f
  crab, 180f
  creation of subjective experience by, 2–3, 222, 223–227
  dorsal parts of, 47, 88
  essential for consciousness, 18t, 176–177t
  evolutionary origin of, 11, 37, 40–49, 46–47f, 71, 87–88, 178, 185, 259n26
  insect, 179f, 181
  lamprey, 87f, 105, 106f, 108, 113, 158–165t
  mammal, 72f, 118, 122–123, 134–135f, 136–137t, 139f, 158–165t
  prevertebrate, 42, 46–47f, 48
  regions of, 43, 70–71, 71f, 73–76t, 178, 186f, 188f
  size of, 118, 122, 125, 176t, 181, 182f, 184, 185
  tunicate, 39, 43, 44f, 77t, 258n11
  ventral parts of, 45, 47
  vertebrate, 46f, 69–79, 71f, 72f, 73–77t, 78f, 87–88, 139f
Brainstem, 70, 71, 75t, 213f
Brain weight vs. body weight, relation, 181, 182f
Burrowing shows behavior in Cambrian explosion, 62–63
Burgess Shale, 55, 91
Butler, Ann B., 88, 118, 126

Cabanac, Michel, 145, 147t, 169
Camazine, S., 21
Cambrian explosion, viii, 51–67
  arms race, 63, 127, 192
  causes and explanation of, 62–67, 261n16
  Darwin's dilemma about, 51, 53, 64
  eve of, 55, 56–57f
  mobility and locomotion, importance in, 64–67, 191
  ocean floor during, 55, 56–57f, 201f
  phyla of, 53, 54f, 56–59f, 61
  vision as component of, 63, 81

Cambrian Period, 51–102, 169, 185, 190, 192, 199–201
Carboniferous Period, 202–203f
Caron, Jean-Bernard, 91
Cartesian theater or framework, 30, 196
Cartilaginous fish. *See* Chondrichthyes
Cave fish, blind, 219
C. elegans (*Caenorhabditis elegans*), 156t, 158t, 175t, 178. *See also* Nematodes
Cells
  diversity of, 27
  as embodied, 18–19
  evolution of, 20f
  as living things, 18
Central canal of spinal cord, 71, 75t
Central nervous system (CNS). *See also* Brain; Nervous system
  amphioxus, 77t
  cephalopod, 188f, 189
  explained, 39
  and eye, 83
  invertebrate (protostome), 171, 173f
  tunicate, 77t
  vertebrate, 73–76t
Central pattern generators, 26, 47, 49, 255n23
Cephalate, 88, 89f
Cephalochordates. *See* Amphioxus
Cephalopods, ix, 187–193, 200
  affective consciousness in, 174, 175t
  arms of, 189–190
  caveats to consciousness, 174, 189–190, 192
  exteroceptive consciousness, 176–177t, 187–191
  nervous system of, 188f
  relation to other animal clades with consciousness, 206, 207f
  vs. vertebrates, 191, 281n45
Cerebellum, 70, 72f, 73t, 78, 79, 210
Cerebral cortex
  affect in, 144 (*see also* Affective consciousness: subcortical, not cortical)
  emergence of, 23, 123, 126–127
  higher mental functions in, 71, 74–75t, 97–98t
  interoceptive pathways in, 133–137
  odor isomorphism in, 96f
  pain pathways in, 135f, 136–137t, 141
  retinotopic mapping in, 33, 93f
  as a site of consciousness, 31–34, 93–96f, 97–98t, 123, 150, 155, 209, 211, 277n32
  as only site of consciousness (*see* Corticothalamic theory)
  somatotopic mapping in, 30, 31f, 94f, 134–135f
  tonotopic isomorphism in, 95f
Cerebral pallium, 71, 74–75t, 78, 79, 112–114, 118–128, 120–121f
  and memory, 114–115, 123–124
  and smell, 112–114, 117
Cerebrum, 71, 79, 92, 118, 150
Chalmers, David, vii, 2, 10, 226, 251n6
Character of experience, 225–226
Chelicerates, 171
Chemoreceptors, 79, 131t, 187
Chemosenses (smell, taste), 131t, 132, 176t, 186f, 187
Chen, Junyuan, 63, 83, 90
Chengjiang Shale, 55, 90–91
Chimpanzees, 14
Chittka, L., 183
Choanoflagellates, 19, 20f
Chondrichthyes, 102f, 104, 113, 158–165t, 167, 182f
Chordates
  brains of, 37, 39, 44f, 46–47f, 48, 71f
  of Cambrian Period, 58f, 60, 88–92
  features of, 37, 52
  neuroanatomy, basic, 39–40, 40f
  relations among, 38f, 102f
  vertebrates as, 37, 70f
Clades, 52
Classical conditioning, 151, 152
C nerve fibers and pathways, 133–137, 141, 149, 158–159t, 167–168
Coleoids, 187, 189–191, 207f

Complex adaptive systems, 22
Complexity of consciousness, 24, 26–27, 176t, 181, 187, 189, 191–192, 197–198, 223, 224
Conditioned place preference, 153, 156–157t, 174t
Connectivity, 18t, 26
Consciousness, general
  as complex, not fundamental, 24, 26–27, 176t, 181, 187, 189, 191–192, 197–198, 223, 224
  continuum of, primary to higher, 215–217
  types: primary vs. higher, 1–2, 215–217, 251n4
Consciousness, sensory (primary, phenomenal)
  in active animals only, 64–67, 191–193, 207, 219–220
  adaptive nature of, 24, 117–118, 128, 145, 169, 192, 217–220, 261n19, 270n54, 284n39, 284n43
  affective (*see* Affective consciousness)
  amphioxus lacking in, 48, 116, 117
  anamniotes' possession of, 104–108, 111, 115, 154–155, 208–211
  ancient, widespread, and diverse nature of, 198–217
  arthropods' possession of, ix, 192–193, 199f, 200, 206, 207f
  attention in relation to, 18t, 35, 108, 118, 124, 177t, 198, 219, 224, 257n47
  birds' advancement of, 118, 122–125, 155, 200
  as by-product (epiphenomenon), 218–219
  centers/sites of, 108–111, 117–118, 122–124, 127–128, 154–155, 183, 190, 209–211, 213f
  cephalopods' possession of, ix, 190–193, 200, 206, 207f
  cerebral cortex's role in, 31–34, 93–96f, 97–98t, 123, 150, 155, 209, 211, 277n32

  commonalities of its diverse forms, 18t, 175–177t, 214–215, 281n4, 282n22, 283n25
  defined, vii, 1–2
  defining features of (criteria for), 17, 18t, 175–177t, 195–198, 281n4
  diversities within, 35–36, 126–128, 130, 131t, 170, 193, 208–217
  as embodied, 18t, 21, 195–196
  emergence approach to, 9–10, 196–197
  energetically expensive, 24, 64, 124–125, 219
  evolution of, 83, 88, 101–128, 168–170, 191–193, 198–206, 215–217, 222–226
  evolutionary loss of, 219–220
  evolvability is high, 117
  exteroceptive (*see* Exteroceptive consciousness)
  genes and, 116–117, 168, 214
  as integrated information (Tononi), 27, 30, 210
  interactions in (*see* Neural interactions)
  interoceptive (*see* Interoceptive consciousness)
  invertebrates' possession of, ix, 12, 48–49, 171–193, 200–207
  irreducibility of, x, 9, 222, 227, 252n11
  isomorphic, 30–34, 110–111, 131t (*see also* Isomorphic sensory pathways; Isomorphism)
  levels of hierarchy in, 98–100, 178, 198
  and life cycle (larvae, infants), 12, 219, 254n33, 284n43
  mammals' advancement of, 118, 122–125, 155, 200, 275n54
  memory's role in, 114–115, 123–124, 128, 152, 177t, 183, 184, 198, 206 (*see also* Memory)
  multiple senses and, 176–177t
  mystery of, 1–15, 195
  as natural, scientific phenomenon, viii–ix, 10, 14, 195–198, 215, 222–225, 227 (*see also* Neurobiological naturalism)

Consciousness, sensory (primary, phenomenal) (cont.)
  neurobiological approach to, 8–10, 14, 195–227
  neurobiological features applying to, 17, 18t, 26–36, 81–88, 105–128, 158–165t, 176–177t, 197–198, 222–225
  neuroevolutionary approach to (domain of), 3, 4f, 11–14, 217–228
  organisms (fossils) first exhibiting, 88, 89f, 90–92, 185, 199f
  origins and dating of, 81–92, 117–118, 168–170, 169f, 185, 190–191, 199f, 200, 222, 226
  partial, 284n43
  philosophical approach to (domain of), 2–8, 19, 220–228
  phylogenetic relations of animals showing evidence of, 207f
  prediction as characteristic of, 117, 158t, 218, 270n53
  as process, 21
  and reality, 219, 284n39
  reciprocal (recurrent, recursive, reentrant) interactions in (*see* Neural interactions)
  reduction approach to, ix–x, 8–9, 14, 220–222, 226–227
  reflexes preceding, 25–26, 197
  in a self-organizing system, 22
  as simulation or representation of world, 18t, 30–33, 64, 83, 107, 118, 127, 198, 218, 265n1, 270n53, 272n88, 284n39
  Step 1: appearance (in vertebrates), 104–117, 198–200
  Step 2: advancement (in vertebrates), 118–127, 192, 199f, 200
  triple approach to, ix–x, 3–14, 4f, 15, 217–222, 226, 227
  types compared (extero-, intero-, affective), 129–131, 211–212, 213f
  vertebrates' possession of, viii–ix, 101–128, 149–170, 169f, 192–193, 198–215, 199f, 219, 265n1

  vision as dawn of, 81–85 (*see also* Vision-first hypothesis)
  as widespread, 206–208
Constraint, in hierarchical systems, 22–24, 22f, 28f, 224
Convergent evolution, 52, 125, 206
Core-self (Damasio), 216
Correlates of consciousness, viii. *See also* Criteria for consciousness; Consciousness, sensory: commonalities
Corticothalamic theory of consciousness, 34, 108, 118, 125, 144, 147t, 209–211, 222, 275n47
Cotterill, Rodney, 118
Craig, A. D. (Bud), 132, 140, 147t
Crick, Francis, 7, 9, 34, 36, 226–227
Criteria for consciousness, 18t, 175–177t, 195–198, 214–215, 281n4, 283n25. *See also* Neurobiological features of consciousness
Crustaceans, 171, 185
Cuttlefish, 171, 174
Cyclopean perception, 6–7, 7f, 224
Cyclostomes, 103, 103f

Damasio, Antonio, 33, 141, 144, 155, 168, 215–216, 251n9, 268n27, 270n54
Darwin, Charles, 51, 253n28
Darwin's dilemma, 51, 53, 64
Decerebrate mammals, behaviors of, 150, 152t
Defining features of consciousness. *See* Criteria for consciousness
Delafield-Butt, Jonathan, 144
Delsuc, Frederic, 43
Dendrites, 25f, 39
Dennett, Daniel, 30
Denton, Derek, 142–143, 146, 147t, 155, 270n54
Descartes, René, 4f, 7f, 252n16
Deuterocerebrum (arthropod), 178, 180f
Deuterostomes, 54f, 60
Dicke, Ursula, 111

Diencephalic locomotor region, 48
Diencephalon, 45, 70–71, 72f, 73–74t, 76f, 83. *See also* Thalamus; Hypothalamus
communication with tectum, 112f, 113
Dinosaurs, 119f, 122, 124, 125, 205f
Distance senses, viii, 63, 81–85, 181, 189, 213
Diversity. *See also* Consciousness, sensory: diversities within
in brain structures for consciousness, 208–214
cellular (neuron types), 27–28
of senses, 27, 79–88, 97–98t, 176–177t, 262n2
in types of consciousness, 1, 2, 131t
Domazet-Loso, Tomislav, 260n26
Dopamine, 158t, 168, 214
Dornbos, Stephen, 63, 83
Dorsal pallium, 112–114, 118, 123–127
Dorsal ventricular ridge (DVR), 121f, 126
Dugas-Ford, Jennifer, 126

Ecdysozoans, 54f, 60, 172f, 206, 207f
Echinoderms, 54f, 60
Ectoderm, 20f, 38, 85f, 86
Ectodermal placodes, 27, 85–88, 167, 263n16, 263n17
Edelman, David, 118
Edelman, Gerald, 33, 114, 218
Ediacaran Period, 53, 55, 56f, 61–62, 61f
Electroreception, 71, 262n2
Embodiment, 18–19, 20f, 195–196, 223, 224, 255n7
Emergence, 9–10, 18t, 19, 21–23, 22f, 28f
consciousness as emergent, 195–197
weak and strong (radical), 9–10, 14, 253n26
Emotions, 130, 138, 142–143, 281n7
Endoderm, 20f, 38
Eph/ephrin genes, 117, 269n52
Epilepsy, 9, 253n23
Epiphenomenon, consciousness is not, 218–219
Episodic memory, 114, 216

Equilibrium (sense of balance), 97t, 103
Eukaryote cells, 19–20
Evolution. *See also* Neuroevolutionary approach to consciousness
basics of, 253n28
of consciousness, 83, 88, 101–128, 168–170, 191–193, 198–206, 215–217, 222–226
convergent, 52, 125, 206
Experience, as basis of consciousness, 1, 2
Explanatory gap(s), 2, 5–11, 19, 26, 190, 221–227
Exteroception, viii, 130
Exteroceptive consciousness, viii, 6, 69–128, 130, 131t, 169f, 174–191, 208–209, 213f
affective consciousness in relation to, 131t, 138, 140, 148, 169–170, 187, 211–214, 213f, 283n25
amphioxus lacks it, 48, 116, 177
arousal and, 115, 212
arthropods and, 178–185
centers/sites of, 108–114, 118, 128, 183, 190, 209–215
cephalopods and, 187–191
defined, viii
gastropods lack it, 185–187
genes and, 116–117
hierarchies of, 98–100, 105–108, 176–177t, 178
insects and, 178–185
interoceptive consciousness in relation to, 130, 131t, 211–212, 213f, 214, 272n24
invertebrates and, 174–191, 176–177t
mammals and birds, 118–128
memory and, 114–115, 123–124, 128, 177t
neural crest and placodes, 85–88
origins and dating of, 81–92, 117–118, 185, 190–191, 199f, 200, 222, 226
Exteroceptive pathways, 92, 93–98, 180f, 186f, 188f
Eyes, 45, 60, 66, 80f, 81–84, 87–91, 93f, 181, 185, 187, 189

Fabbro, F., 265n1
Fauria, Karine, 184
Fear, 145–146
Feelings. *See* Affective consciousness; Sentience
Feinberg, Todd, 223, 224
Filter-feeding, 42, 43, 55, 66, 101–103
　in the first vertebrates, 66, 101
First-order multipolar neurons, 92, 93–96f, 97–98t
First-order representational theories, 265n1
Fish, 64, 66–67, 90–91, 101–105, 127, 156–165t, 167–168, 201–204f, 208. *See also* Anamniotes
　advanced behaviors in, 209, 282n12
　and pain, 151, 166–168
Flatworms. *See* Platyhelminthes
Forebrain, 23, 70–71, 73–75t
Fossil animals/organisms, 19–20, 51–67, 88–92, 101–103, 118–123, 185, 198–208
Fossil record, 51, 53, 83
Frustration, 152–153, 156–157t, 175–177t
Fuxianhuia (arthropod), 65f, 66, 185

Gamma oscillations. *See* Oscillatory patterns of neuronal communication
Gastropods, 171, 174, 175t, 185, 186f, 187
General biological features of living things, 17–24, 18t, 195–197, 222–225
Genes
　and affective consciousness, 168, 214
　of bilaterians, 61–62, 261n13
　and exteroceptive consciousness, 116–117
Genomic complexity and duplication, 67, 262n26, 280n33
Global organismic state (LeDoux), 146
Globus, Gordon G., 220, 251n6
Gnathostomes, 70f, 102f, 104
Gottfried, Jay, 265n31
Grain argument, 6

Grandmother (pontifical) neurons, 29, 30
Grantham, Todd, 24, 218
Greater limbic system, 76t, 78
Gutnick, Tamar, 190

Habenula, 74t, 78f, 139f, 162t, 213f, 277n33
Hagfish, 103, 103f, 104, 127, 202f
Haikouella (fossil chordate), 58f, 70f, 89f, 90–91, 201f, 264n24, 264n26
Haikouichthys (fossil fish), 4, 57f, 58f, 65f, 66, 70f, 89f, 90–91
Hall, Margaret, 122
Hard problem of consciousness, 2–15, 220–227
　animal cognition and, 12
　defined, vii, 2, 251n6, 251n7
　as marker for origins of sensory consciousness, 11, 222
　mind–body problem and, 14–15
　solution, 220–227
Hearing, 33, 86–88, 95f, 97t, 189
Hearing pathway, 95f, 97–98t, 180f
Hemichordates, 58f, 60
Hierarchy. *See also* Neural hierarchies
　biological, 18t, 22–24, 22f
　constraints and, 22–24, 22f
　emergence and, 22–23, 22f
　nested and non-nested, 28–30, 28f
Higher consciousness, 1, 2, 215–217, 251n4
Hindbrain, 70, 71f, 73t
Hippocampus, 75–76t, 112, 114–115, 124, 128, 139f, 162t, 214. *See also* Memory; Medial pallium
Holland, Linda, 42, 43, 258n17, 259n20
Holland, Nicholas, 116
Holland, Peter, 38
Homeostasis/homeostatic functions, 45, 47, 132, 142, 143, 210, 222, 274n22
Homeothermy, 122
Homology, 52
Homunculus, 31f
How the brain created experience, 222, 223–226

Humans
    auditory pathway in, 95f, 97–98t
    first modern, 147, 199f, 206, 208
    interoceptive pathways in, 134–135f
    olfactory pathway in, 96f, 97–98t
    as the only conscious organisms, 146–147, 254n34
    overemphasized in consciousness studies, 11, 108, 143, 209–211
    somatosensory-touch pathway in, 94f, 97–98t
    visual pathway in, 93f, 97–98t
Hydranencephaly, 144, 268n27
Hyperpallium (Wulst, in birds), 120–121f, 126
Hypothalamus, 45, 46f, 47–48, 71, 74t, 78f, 133, 134–135f, 139f, 142, 164t

Images. *See* Mental images
Infundibular organ, 45, 46f
Insects, 171, 175–177t, 178–185, 200, 205f
Instrumental learning. *See* Operant conditioning
Insula, 134–142, 134–135f, 139f, 162–163t, 213f
Integrated information in consciousness (Tononi), 27, 30, 210
Intelligence, animal, 12, 217
Interactions, neural. *See* Neural interactions
Internal milieu, 138, 274n22
Interoception and interoceptive consciousness, 6, 130–138, 131t
    affective consciousness in relation to, 130–133, 138, 140, 211–212, 213f
    exteroceptive consciousness in relation to, 130–133, 211–212, 213f, 214, 272n24
    and isomorphism, 130–135, 138, 167, 212
    neuroanatomic structures, 131t, 134–137
    origin, date, and basis of, 140–142, 166–170, 206

Interoceptive pathways, 133, 134–135f, 136–137t, 138
Invertebrates, 52, 54f. *See also* Protostome invertebrates; Amphioxus; Tunicates; Sponges
Irreducibility of consciousness and subjectivity, x, 9, 217, 220–227, 252n11
Isomorphic (topographic) sensory pathways, 30, 31f, 33–34, 88, 93–96f, 105, 106f, 107–108, 110, 113, 116, 117, 128, 131t, 176t, 182, 189–190, 223, 225, 272n88
Isomorphism (defined), 30, 33. *See also* Maps; Isomorphic; Retinotopy; Somatotopy; Tonotopy
Isthmus nuclei, 109f, 110, 111

James, William, 21, 35, 140, 218
James–Lange theory, 140, 274n23, 274n24
Jarvis, Erich, 126
Jawless fish, 58f, 66, 91, 101–104. *See also* Lampreys; Hagfish
Jaws, evolution of, 104
Jayaraman, Vivek, 182
Jonkisz, J., 270n54, 281n4

Kaas, Jon, 110
Karten, Harvey, 126
Keller, A., 270n54
Kielan-Jaworowska, Zofia, 124
Kim, Jaegwon, 8, 22–23
Koch, Christof, 7, 27, 34, 36, 210

Lacalli, Thurston, 43–48, 87–88, 258n17, 259n20, 259n26
Lamellar body, 45, 46f
Lampreys, 4, 66, 87f, 101–105, 103f, 106f, 113, 115, 127, 158–165t, 166, 182f, 202f, 209
    brain as representative of first vertebrates, 104–105, 266n9, 266n10, 266n11
    larval, 65f, 66, 264n24
    optic tectum of, 106f, 108

Lancelets. *See* Amphioxus
Language, 206
Larvaceans, 41f, 42, 44f, 258n11
Lateral line, 103
Lateral septum. *See* Septum
Laterodorsal tegmental nucleus, 160t, 214
LeDoux, Joseph, 129, 145–147, 147t
Lee, Michael, 124, 125
Levels, neuronal, sufficient for consciousness, 98–100, 178, 198
Levine, Joseph, 2
Life, living organisms
  ancestors of, 20f, 54f
  consciousness in relation to, 17, 18t, 195–196
  defining, 17–18, 254n2
  embodiment of, 18,19, 20f, 195–196, 223, 224
  general features of, 17–24, 18t, 195–196, 222–225
  ontological subjectivity grounded in, 19, 195–196
  as process, 21, 223
  as self-organizing systems, 21–22
Life stages and consciousness. *See* Consciousness, sensory: and life cycle
Light-Switch hypothesis, 81
Limbic system, 47–48, 76t, 78, 78f, 133, 138, 139f, 140, 154–155, 212, 213f
  definition used here, 78
Llinás, R. R., 270n53
Lobopodian worms (and arthropods), 58f, 60, 262n22
Lophotrochozoans, 54f, 60, 206, 207f
Lowe, Christopher, 258n20

Mallatt, Jon, 90
Mammal-like reptiles, 118, 119f, 122, 203f, 205f. *See also* Synapsids
Mammals
  brains of, 72f, 118–127, 136–137t, 139f, 158–165t
  consciousness advancing in, 118–127, 155, 199f, 200, 275n54
  early environment of, 124–125
  interoceptive and pain pathways in, 133–138
  as the only conscious organisms (with birds), 101, 118, 125, 209–211, 254n34 (*see also* Corticothalamic theory of consciousness)
  overemphasized in consciousness studies (*see* Humans)
  smell vs. vision in, 122–123
Maps (in nervous system), 30, 31f, 33–34, 63–64, 92–98, 108, 110, 130–132, 176–177t, 182, 190, 198. *See also* Isomorphic sensory pathways; Isomorphism; Mental images
Masaccio, *The Expulsion from the Garden of Eden*, 31, 32f
Mashour, G. A., 265n1
Mats. *See* Microbial mats
Mayr, Ernst, 18, 21, 225, 254n2, 255n21
Mechanoreceptors (mechanoreception, mechanosenses, mechanical senses), 79, 97t, 131t, 176t, 181, 186f, 187, 189
Medial pallium, 75t, 78f, 112f, 114–115, 139f. *See also* Hippocampus
Medulla oblongata, 70, 71f, 72f, 73t
Mehta, N., 265n1
Melanocytes, 86
Memory
  forms (Damasio and Tulving), 215–217
  in invertebrates, 177t, 183–184
  in mammals and birds, 123–124
  in mental causation, 225
  necessary criterion for consciousness, 114, 152, 177t, 198, 206
  role of, in affective consciousness, 152
  role of, in advancement of consciousness, 124, 128, 152, 200, 215
  role of, in sensory consciousness, 114–115
  smell in relation to, 84, 115
Mental causation, ix, 8, 170, 218, 224–225

Index

Mental images (sensory), viii, 6, 18t, 33, 83, 92–100, 107, 117, 131t, 184, 198, 200, 209, 211–213, 219, 222
Mental states, 34
Mental unity, ix
 adaptive function of, 217–218, 224
 defined, 6–7
 multisensory convergence contributes to, 110–111, 183, 189–190, 198
 neural interaction as contributing to, 34
 optic tectum's role in, 110–111
 transition to, 224
Merker, Bjorn, 144, 272n88
Mesoderm, 20f, 38
Mesolimbic reward system, 155, 160–163t, 166, 212, 213f
Mesozoic Era, 13f
 jump in consciousness, 127
Metaspriggina (fossil fish), 58f, 70f, 89f, 91
Metencephalon, 71f, 73t, 76f, 79
Microbial mats, 55, 56f, 62, 66, 83
Midbrain, 70, 71f, 73t, 76f
Mind–body problem, 14–15, 19, 196
Mind–brain problem, ix–x, 222
Minimum number of neuronal levels for consciousness, 98–100, 178, 198
Mitochondria, in evolution of cell complexity, 19, 255n6
Mobility, locomotion, high activity, and consciousness, 64–67, 191–193, 207, 219–220
Molluscs, 58, 64. *See also* Protostome invertebrates
 and affective consciousness, 174, 175t, 187
 and exteroceptive consciousness, 176–177t, 185–191
 overview of, 171, 173
 relation to other animal clades with consciousness, 206, 207f
Molting in arthropods, 185
Monod, Jacques, 24
Morris, Simon Conway, 91

Motor origin of consciousness, 270n53, 274n41
Motor programs. *See* Basic motor programs
Multiple sensory hierarchies, 176t, 181, 189, 198
Multisensory convergence/multisensory mapping, 88, 92, 98t, 99, 110–111, 114, 117, 134f, 183, 176–177t, 206
Myomeres, 65f, 66–67
Myriapods, 171, 220

Nagel, Thomas, vii, 1, 252n11
National Research Council, 149, 151, 153
Natural selection. *See also* Adaptation; Consciousness, sensory: adaptive nature of
 and consciousness, 117, 217, 219
 defined, 253n28
 general, 18, 22, 51, 52
Nautilus, 172f, 187, 191
Negative contrast. *See* Behavioral negative contrast
Nematodes (roundworms), 154, 156t, 172f, 173–175, 178, 191, 207f
Neoteny, 42
Nerve fiber, defined, 39
Nervous system. *See also* Central nervous system; Peripheral nervous system
 of cephalopods, 188f
 of chordates, 39–40, 40f, 73–76t
 complexity greater than other organ systems, 14, 28
 of gastropods, 186f
 of insects and arthropods, 179–180f
 necessary for consciousness, 1–15, 18t, 25–26, 197–198
Nested hierarchies, 18t, 28–30, 28f, 224
Neural correlates of consciousness. *See* Criteria; Consciousness, sensory: commonalities of its diverse forms
Neural crest, 27, 85–88, 85f, 263n17
Neural hierarchies, 26–34, 105
 complex, 18t, 26–28, 67
 consciousness and, 197–198, 210, 214–215, 223, 224

Neural hierarchies (cont.)
   levels of, sufficient for sensory consciousness, 98–100, 178, 198
   mapping in, 30, 31f, 33–34, 63–64, 92–98, 108, 110, 130–132, 176–177t, 182, 190, 198
   nested and non-nested, 28–30, 28f, 224
   non-nested as special, 28–30, 28f
   neural-neural interactions created in (*see* Neural interactions)
   sensory, 93–100, 105, 106f, 107–108, 107f, 134–135f, 176–177t, 178, 180f, 181, 188f
   unconscious, 210
Neural interactions (reciprocal, recurrent, recursive, reentrant), 18t, 27, 34, 105–108, 177, 183, 196–198, 214, 223, 225
Neural plate, 85f, 86
Neural tube, 85f, 86
Neuraxis, 23
Neurobiological approach to (domain of) consciousness, ix, 3, 4f, 11–14, 217–218
Neurobiological features of consciousness, 17, 18t, 26–36, 81–88, 105–128, 158–165t, 177–175t, 197–198, 222–225, 227
Neurobiological naturalism, viii, ix, 14, 195–227
   sensory consciousness as ancient, widespread, and diverse, 198–217
   sensory consciousness explained by neurobiological principles, 195–198
   three postulates of, 195–222, 196t
   transition to consciousness, 222–225
   solution to the problems of consciousness, 222–227
Neurochemicals. *See* Neurotransmitters; Neuromodulators
Neuroevolutionary approach to (domain of) consciousness, 3, 4f, 11–14, 37–170, 191–193, 198–208, 217–228
Neurohierarchies. *See* Neural hierarchies
Neurohypophysis, 45, 46f, 74t, 77t

Neuromodulators, 48, 115–116, 168, 269n46
Neuron (nerve cell)
   basic anatomy, 25, 40f
   explained, 39
   functions of, 27, 39
   of invertebrates, 184, 256n28
   nucleus, 39
   numbers of, and consciousness, 175–177t
   types of, 27, 40f
Neuroontologically irreducible features of consciousness (NOIFs), 252n12. *See also* Neuroontologically subjective features of consciousness
Neuroontologically subjective features of consciousness (NSFCs), 5–8, 5t, 217–218, 223–225
Neuropile, 46f, 47, 48
Neurotransmitters, 39, 168, 215
Nichols, Shaun, 24, 218
Nieuwenhuys, Rudolph, 47–48, 78
Nilsson, Dan-Eric, 80, 81
Niven, J., 183
Nociception (vs. pain), 150
Nociceptive pathways for pain. *See* Pain pathways
Nociceptors, 79
Noetic consciousness (Tulving), 216
Nonconscious clades of animals, 13, 48–49, 64, 116, 169f, 187, 191–192, 207, 261n19
Nonconscious perception, 209–210
Norepinephrine, 115, 116, 269n46
Northcutt, R. Glenn, 54, 61
Northmore, David, 110
Northoff, Georg, 267n13, 274n24, 284n46
Notochord, 37, 39, 41f, 85f, 89f
Nucleus accumbens, 139f, 160t, 213f, 277n33

Objective–subjective divide, ix, 2, 5–8, 14, 33, 220–223, 226, 227, 284n47

Ocean floor (as important place for evolution), 55, 56–57f, 62, 201–202f, 204f
Ocelli, 41f, 61, 66, 82
Octopuses, 14, 174, 188f, 189–190, 280n33
Olfactory bulb, 71, 74t, 96f, 106f
Olfactory consciousness, 127, 128, 211. *See also* Smell
Olfactory pathway
  description, 96f, 97–98t, 106f, 180f
  thalamus bypassed, 98, 211, 265n31, 282n19
Ontological gap (at objective–subjective divide), 227
Ontological reduction, 8–9. *See also* Irreducibility of consciousness and subjectivity
Ontological subjectivity and irreducibility, vii, 4f, 5–12, 15, 195–196, 217, 220–227, 284n46, 285n47
Ontology, defined, 5
Operant conditioning and learning, 144, 151–153, 156–157t, 175t, 187
Optic tectum, 70, 72f, 73t, 78, 99, 108–115, 109f, 123, 209, 272n88
  of anamniotes, 106f, 108–115
  of birds, 111, 123, 125
  communication with other brain regions, 110–114
  and consciousness, 108–111, 127, 209, 268n27
  evolution of, 79, 88, 117
  functions, 108–111
  and gamma oscillations, 111
  and isthmus nuclei, 110
  of lamprey, 108
  of mammals (*see* Superior colliculus)
  for multisensory, isomorphic convergence, 108, 110
  in visual pathway, 93f, 97t, 106f
Oscillatory patterns of neuronal communication, 107–108, 111, 183–184, 198, 211, 267n13, 283n25
Osteichthyes, defined, 104

Packard, Andrew, 144
Pain, 132, 167. *See also* C nerve fibers; Aδ nerve fibers; Nociception
  agonizing (burning, slow, dull, prolonged, second, suffering), 132, 133, 137t, 149, 167–168
  in animals, 149–151, 167–168, 174
  behavioral evidence for, 149–151, 152t, 153t
  conscious experience of, 150–151
  defined, 132, 149
  in fish, 151, 166–168
  and isomorphism (somatotopy), 132, 137t
  origin and date of, 167–168
  sharp (fast), 132
  theories of, 141–142, 167–168
Pain pathways, 133, 135f, 136–137t, 138
Paleobiological domain. *See also* Fossil animals
  consciousness in, 4f, 11–13
  dates of evolution of consciousness, 198–206, 199f
  timeline of, 13f
Pallidum, 74t, 78f
Pallium. *See also* Cerebral pallium; Cerebral cortex; Medial pallium
  anamniote, 112–114, 127
  basic parts, 74–75f, 112–114
  of birds vs. mammals, 118–127, 120–121f
  as site for all sensory consciousness, 118, 122–124, 209 (*see also* Corticothalamic theory of consciousness)
  as site for olfactory consciousness in all vertebrates, 113, 117, 127
Pani, Ariel, 258n20
Panksepp, Jaak, 143–144, 146, 147t, 155, 274n41, 281n7
Parabrachial nucleus, 133, 134–135f, 139f
Paracrine core of brain, 47–48
Parasites and consciousness, 220
Parker, Andrew, 81

Partial consciousness, 284n43
Pattee, Howard H., 23
Pavlovian conditioning. *See* Classical conditioning
Periaqueductal gray of midbrain, 73t, 76t, 134–135f, 139t, 158–159t
Peripheral nervous system (PNS), 39, 40f, 83, 86
Persistence, 153
Phasically firing axons, 27
Phenomenal consciousness. *See* Consciousness, sensory
Philippe, Herve, 43
Philosophical approach to (domain of) consciousness, vii–x, 2–8, 19, 217, 220–228
Photoreceptors, 45, 79, 80f, 82–83, 93f, 97t, 106f
Phyla, 52, 54f
Phylogenetics, 52
Pikaia, 58f, 60
Pineal organ, 45, 46f, 72f, 74t, 252n16
Placodes. *See* Ectodermal placodes
Platyhelminthes (flatworms), 54f, 154, 174, 175, 207f
Play, 143, 145, 153, 174, 175t
Plotnick, Roy, 63, 83
PNS. *See* Peripheral nervous system
Pons, 70, 71–72f, 73t, 79
Posterior tuberculum, 78f, 235–236, 277n33
Predation
 as cause of Cambrian explosion, 62–64, 81, 198
 first vertebrates were not predators, 66–67, 104
 jaws and, 104
 limiting arthropod and cephalopod evolutionary potential, 185, 191
 qualia and, 218
 vision and, 82, 84
Predatory arms race. *See* Arms race, Cambrian
Prediction, 117, 158, 218, 270n53
Premotor center, 178

Prevertebrates, early
 amphioxus similar to, 43, 46–47f, 48, 67, 79
 brains of, 42, 48, 79
 as filter feeders, 66
 senses of, 79
Prevertebrates, later
 candidates, actual and hypothetical, 88, 90
 sensory and brain revolution in, 83–88
 vision of, 81–83
Primal emotions (primary, basic) (Panksepp), 143
Primary consciousness. *See* Consciousness, sensory
Primates, 142, 199f, 206, 209
Primordial emotions (Denton), 142
Procedural memory, 216
Process (consciousness as, life as), 18t, 21
Processes (projections), of neurons, 39–40
Projicience, 6, 13t
Proprioception, 210
Protocerebrum (arthropod), 178, 179–180f, 183, 279n14
Proto-self (Damasio), 215–216
Protostome invertebrates. *See also* Arthropods; Molluscs; Nematodes; Platyheliminthes
 and affective consciousness, 174, 175t
 brain not homologous to that of vertebrates, 154–155, 171
 central nervous system, basic plan, 171, 173f
 and consciousness, 171–193, 199f, 200, 206, 207f
 defined, 54f, 60
 and exteroceptive consciousness, 174, 176–193
 living animal groups of, 172f
 neurons and neuron numbers, 175t, 176t, 184, 256n28
Pupillary-light reflex, 25
Purposive actions, as conscious and ancient, 169–170

Qualia (images and affects)
  adaptive character of, 218
  defined, viii, 1, 7
  diversity of images vs. affects, 215
  neural interaction as contributing to, 34, 225
  origins of (over 520 mya), viii, 99, 169, 185, 222
  transition to, 225

Rates, fast (of neural communication), 18t, 25–26, 124, 197, 275n75
Reality and consciousness, 219, 284n39
Receptors, sensory, 27, 79, 97–98, 130–132
Reciprocal interactions (recurrent, recursive, reentrant). *See* Neural interactions
Reduction, ix–x, 8–9, 14. *See also* Irreducibility
Referral, ix, 6, 26, 218, 223
Reflexes, 17, 18t, 25–26, 25f, 152, 197, 222, 223, 224
  vs. affective behavior, 152, 153
  in amphioxus and early prevertebrates, 37, 49, 79
  in bilaterian ancestor, 62
  as inefficient, 145, 152
  and mental causation, 224
  monosynaptic and polysynaptic, 25
  and nociception, 150
  and referral, 26, 223
  as royal road to consciousness, through elaboration, 25–26, 92, 100, 142, 197, 222
  in young or larval animals, 219
Representations vs. images, 33
Reptiles, 52, 102f, 121f, 126–127, 156–165t, 272n83
  as poorly studied, 127, 272n81
Reticular activating system, 75t, 115, 212, 213f, 214
Reticular formation, 45, 48, 75t, 77t, 78f, 115–116, 133–138, 213f

Retina, 45, 46f, 80f, 83, 93f, 106f, 179–180f
Retinotopy, 33, 93f, 182–183, 190, 192, 246
Revonsuo, Antti, 1, 8
Rhizocephalan barnacles, 220
Rose, James D., 167
Roth, Gerhard, 111
Roundworms. *See* Nematodes
Rowe, Timothy, 122

Salience, 110, 145, 155, 166, 267n19
Sanctacaris (arthropod), 65f, 66
Sauropsids (Sauropsida), 102f, 118, 119f, 120f
  include all modern reptiles and birds, 126
Schrödinger, Erwin, 10
Schultz, Wolfram, 170
Searle, John, vii, 5, 8–9, 14
Sea squirts. *See* Tunicates
Seeking behavior, 143, 153
Seelig, Johannes, 182
Selective attention, 18t, 35, 108, 118, 124, 177t, 183, 187, 190, 198, 211, 219, 223, 224
Selective pressure, 52
Self, 138, 140–141, 215–216
Self-awareness/consciousness, 2, 206, 215–217
Self-delivery of analgesics or rewards, 151, 153, 156–157t, 175t
Self-organization, 21–22, 225
Semantic memory, 216
Senses
  amphioxus, 40, 42, 45, 79, 86, 87
  common pattern of pathways for all vertebrate senses, 92–98, 93–96f, 97–98t, 134f
  convergence of different senses, 88, 99, 110–111, 114, 117, 177t, 183, 206
  evolution of, in Cambrian explosion, 63–64, 67, 79–92, 187
  of fish, 103
  of gastropods, 187

Senses (cont.)
  of insects and other arthropods, 176t, 180f, 181
  of lampreys, 105
  prevertebrate, 79–92
  tunicate, 40, 42, 44f, 80f, 86–87
  vertebrate, 79–89, 92, 93–96f, 97–98t, 98, 103, 176t
Sensory consciousness. *See* Consciousness, sensory
Sensory neural hierarchies. *See* Neural hierarchies: sensory
Sensory pathways. *See* Neural hierarchies: sensory
Sensory receptors, 27, 79, 97–98t, 130–132
Sentience, 129, 272n1. *See also* Affective consciousness
Sentient self (Craig), 140–141
Septum (septal nuclei, and lateral septum), 74t, 158t, 213f
Šestak, Martin, 84, 260n26
Seth, Anil, 118
Sherrington, Charles, 6, 33, 150
Shu, Degan, 91
Simon, Herbert, 27
Sister groups, 43, 52, 254
Slugs, 185, 187. *See also* Gastropods
Smell, 63, 83–84, 96f, 98t, 112–113, 122–123, 130–133, 265n31. *See also* Olfactory consciousness; Olfactory pathway
  in chemoreception, 79, 131
  and memory, 84, 115
Smell-brain concept, 112–113
Smell-first hypothesis, 83–85, 263n12, 265n31
Snails, 185, 187. *See also* Gastropods
Social behavior network, 155. *See also* Adaptive behavior network
Solitary tract and nucleus, 98t, 133–139
Solms, Mark, 143, 147t, 155
Somatosensory cortex (primary) (SI), 31f, 94f, 97t, 120f, 135f, 136–137t, 141
Somatosensory pathway, 94f, 97–98t, 99, 106f, 180f, 234n48

Somatotopy, 30, 31f, 33, 94f, 134–138, 182, 187, 189–190
Special neurobiological features of consciousness. *See* Neurobiological features of consciousness
Sperry, Roger, 10
Spinal cord, 39, 71, 73t
  learning by, 152t
  pathways in, 94t, 133–138
Spinal trigeminal nucleus, 136, 138
Spinothalamic tract, 131t, 133–138
  includes spinohypothalamic, spinomesencephalic, spinoparabrachial, spinoreticular, 136–137t
Sponges, 19, 20f, 54f, 55, 201f, 224, 260n7
Strausfeld, Nicholas J., 185
Striatum (corpus striatum, basal ganglia), 74t, 76t, 78f, 112, 160t, 178
Structural complexity, 24
Subcortical limbic brain regions, 139f, 144, 155, 166, 170, 211–214
Subjectivity. *See* Consciousness; Objective–subjective divide; Ontological subjectivity
Suffering. *See* Pain: agonizing
Superior colliculus, 73t, 92, 93–95f, 108, 121f, 268n27, 282n11. *See also* Optic tectum
Suprachiasmatic nucleus, 45
Survival role of consciousness. *See* Sensory consciousness: adaptive nature of
Synapses, 39
Synapsids (Synapsida: mammal-like reptiles and mammals), 102f, 118, 119f, 120f
System and systems theory, 21–24, 254n1

Taste, 63, 97–98t, 130–134, 176t, 180f
  in chemoreception, 79, 131t
Taste pathway, 97–98t, 134f, 138, 180f
Tectum. *See* Optic tectum

Teeth, 104
Tegmentum, 70, 73t
Telencephalon, 71, 74–75t, 88, 112–114
 amphioxus probably lacks, 45, 77t, 258n17
 and smell, 88, 112–114, 122–123
Teleological thinking, 24
Teleonomy, 18t, 24
 and adaptation, 255n21
Teleost fish, 104, 127, 156, 167, 208
Thalamus, 31f, 34, 71, 73–74t, 78f, 79, 213f
 relays to cerebral cortex/pallium, 73t, 92–95f, 97–98t, 106f, 133–137
 and optic tectum, 112f, 113
 smell consciousness, not essential for, 98, 211, 265n31, 282n19
 in vertebrates, not amphioxus, 77t, 259n20
Thaliaceans, 41f, 42, 44f
Thompson, Evan, 19, 196
Timelines
 of Earth's history, 13f, 53f
 of history of consciousness, 198–206, 199f
Tonically firing axons, 27
Tononi, Giulio, 27, 210
Tonotopy, 95f, 182, 246
Topical convergence, 29
Topographic maps, 30, 33. *See also* Maps; Isomorphic; Isomorphism
Touch, 30–32, 40, 48, 94f, 97t, 122–123, 131, 176t, 180f, 181. *See also* Mechanoreceptors; Somatosensory pathway; Somatotopy
Trestman, Michael, 81
Triassic Period, 200, 204–205f, 207–208
Trigeminal nucleus, 106f, 135f. *See also* Spinal trigeminal nucleus
Trigeminothalamic tract, 133, 134–135f
Tulving, Endel, 216–217
Tunicates
 brains of, 39, 43, 44f, 258n11
 of Cambrian Period, 58f, 60
 central nervous system of, 77t

 overview of, 41f, 42–43
 senses of, 40, 42, 44f, 80f, 86–87

Unconscious hierarchies and senses, 210
Unity. *See* Mental unity
Urochordates. *See* Tunicates

Valence and affects, 129
Ventral pallidum, 139f, 160t, 277n33
Ventral tegmental area, 133, 134–135f, 213f, 277n33
 and salience, 160t
Ventricles of brain, 71, 75t
Vernier, Philippe, 111
Vertebrates, 66–170, 192–227
 affective consciousness possessed by all, 154, 155, 168–170
 affective neurostructures of, 155, 158–165t, 277n33
 arthropods compared to, 64–67, 104, 192–193, 185, 206–207
 brains of, 46f, 69–79, 71f, 72f, 73–77t, 87–88, 87f, 139f
 of Cambrian Period, 58f, 64–67, 89
 central nervous system of, 69–79
 cephalopods compared to, 191–193, 206–207
 as chordates, 37–38, 54f
 clades and phylogenetic relationships of, 38f, 102f
 consciousness possessed by all, viii–ix, 101–128, 149–170, 169f, 192–193, 198–215, 199f, 219, 265n1
 defining features of, 52
 evolution of, earliest (Cambrian), 66–67, 71, 88–92, 199–200
 evolution of, later (post-Cambrian), 101–104, 118–128, 192–193, 200–206
 evolutionary success of, 67, 192–193, 262n27
 eyes of, 80f, 81–83
 filter feeding as ancestral, 66, 101–104
 genome of, 67, 262n26
 interoceptive consciousness possessed by all, 131t, 166–167

Vertebrates (cont.)
  interoceptive consciousness and
    pathways in, 130–138
  neural crest and placodes of, 85–88, 85f
  neuron numbers of, 176t
  overview of all, 101–104
  pain (agonizing) only in tetrapods,
    167–168
  senses of, 79–89, 92, 93–96f, 97–98t, 98,
    103, 176t
Vetulicolians, 59f, 60
Vierck, Charles, 141–142, 147t
Vision, 81–84, 93f, 97t, 106f, 108–111,
  122–124, 176t, 182–183, 189. *See also*
  Eyes
Vision-first hypothesis, 81–85, 110–111
Visual pathway, 93f, 97–98t, 106f, 180f,
  188f
Von Economo neurons, 141

Watson, James, 9
Watt, Douglas, 143
Wicht, Helmut, 45
Wilczynski, Walter, 113
Worms. *See also* Lobopodian worms,
    Nematodes; Platyhelminthes
  as ancestral bilaterians, 20f, 53, 54f, 55,
    56f, 57, 61–62, 61f, 145, 200
  Cambrian, 57f, 58–59f, 60, 63, 66
  Modern, 54f, 156t, 172–175
Wullimann, Mario, 111

Yu, Jr-Kai, 116
Yunnanozoans, 89f, 90, 264n26. *See also*
  Haikouella

Printed in Poland
by Amazon Fulfillment
Poland Sp. z o.o., Wrocław

94961794R00228